吃透
Ansible
核心源码剖析
与项目实战

沈聪◎编著

机械工业出版社
China Machine Press

图书在版编目（CIP）数据

吃透Ansible：核心源码剖析与项目实战 / 沈聪编著. —北京：机械工业出版社，2021.6

ISBN 978-7-111-68580-7

Ⅰ. ①吃… Ⅱ. ①沈… Ⅲ. ①程序开发工具 Ⅳ. ①TP311.561

中国版本图书馆CIP数据核字（2021）第126378号

吃透 Ansible：核心源码剖析与项目实战

出版发行：机械工业出版社（北京市西城区百万庄大街 22 号　邮政编码：100037）	
责任编辑：刘立卿	责任校对：姚志娟
印　　刷：中国电影出版社印刷厂	版　　次：2021 年 7 月第 1 版第 1 次印刷
开　　本：186mm×240mm　1/16	印　　张：30.5
书　　号：ISBN 978-7-111-68580-7	定　　价：149.00 元

客服电话：（010）88361066　88379833　68326294　　　投稿热线：（010）88379604
华章网站：www.hzbook.com　　　　　　　　　　　　　　读者信箱：hzit@hzbook.com

|前言|

　　Python 是目前自动化运维领域应用最多的语言之一，而基于 Python 开发的 Ansible 则是自动化运维管理的首选工具。对 Python 程序员来说，学习 Ansible 源码一方面能从中学到不少 Python 的高级用法，另一方面也能熟练掌握这款流行工具并可以随时对其进行深度改造和定制，这对他的个人成长至关重要。此外，Ansible 源码中的很多函数与类均可以在简单改造后成为运维人员的常用脚本。

　　初识 Ansible 2.8 的源码时，笔者感受到了它的复杂与宏大。单纯地分析整个 Ansible 的源码及其所有细节并让 Python 初学者接受是一件非常困难的事情，笔者一开始也不打算这么做，直到有一天，笔者在 Ansible 的官方代码库中看到了最早发布的 Ansible 1.1，随后几天迅速阅读和调试完该版本的源码，最终发现早期的 Ansible 1.1 与当前的 Ansible 2.8 有着共同的核心理念，即相同的模块运行机制及剧本集概念。Ansible 1.1 只用非常少的代码就实现了一个在当时看来十分不错的运维工具，其核心设计思想一直延续至今，从未改变。笔者意识到，对于想要深入学习 Ansible 源码的运维人员而言，Ansible 1.1 的源码是一份绝佳的入门材料。由浅入深是笔者最喜欢用的研究源码的方式，这样的理念也促成了本书的形成。此外，为了帮助读者理解基于 Ansible 的 Playbook 项目，笔者精选了自动部署工具 ceph-ansible 作为实战对象，它也是笔者完成各种 Playbook 项目的首选参考。

　　本书融合了笔者在 Ansible 源码方面积累的一些学习心得，从小版本的 Ansible 源码，到当前流行的 Ansible 2.8 源码，乃至从模块运行原理，到 Playbook 的解析与执行，其间都有笔者总结的一些经验。

本书特色

　　本书的主要特色有：
- 按照由浅入深的方式剖析三个经典版本的 Ansible 源码，每个版本的源码剖析重点各不相同。
- 对 Ansible 1 和 Ansible 2 中的核心源码做了大量的测试与实践，从而辅助读者更好地理解相关知识。
- 通过完整对比小版本的 Ansible 源码及对 GitHub 上相关源码的 Bug 进行溯源两种方式，帮助读者深入学习相关知识，从而吃透 Ansible。
- 完整剖析 ceph-ansible 项目并深入实践与改造，帮助读者更好地理解基于 Ansible 开发的 Playbook 项目。

本书内容

本书主要介绍 Ansible 的核心源码与 Playbook 项目实战，共分为 6 章。

第 1 章剖析 Ansible 1.1 的源码，重点分析 Ansible 1.1 的底层通信机制及其模块运行原理。

第 2 章剖析 Ansible 1.9.6 的源码，并对比它与 Ansible 1.1 源码的不同之处，另外还会重点分析 Playbook 的执行逻辑。

第 3 章介绍 Ansible 2.8 中部分核心模块的使用并剖析其源码，同时还会介绍 Playbook 的常用语法。

第 4 章剖析 Ansible 2.8 的核心源码。首先进行基础功能模块的源码分析，如常量值定义、inventory 文件解析和模板渲染等，接着对 Ansible 2.8 中模块运行的全流程及剧本集的运行源码进行跟踪和分析。

第 5 章深入学习 Ansible 源码，主要通过小版本的源码对比和 GitHub 上相关源码的 Bug 溯源两种方式，帮助读者熟悉深入学习 Ansible 源码的方式。

第 6 章主要介绍基于 Ansible 的经典项目——ceph-ansible，并对该项目进行改造和优化。

读者对象

顶级开源项目的源码本身既复杂又难懂，想要掌握本书的内容，领会笔者的分析思路，读者需要具备一定的 Python 基础，同时要有坚持学习的信念。本书不是帮助 Python 初学者入门，而是深度剖析 Python 顶级开源项目的源码，主要适合以下读者阅读：

- 拥有一定 Python 基础的运维人员；
- 熟练使用 Ansible 并想深入理解其核心源码的后端开发人员；
- 渴望进入开源世界，探索一个顶级开源项目全貌的初、中级程序员；
- 高等院校的老师和学生；
- 相关培训机构的学员。

版本说明

在本书中，Ansible 1、Ansible 2 和 Ansible 2.8 代表的是 Ansible 的几个大版本，但并非具体的发行版本，而 Ansible 1.1、Ansible 1.9.6 和 Ansible 2.8.16 分别是各大版本下的具体发行版本，笔者主要基于这三个发行版本剖析 Ansible 的源码，当然也会提及其他发行版本。

Ansible 源码获取方式

本书涉及的所有源码均已开源，读者可以通过 https://releases.ansible.com/ansible 网站进行下载。也可以在华章公司的网站（www.hzbook.com）上搜索到本书，然后单击"资料下载"按钮，即可在本书页面上找到下载链接。

售后支持

Ansible 从版本 2 开始已经变得庞大且臃肿，相关的小版本迭代也在飞速进行。尽管本书剖析的 Ansible 源码不是最新的 2.10 或者更高的版本，但其核心思想一直未曾改变，这是读者在学习 Ansible 源码时要重点把握的。笔者完成本书的主要资料正是 Ansible 各个版本的源码，许多关于 Ansible 的术语，如动作插件、通信插件等，都是直接翻译英文注释而来的，可能会有不当之处，加之笔者水平和精力所限，书中可能还存在一些疏漏与错误，敬请各位读者不吝指正。读者在阅读本书时若有疑问，可以发电子邮件到hzbook2017@163.com，笔者将尽量解答相关疑问。此外，读者还可以关注"源码探索之旅"微信公众号，笔者会在公众号上定期对读者的一些疑惑进行文字或视频解答，与读者共同成长与进步。

致谢

首先感谢我的家人！没有父母和姐姐给我的鼓励与支持，就没有我的今天，我也不会有机会进入计算机行业，更不会有机会完成本书。

感谢我的女朋友黄宝宝！她花费了很多时间和精力帮我整理书稿，并改正书稿中的不当语句和错别字。

感谢天翼云科技有限公司的领导黄润怀副总经理和我的直接领导林洁琬组长！他们给我们组创造了一个难得的机会，让我们能全身心投入分布式存储领域的研发中。正是这次机会，让我接触了 Ansible 及部署分布式存储集群的开源工具 ceph-deploy 和 ceph-ansible，从此踏上了 Python 工具的源码分析之旅。

感谢媒体存储小分队的黄鹄、陈涛、毛廷鸿、谭伟杰、吴文峰和杨佑！他们拥有的丰富的互联网产品研发经验给了我很多帮助，让本书的写作比预期顺利很多。

感谢本书的编辑欧振旭！他给了我出版本书的机会与动力，让我得以完成这本心中想了无数遍的书。

还要感谢为本书的出版提供过帮助的其他编辑和朋友！没有他们的大力支持，本书也很难与读者见面。

最后感谢本书读者！你们的肯定与支持会激励笔者继续前行。

沈聪

|目录|

第 1 章 Ansible 1.1 源码剖析

本章将完整剖析 Ansible 1.1 的源码,包括 Ansible 模块的执行流程和底层通信机制,以帮助读者彻底掌握 Ansible 1.1。Ansible 1.1 类似于婴儿期的 Ansible,虽然功能不强,但胜在"五脏俱全"。通过学习 Ansible 1.1 源码,读者就能掌握 Ansible 工具的实现原理,而这一原理依旧是 Ansible 2 源码的核心,只不过后者的实现过程更为复杂。

1.1 Ansible 1.1 的环境搭建

本节将介绍如何搭建 Ansible 1.1 代码的测试与调试环境。由于 Ansible 1.1 形成于 2013 年年中,只支持 Python 2,为了后续能更好地学习其他版本的 Ansible,笔者使用虚拟环境来安装 Ansible 的各个版本。此外,笔者还使用了一台安装 CentOS 7 桌面版的虚拟主机,并在其中使用 VSCode 演示如何调试 Ansible 代码。

1.1.1 Ansible 1.1 测试环境搭建

这里笔者使用 VMware 软件创建了 4 台装有 CentOS 7 系统的虚拟主机,并配置好了相应的静态 IP 以及网络,确保能正常连通外网。4 台主机的 IP 及其设定的主机名如下:

```
[root@master ~]# cat /etc/hosts
127.0.0.1   localhost localhost.localdomain localhost4 localhost4.localdomain4
::1         localhost localhost.localdomain localhost6 localhost6.localdomain6


192.168.26.110 master
192.168.26.120 ceph-1
192.168.26.121 ceph-2
192.168.26.122 ceph-3
```

然后安装一个好用的虚拟环境管理工具——pyenv。其手工安装过程非常简单,保证网络畅通即可,具体操作命令如下:

```
[root@master ~]# yum install git -y
[root@master ~]# git clone https://github.com/pyenv/pyenv.git ~/.pyenv
[root@master ~]# git clone https://github.com/pyenv/pyenv-virtualenv.git
~/.pyenv/plugins/pyenv-virtualenv
```

直接从 GitHub 上下载 pyenv 项目源码及创建虚拟环境的插件（pyenv-virtualenv）并将其放到对应的目录下，接着在~/.bashrc 文件中配置 pyenv 的命令路径即可使用 pyenv 工具。具体配置命令如下：

```
[root@master ~]# cat ~/.bashrc
# .bashrc

# User specific aliases and functions

alias rm='rm -i'
alias cp='cp -i'
alias mv='mv -i'

# Source global definitions
if [ -f /etc/bashrc ]; then
. /etc/bashrc
fi

# 添加 pyenv 命令路径
export PATH="~/.pyenv/bin:$PATH"
eval "$(pyenv init -)"
eval "$(pyenv virtualenv-init -)"
[root@master ~]# source ~/.bashrc
[root@master ~]# pyenv --version
pyenv 1.2.21
```

此时便可以使用 pyenv 来创建相应的虚拟环境。在使用 pyenv install 命令安装 Python 版本时，默认从 python.org 网站上下载指定的版本，国内用户访问速度通常非常慢。此时可以先从国内的 Python 镜像源中下载指定的 Python 版本到 pyenv 的插件缓存目录中，然后再执行安装命令。具体操作如下：

```
[root@master ~]# yum -y install gcc gcc-c++ zlib zlib-devel bzip2-devel
openssl-devel ncurses-devel sqlite-devel readline-devel tk-devel gdbm-
devel db4-devel libpcap-devel libffi-devel xz-devel
[root@master ~]# v=2.7.18;wget http://mirrors.sohu.com/python/$v/Python-
$v.tar.xz -P ~/.pyenv/cache/;pyenv install $v
```

以上安装的是 Python 2 的最新版本，在此之前需要安装一些依赖库。由于 Python 2 在 2020 年之后不再进行维护，许多大型 Python 项目也不再支持 Python 2，例如 Django 项目从 Django 2 开始就不再支持 Python 2，因此对于刚开始接触 Python 语言的读者，建议直接使用 Python 3 进行学习。

Ansible 在兼容性方面很不错，即使到了如今的 Ansible 2.9，也依然支持 Python 2。目前 CentOS 的 Minimal 版本中只有 Python 2，Python 3 通常需要额外安装。许多旧的生产系统上依旧运行着各种 Python 2 脚本，多种自动化工具依旧是基于 Python 2 开发的，预计这一现象会在 3~5 年后有所改变。

完成 Python 2.7 的安装后，就可以使用 pyenv virtualenv 命令创建对应 Python 版本的虚拟环境了。具体操作如下：

```
[root@master ~]# pyenv versions
* system (set by /root/.pyenv/version)
  2.7.18
[root@master ~]# pyenv virtualenv 2.7.18 ansible1.1
[root@master ~]# pyenv versions
* system (set by /root/.pyenv/version)
  2.7.18
  2.7.18/envs/ansible1.1
  ansible1.1
```

以上操作创建了一个名为 ansible1.1 的虚拟环境，使用 pyenv activate ansible1.1 命令就可以激活该虚拟环境：

```
[root@master ~]# pyenv activate ansible1.1
(ansible1.1) [root@master ~]#
```

接下来在虚拟环境中安装 Ansible 1.1 工具，直接使用 pip 命令安装 Ansible 并指定 1.1 的版本即可。如果下载速度过慢，可以指定使用清华源或者阿里源。

```
(ansible1.1) [root@master ~]# pip list
Package     Version
---------- -------
pip         20.2.3
setuptools  44.1.1
wheel       0.35.1
(ansible1.1) [root@master ~]# pip install ansible==1.1 -i https://pypi.
tuna.tsinghua.edu.cn/simple
(ansible1.1) [root@master ~]# ansible --version
ansible 1.1
(ansible1.1) [root@master ~]# pip list
Package      Version
------------ -------
ansible      1.1
bcrypt       3.1.7
cffi         1.14.3
cryptography 3.1.1
enum34       1.1.10
ipaddress    1.0.23
Jinja2       2.11.2
MarkupSafe   1.1.1
paramiko     2.7.2
pip          20.2.3
pycparser    2.20
PyNaCl       1.4.0
PyYAML       5.3.1
setuptools   44.1.1
six          1.15.0
wheel        0.35.1
```

这样就在虚拟环境 ansible1.1 中成功安装了 Ansible 1.1 工具。对比安装前后的 pip list 命令结果可知，Ansible 依赖众多的 Python 第三方模块。如 paramiko 模块用于远程 SSH 通信（通信方式可选）、PyYAML 模块用于解析 YAML 文件、Jinja2 模块用于实现 Ansible 中模板文件或者模板变量的渲染工作等。

此外，还需要对刚刚安装好的 Ansible 工具进行测试，确保能正常使用。首先准备一个 hosts 文件并按照组归类 Ansible 目标主机，同时给组变量添加主机的 SSH 登录账号、密码及端口（用户和端口均有默认值，可以忽略），具体操作命令如下：

```
(ansible1.1) [root@master ~]# cat hosts
[nodes]
ceph-[1:3]

[nodes:vars]
ansible_ssh_user=root
ansible_ssh_pass=@SHENcong19920522
ansible_ssh_port=22

(ansible1.1) [root@master ~]# ansible all -i hosts -m ping
ceph-2 | FAILED => module ping not found in /root/library:/root/.pyenv
/versions/2.7.18/envs/ansible1.1/share/ansible/
ceph-3 | FAILED => module ping not found in /root/library:/root/.pyenv
/versions/2.7.18/envs/ansible1.1/share/ansible/
ceph-1 | FAILED => module ping not found in
/root/library:/root/.pyenv/versions/2.7.18/envs/ansible1.1/share/ansible/
```

上面的结果显示，当前环境中没有找到 Ansible 的 ping 模块，查找的路径有两个：/root/library 和/root/.pyenv/versions/2.7.18/envs/ansible1.1/share/ansible/。这个问题也容易解决，只需要把 Ansible 1.1 内置的模块代码放到提示的任意一个目录下即可，这些模块文件在 Ansible 的源码中可以找到。

首先从官方的 Ansible 代码库中下载 Ansible 1.1 源码包，然后复制源码包下的模块并存放到指定位置即可。具体操作命令如下：

```
(ansible1.1) [root@master ~]# wget https://releases.ansible.com/ansible/
ansible-1.1.tar.gz
(ansible1.1) [root@master ~]# tar -xzf ansible-1.1.tar.gz
(ansible1.1) [root@master ~]# mkdir -p /root/.pyenv/versions/2.7.18/envs/
ansible1.1/share/ansible/
(ansible1.1) [root@master ~]# cp ansible-1.1/library/* /root/.pyenv/
versions/2.7.18/envs/ansible1.1/share/ansible/
```

最后，再次运行前面的测试命令，运行结果如下：

```
(ansible1.1) [root@master ~]# ansible all -i hosts -m ping
ceph-3 | success >> {
    "changed": false,
    "ping": "pong"
}

ceph-2 | success >> {
    "changed": false,
    "ping": "pong"
}

ceph-1 | success >> {
    "changed": false,
    "ping": "pong"
}
```

至此，Ansible 工具已经在虚拟机上安装完毕并且能正常使用了。在后面的章节中会简单使用 Ansible 1.1 的其他模块并编写 Playbook 进行测试。

Ansible 1.1 源码全部位于~/.pyenv/versions/ansible1.1/lib/python2.7/site-packages/ansible 目录下，后面的章节中将解析这些代码。在测试时，读者可以任意修改该目录下的代码文件，打印 ansible 命令运行的结果，这是一种比较"笨"但是非常实用的调试方法。

1.1.2　Ansible 1.1 调试环境搭建

接下来，笔者将在 CentOS 7 中搭建 Ansible 的源码调试环境。这里不再使用虚拟环境，而是直接使用 CentOS 7 自带的 Python 2.7。本节不会介绍 CentOS 7 桌面版的 VSCode 安装过程，具体步骤请参考网上教程。

为了能运行和调试 Ansible 1.1 源码，需要先安装 Ansible 1.1 依赖包。在虚拟环境中安装 Ansible 1.1 时可以看到 Ansible 所依赖的第三方模块，首先将其全部模块导出得到 requirements.txt 文件如下：

```
(ansible1.1) [root@master ~]# pip freeze > requirements.txt
(ansible1.1) [root@master ~]# cat requirements.txt
ansible==1.1
bcrypt==3.1.7
cffi==1.14.3
cryptography==3.1.1
enum34==1.1.10
ipaddress==1.0.23
Jinja2==2.11.2
MarkupSafe==1.1.1
paramiko==2.7.2
pycparser==2.20
PyNaCl==1.4.0
PyYAML==5.3.1
six==1.15.0
```

上面的 ansible==1.1 需要去掉，在调试环境下不要通过 pip 安装 Ansible，而是通过源码方式来安装。将依赖包内容（删除了第一行）复制到 CentOS 桌面版的虚拟机中，然后执行如下命令：

```
[shen@localhost ~]$ cat ~/Desktop/requirements.txt
bcrypt==3.1.7
cffi==1.14.3
cryptography==3.1.1
enum34==1.1.10
ipaddress==1.0.23
Jinja2==2.11.2
MarkupSafe==1.1.1
paramiko==2.7.2
pycparser==2.20
PyNaCl==1.4.0
PyYAML==5.3.1
six==1.15.0
```

```
[shen@localhost ~]$ sudo pip install -r ~/Desktop/requirements.txt -i
https://pypi.tuna.tsinghua.edu.cn/simple
```

注意：在使用 pip 安装 Ansible 1.1 依赖包时，如果遇到类似 Python.h: No such file or directory 的报错信息，需要手动安装 python-devel 包。

```
[shen@localhost ~]$ sudo yum install python-devel.x86_64
```

安装好上述依赖的第三方模块后，就可以下载 Ansible 1.1 源码并用 VSCode 打开，命令如下：

```
[shen@localhost python]$ pwd
/home/shen/shencong/code/python
[shen@localhost python]$ wget https://releases.ansible.com/ansible/ansible-1.1.tar.gz
[shen@localhost python]$ tar -xzf ansible-1.1.tar.gz
```

在 VSCode 中导入 Ansible 1.1 源码，结果如图 1-1 所示。

图 1-1　在 VSCode 中导入 Ansible 1.1 源码

在使用 VSCode 调试 Ansible 源码之前，最重要的一步就是安装 Python 插件，然后设置 python.pythonPath 的值，指定为系统的 Python 2.7 路径。此外，在最左边的竖条工具栏中，单击 Run 图标后会出现相应的调试设置区，如图 1-2 所示。

在图 1-2 中，单击 create a launch.json file 文本并在弹出的配置框中依次选择 Python 选项和 Python File 选项，便可创建一个 DEBUG 模式的配置文件——launch.json。默认创建的文本（launch.json）内容如下（位于源码目录下的.vscode 目录中）：

```
{
    "version": "0.2.0",
    "configurations": [
        {
            "name": "Python: Current File",
            "type": "python",
            "request": "launch",
            "program": "${file}",
            "console": "integratedTerminal"
        }
    ]
}
```

图 1-2　单击 Run 图标后出现的面板

下面对配置文件中的几个重要参数进行说明（其中除了已有的参数外，还有几个非常重要的参数需要手动添加）。

（1）program：提供 Python 程序入口模块的完全限定路径，推荐值为${file}。当在某个 Python 文件下按 F5 键或者单击 Run 菜单下的 Start Debugging 选项时，${file}就是该 Python 文件的全路径。

（2）args：指定传递给 Python 程序的参数。

（3）env：在这里非常重要，没有这个配置就无法搭建源码的调试环境。在 Ansible 的源码中出现的源码导入语句，如 from ansible 等，需要能将其定位至源码文件所在目录。因此调试 Python 项目源码时会使用该字段设定 PYTHONPATH 值。

接着在桌面（~/Desktop/）建立一个 hosts 文件，里面包含笔者的一台阿里云主机地址，当然这里随便使用一台虚拟机进行测试也是可以的。

```
[shen@localhost ~]$ cat ~/Desktop/hosts
[nodes]
47.115.61.209

[nodes:vars]
ansible_ssh_user=root
ansible_ssh_pass=你的云主机密码
ansible_ssh_port=22
```

接下来设置新的 launch.json 内容，具体如下：

```
{
    "version": "0.2.0",
    "configurations": [
        {
            "name": "Python: Current File",
            "type": "python",
            "request": "launch",
            # 也可以直接指定 ansible 命令文件: ${workspaceRoot}/bin/ansible
            "program": "${file}",
            # 设置 Python 环境路径
            "env": {"PYTHONPATH":"${workspaceRoot}/lib"},
            "args": [
              "all",
```

```
            "-i", "~/Desktop/hosts",
            "-m", "ping",
            "-vvv"
        ],
        "console": "integratedTerminal"
    }
  ]
}
```

说明：笔者希望能调试 ansible all -i ~/Desktop/hosts -m ping -vvv 这个命令。其中，program 参数代表的是当前调试的文件，即想要执行的代码文件位于 Ansible 源码目录下的 bin 目录中。另外，Ansible 的源码均位于 lib/ansible 目录下，变量${workspaceRoot} 是项目的根目录。而 Ansible 源码的根目录希望从 lib 目录开始，这样导入语句 from ansible.runner import Runner 定位的其实就是源码目录 lib/ansible/runner 下的 Runner 类（位于 __init__.py 文件中）。

为了验证上述调试配置是否正确，可以在源码目录 lib/ansible/runner 下 __init__.py 文件中 Runner 类的初始化方法中打上断点，如果能断点至此处且释放后命令正常运行，则说明调试环境搭建成功。打开源码中 bin 目录下的 Ansible 文件，在该文件上按 F5 键后进入调试状态，此时程序成功断点在预定源码位置，如图 1-3 所示。

图 1-3　调试断点结果

此外，从图 1-3 中可以看到 VSCode 的控制台输出结果，具体如下：

```
[shen@localhost ansible-1.1]$ cd /home/shen/shencong/code/python/ansible-1.1 ;
```

```
/usr/bin/env /usr/bin/python /home/shen/.vscode/extensions/ms-python.
python-2020.12.424452561/pythonFiles/lib/python/debugpy/launcher 40953
-- /home/shen/shencong/code/python/ansible-1.1/bin/ansible all -i ~/Desktop/
hosts -m ping -vvv
```

接着释放断点，从控制台中可以看到最终命令成功执行，结果如图 1-4 所示。

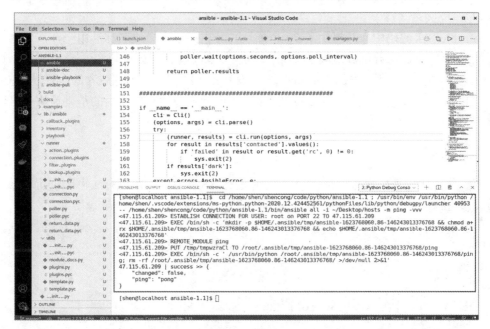

图 1-4　Ansible 源码调试结果

至此，整个 Ansible 的测试及调试环境搭建完毕，接下来将正式进入 Ansible 1.1 的使用和源码剖析阶段。

1.2　Ansible 1.1 的基本使用

1.1 节中介绍了如何在 Linux 环境下搭建 Ansible 1.1 的测试和调试环境，本节将进一步熟悉和掌握 Ansible 1.1 的基本功能，为后续的源码剖析打好基础。

1.2.1　Ansible 1.1 模块说明

在 1.1 节中提到过，Ansible 1.1 的模块文件均在源码的 library 目录下，但这些都是不完整的 Python 代码，无法直接运行。在 ansible 命令执行过程中，会将对应模块文件中的 Python 代码和一个公共的代码合并，最后形成可执行的 Python 代码并上传到远端主机执行。通过阅读 library 目录下的模块代码，可以很清楚地了解相应 Ansible 模块所支持的参数及该模块在远端主机上的执行动作，这是掌握该模块用法的最佳途径。在后续 Ansible 2 的模块学习中，也将沿着这一模式进行。

1.2.2　Ansible 1.1 的常用模块

Ansible 1.1 源码中一共提供了 79 个模块，其中的许多模块名称沿用至今，例如常用的 ping 模块、shell/command 模块、copy 模块、cron 模块、file 模块和 user 模块等。这些模块的基本功能和使用方法与 Ansible 2.8 并没有太大区别，只不过后者的内置模块更多，模块功能更为强大，且支持更多的参数控制。下面简单介绍一些常用的基础模块及这些模块的使用。

1．ping模块

ping 模块用于测试远端主机是否能正常进行 SSH 通信，使用方法也比较简单，只有一个 data 参数，用于填充返回 ping 字段的结果。

```
(ansible1.1) [root@master ~]# cat hosts
[nodes]
ceph-[1:3]

[nodes:vars]
ansible_ssh_user=root
ansible_ssh_pass=@SHENcong19920522
ansible_ssh_port=22

(ansible1.1) [root@master ~]# ansible ceph-1 -i hosts -m ping
ceph-1 | success >> {
    "changed": false,
    "ping": "pong"
}

(ansible1.1) [root@master ~]# ansible ceph-1 -i hosts -m ping -a "data=
'hello, ansible'"
ceph-1 | success >> {
    "changed": false,
    "ping": "hello, ansible"
}
```

说明：ansible 命令的使用非常简单。-i 选项表示指定目标主机的文件，此外，在该文件中也可以设置目标主机的一些变量信息等，该文件的默认路径为/etc/ansible/hosts；-m 选项后面跟使用的模块，如 ping 模块、copy 模块等；-a 选项后面跟模块参数，各参数之间用空格隔开，使用 key=value 形式赋值，如 copy 模块的参数书写为-a "src=/ect/hosts dest=/tmp/hosts"。ansible 命令的格式如下：

```
ansible <目标主机> [-i 指定目标主机的文件] [-m 模块] [-a 参数] [其他选项]
```

以下是 ping 模块的代码，该部分代码位于源码目录的 library 下的 ping 文件中。

```
# 源码位置: library/ping
# ...

def main():
    module = AnsibleModule(
        # 定义模块支持的参数
        argument_spec = dict(
            data=dict(required=False, default=None),
        ),
        supports_check_mode = True
    )
    # 初始化返回结果
    result = dict(ping='pong')
    if module.params['data']:
        result['ping'] = module.params['data']
    # 带上结果，退出
    module.exit_json(**result)

# this is magic, see lib/ansible/module_common.py
#<<INCLUDE_ANSIBLE_MODULE_COMMON>>
main()
```

ping 模块在远端主机上的执行动作为：从 ping 模块中取出 data 参数的值（如果存在）并放到结果的 ping 字段中，最后再将结果返回。ping 模块代码中涉及的 AnsibleModule 类将由 Ansible 1.1 中的公共代码（module_common.py）提供，这里不用过于深究。

2. shell模块

shell 模块可以说是 Ansible 中最常用的模块了，也是 Ansible 中不可缺少的模块。它用于对远端主机批量执行 shell 命令，支持管道写法。shell 模块提供的重要参数如下：

- creates：该参数值需要传入一个文件路径，如果指定的文件存在，则不执行 shell 模块；否则执行 shell 模块。
- removes：和 creates 参数正好相反，如果指定的文件存在，则执行 shell 模块；否则不执行 shell 模块。
- chdir：先进入该目录然后再执行 shell 命令。
- executable：选择执行 shell 命令的解释器，如/bin/sh 等。一般不用指定，使用默认值即可。

以下是关于 shell 模块的若干使用案例：

```
(ansible1.1) [root@master ~]# ansible ceph-1 -i hosts -m shell -a "hostname"
ceph-1 | success | rc=0 >>
ceph-1

(ansible1.1) [root@master ~]# ansible ceph-1 -i hosts -m shell -a "hostname
creates=/etc/hosts"
ceph-1 | skipped
(ansible1.1) [root@master ~]# ansible ceph-1 -i hosts -m shell -a "hostname
```

```
creates=/etc/not-exists"
ceph-1 | success | rc=0 >>
ceph-1

(ansible1.1) [root@master ~]# ansible ceph-1 -i hosts -m shell -a "hostname
removes=/etc/hosts"
ceph-1 | success | rc=0 >>
ceph-1

(ansible1.1) [root@master ~]# ansible ceph-1 -i hosts -m shell -a "hostname
removes=/etc/not-exist"
ceph-1 | skipped

(ansible1.1) [root@master ~]# ansible ceph-1 -i hosts -m shell -a "pwd"
ceph-1 | success | rc=0 >>
/root

(ansible1.1) [root@master ~]# ansible ceph-1 -i hosts -m shell -a "pwd
chdir=/tmp"
ceph-1 | success | rc=0 >>
/tmp
```

Ansible 中还有一个 command 模块，它和 shell 模块类似但不如 shell 模块的功能强大。command 模块在远程主机上执行命令时不会经过远程主机的 shell 处理，即在使用 command 模块时，如果待执行的命令中出现管道符（|）或重定向操作等，都会导致命令运行失败。例如以下操作：

```
(ansible1.1) [root@master ~]# ansible ceph-1 -m command -a "ps -ef | grep
sshd"
ERROR: Unable to find an inventory file, specify one with -i ?
(ansible1.1) [root@master ~]# ansible ceph-1 -i hosts -m command -a "ps -ef
| grep sshd"
ceph-1 | FAILED | rc=1 >>
error: garbage option

Usage:
 ps [options]

 Try 'ps --help <simple|list|output|threads|misc|all>'
  or 'ps --help <s|l|o|t|m|a>'
 for additional help text.

For more details see ps(1).
```

那究竟是什么原因造成 shell 和 command 模块的运行差别呢？通过查找 shell 模块的代码发现，shell 模块文件中仅有相关的模块说明而没有相关代码。通过后续的学习可知：shell 模块复用了 command 模块中的代码，但 Ansible 会在 shell 模块的参数中添加一个 #USE_SHELL 标识，用于标明本次执行的是 shell 模块；如果没有该标识，则会被认为是 command 模块。command 模块中的代码内容如下：

```
# 源码位置：library/command
# ...
```

```
DOCUMENTATION = '''

忽略文档说明，关于 command 的使用注意事项有详细的说明

'''

def main():

    # command 模块的参数提取比较特殊，不能直接通过 AnsibleModule 对象获取
    module = CommandModule(argument_spec=dict())

    # 判断是 shell 还是 command 模块
    shell = module.params['shell']
    chdir = module.params['chdir']
    executable = module.params['executable']
    args = module.params['args']
    creates = module.params['creates']
    removes = module.params['removes']

    if args.strip() == '':
        module.fail_json(rc=256, msg="no command given")

    if chdir:
        # 如果存在该参数，则调用 os.chdir()方法进入该目录
        os.chdir(os.path.expanduser(chdir))

    if creates:
        v = os.path.expanduser(creates)
        # 如果文件存在，则模块不执行，直接退出
        if os.path.exists(v):
            module.exit_json(...)

    if removes:
        v = os.path.expanduser(removes)
        # 如果文件不存在，则模块不执行，直接退出
        if not os.path.exists(v):
            module.exit_json(...)

    # shell 和 command 的区别：shell 信息会保存在 args 中并被传递到 module.run_
      command()方法
    if not shell:
        args = shlex.split(args)
    startd = datetime.datetime.now()

    rc, out, err = module.run_command(args, executable=executable)

    endd = datetime.datetime.now()
    delta = endd - startd

    if out is None:
        out = ''
    if err is None:
        err = ''
```

```
    module.exit_json(...)

# include magic from lib/ansible/module_common.py
#<<INCLUDE_ANSIBLE_MODULE_COMMON>>

# 只有 command 模块的参数需要这样单独处理，其他模块参数必须是 key=value 的形式
class CommandModule(AnsibleModule):

    # ...

    def _load_params(self):
        # 读取输入，返回字典信息及参数字符串
        args = MODULE_ARGS
        params = {}
        params['chdir'] = None
        params['creates'] = None
        params['removes'] = None
        params['shell'] = False
        params['executable'] = None
        if args.find("#USE_SHELL") != -1:
            args = args.replace("#USE_SHELL", "")
            params['shell'] = True

        # 忽略一些处理输入的代码
        # ...

        return (params, params['args'])

main()
```

笔者简化了 command 模块中的一些行并翻译了部分英文注释。对于有 Python 基础的读者而言，上述代码非常简单和清晰。从这里可以看到 chdir、creates 和 removes 三个参数的作用，具体可参考注释，这里不再说明。至于模块中得到的 shell 和 executable 参数，有如下关键代码：

```
if not shell:
    args = shlex.split(args)
rc, out, err = module.run_command(args, executable=executable)
```

上面的 args 就是模块参数中的 shell 命令。对于 command 模块，如 hostname -s 命令，会被 Python 内置的 shlex 模块切割处理，得到['hostname', '-s']这样的数组形式；而对于 shell 模块而言，输入的命令可以看作一个长字符串，如"ps -ef | grep sshd"。最后，args 值连同 executable 参数值一起传给 AnsibleModule 对象的 run_command()方法，并在当前主机上执行。下面继续深入学习 run_command()方法，它位于源码的 lib/ansible/module_common.py 中，具体代码如下：

```
# 源码路径: lib/ansible/module_common.py
# ...

MODULE_COMMON = """
```

```
# 忽略部分代码

class AnsibleModule(object):

    # ...

    def run_command(self, args, check_rc=False, close_fds=False, executable=
    None, data=None):
        # 处理 args 参数类型
        if isinstance(args, list):
            shell = False
        elif isinstance(args, basestring):
            shell = True
        else:
            msg = "Argument 'args' to run_command must be list or string"
            self.fail_json(rc=257, cmd=args, msg=msg)
        rc = 0
        msg = None
        st_in = None
        if data:
            st_in = subprocess.PIPE
        try:
            # 调用 Python 内置的 subprocess 模块去执行 shell 命令
            cmd = subprocess.Popen(args,
                                executable=executable,      # 执行 shell 命令的
                                                             解释器
                                shell=shell,                # shell 命令
                                close_fds=close_fds,
                                stdin=st_in,                # 输入
                                stdout=subprocess.PIPE,     # 输出
                                stderr=subprocess.PIPE)     # 错误输出
            if data:
                # 如果有 data 数据，则通过 stdin 写入
                cmd.stdin.write(data)
                cmd.stdin.write('\\n')
            # 执行命令并获取结果，rc 为本次命令执行返回的状态码
            out, err = cmd.communicate()
            rc = cmd.returncode
        except (OSError, IOError), e:
            self.fail_json(rc=e.errno, msg=str(e), cmd=args)
        except:
            self.fail_json(rc=257, msg=traceback.format_exc(), cmd=args)
        if rc != 0 and check_rc:
            msg = err.rstrip()
            self.fail_json(cmd=args, rc=rc, stdout=out, stderr=err, msg=msg)
        return (rc, out, err)
    # ...
"""
```

这里的 run_command() 方法是通过 subprocess 模块执行 shell 命令的，该方法通过传入的命令参数（args）判断是否需要 shell。如果 args 为数组形式，则为 command 模块，设置 shell 参数为 False；否则为 shell 模块，设置 shell 参数为 True。subprocess 模块能否支

持 ps -ef | grep sshd 这样带管道或重定向符的命令，依赖于 subprocess.Popen()方法中输入的 shell 参数。关于这些参数的更详细的介绍，可以直接参考 Python 的官方文档，因为 subprocess 模块是 Python 内置的标准模块。

此外，可以在 command 模块的 main()方法的最后一个 module.exit_json()中添加额外的输出信息，用于检验上面分析过程的正确性，具体代码如下：

```python
# 源码位置：library/command
# ...

def main():
    # ...

    module.exit_json(
        cmd     = args,
        stdout  = out.rstrip("\r\n") + "\nshell={}, args={}".format(shell, args),
        stderr  = err.rstrip("\r\n"),
        rc      = rc,
        start   = str(startd),
        end     = str(endd),
        delta   = str(delta),
        changed = True
    )
```

对应修改/root/.pyenv/versions/2.7.18/envs/ansible1.1/share/ansible/command（1.1 节中介绍过 Ansible 模块的位置）中的相应代码，然后再次运行 ansible 命令，具体代码如下：

```
(ansible1.1) [root@master ~]# ansible ceph-1 -i hosts -m command -a "hostname
-s  removes=/etc/hosts"
ceph-1 | success | rc=0 >>
ceph-1
shell=False, args=['hostname', '-s']

(ansible1.1) [root@master ~]# ansible ceph-1 -i hosts -m shell -a "ps -ef
| grep ssh  removes=/etc/hosts"
ceph-1 | success | rc=0 >>
root      1155       1  0 04:37 ?  00:00:00 /usr/sbin/sshd -D
root     10728    1155  0 15:33 ?  00:00:00 sshd: store [priv]
store    10735   10728  0 15:33 ?  00:00:00 sshd: store@pts/0
root     15087    1155  0 20:39 ?  00:00:00 sshd: root@notty
root     15099   15087  0 20:39 ?  00:00:00 /usr/libexec/openssh/sftp-server
root     15118   15117  0 20:39 ?  00:00:00 /bin/sh -c ps -ef | grep ssh
root     15120   15118  0 20:39 ?  00:00:00 grep ssh
shell=True, args=ps -ef | grep ssh
```

从上面的打印结果中可以看到 command 模块与 shell 模块代码中有关 shell 参数的值，该参数决定了这两个模块的不同之处。

3. copy模块

copy 模块也是 Ansible 中的一个常用模块，使用频率仅次于 shell 模块。通常，copy 模块用于向多个远端主机下发文件，支持强制覆盖、文件备份、修改文件权限及属主等。

下面是 copy 模块的一些使用案例。

```
# 使用 copy 模块之前的 ceph-1 主机
[root@ceph-1 ~]# ls
anaconda-ks.cfg

# 在 master 主机上新建一个测试文件
(ansible1.1) [root@master ~]# cat test_copy.txt
测试 Ansible 的 copy 模块，上传

(ansible1.1) [root@master ~]# ansible ceph-1 -i hosts -m copy -a "src=
test_copy.txt dest=~/test_copy.txt"
ceph-1 | success >> {
    "changed": true,
    "dest": "/root/test_copy.txt",
    "group": "root",
    "md5sum": "8b8437beb7107dee441dbb9cb4ca9520",
    "mode": "0644",
    "owner": "root",
    "size": 36,
    "src": "/root/.ansible/tmp/ansible-1602680244.85-189392735288190/source",
    "state": "file"
}

# 执行 copy 模块后，再次查看 ceph-1 主机
[root@ceph-1 ~]# ls
anaconda-ks.cfg  test_copy.txt
[root@ceph-1 ~]# cat test_copy.txt
测试 Ansible 的 copy 模块，上传
```

在 copy 模块中，src 参数表示本地文件路径，dest 参数表示上传文件到目标主机的位置，该值可以是文件全路径，也可以是目录。如果 dest 是目录，此时上传的文件名默认为 src 中的文件名。另外，backup 和 force 参数则表示是否进行文件备份和是否强制覆盖原文件。下面是关于 force 参数的一个示例。

```
# 目标主机上存在 test_copy.txt 文件
[root@ceph-1 ~]# ls
anaconda-ks.cfg  test_copy.txt
[root@ceph-1 ~]#

# master 节点
(ansible1.1) [root@master ~]# ansible ceph-1 -i hosts -m copy -a "src=
test_copy.txt dest=~ force=no"
ceph-1 | success >> {
    "changed": false,
    "dest": "/root",
    "group": "root",
    "mode": "0550",
    "msg": "file already exists",
    "owner": "root",
    "size": 220,
    "src": "/root/.ansible/tmp/ansible-1602682077.88-25716069795499/source",
    "state": "directory"
}
```

在 Ansible 的公共代码中有如下设置：

```
MODULE_COMMON = """
# ...

BOOLEANS_TRUE = ['yes', 'on', '1', 'true', 1]
BOOLEANS_FALSE = ['no', 'off', '0', 'fyesalse', 0]

# ...

"""
```

即字面值'yes'、'on'、'1'、'true'、1 都会被解析成 Python 中的 True，而'no'、'off'、'0'、'fyesalse'、0 则都会被解析成 Python 中的 False，这些技巧在编写 Playbook 文件时非常有用。

再回到 copy 模块，当 force 值设置为字面值的 False 后，copy 模块会返回文件已经存在的信息且不会强制替换。force 参数默认为 yes。以下是关于 backup 参数的一个例子。

```
(ansible1.1) [root@master ~]# ansible ceph-1 -i hosts -m copy -a "src=
test_copy.txt dest=/root/test_copy.txt backup=yes"
ceph-1 | success >> {
    "changed": false,
    "group": "root",
    "mode": "0644",
    "owner": "root",
    "path": "/root/test_copy.txt",
    "size": 36,
    "state": "file"
}

# 主机 ceph-1 上并没有备份文件
[root@ceph-1 ~]# ls
anaconda-ks.cfg  test_copy.txt
```

可以看到上面的 backup 参数并没有生效，这是什么原因呢？修改 test_copy.txt 文件中的内容再试一次：

```
(ansible1.1) [root@master ~]# cat test_copy.txt
测试 Ansible 的 copy 模块，上传

新的版本

(ansible1.1) [root@master ~]# ansible ceph-1 -i hosts -m copy -a "src=
test_copy.txt dest=/root/test_copy.txt backup=yes"
ceph-1 | success >> {
    "backup_file": "/root/test_copy.txt.2020-10-20@23:24~",
    "changed": true,
    "dest": "/root/test_copy.txt",
    "group": "root",
    "md5sum": "b98cff05b477c23067bf6eb2c2ddab4f",
    "mode": "0644",
    "owner": "root",
    "size": 50,
```

```
        "src": "/root/.ansible/tmp/ansible-1602682811.7-79801427392161/source",
        "state": "file"
}

# 查看 ceph-1 主机，备份文件已经生成
[root@ceph-1 ~]# ls
anaconda-ks.cfg  test_copy.txt  test_copy.txt.2020-10-20@23:24~
```

从上面的演示结果中可以看到，修改源文件后再执行 copy 模块，backup 参数生效。这一现象说明，Ansible 会对比本地文件和远端文件，如果文件相同，则不会做任何操作便直接返回。后面在 copy 模块的源码中可以清楚地看到这个比较过程。

接下来看 copy 的另一个参数 content，该参数和 src 参数是互斥的。content 参数的内容将写入 dest 参数指定的文件中，请看如下示例代码：

```
(ansible1.1) [root@master ~]# ansible ceph-1 -i hosts -m copy -a "dest=
/tmp/test_content.txt content=xxxxxxx"
ceph-1 | success >> {
    "changed": true,
    "dest": "/tmp/test_content.txt",
    "group": "root",
    "md5sum": "04adb4e2f055c978c9bb101ee1bc5cd4",
    "mode": "0644",
    "owner": "root",
    "size": 7,
    "src": "/root/.ansible/tmp/ansible-1602681193.07-232005738125574/source",
    "state": "file"
}

# 在 ceph-1 主机上查看上传文件结果
[root@ceph-1 ~]# ls /tmp/
test_content.txt  test_copy.txt
[root@ceph-1 ~]# cat /tmp/test_content.txt
xxxxxxx[root@ceph-1 ~]#
```

接下来看 content 参数的另一个例子，此时的 copy 模块中隐藏着一个不能满足需求的情况，具体代码如下：

```
(ansible1.1) [root@master ~]# ansible ceph-1 -i hosts -m copy -a 'dest=
/tmp/test_content.txt content="xxxxxxx\n"'
ceph-1 | success >> {
    "changed": false,
    "group": "root",
    "mode": "0644",
    "owner": "root",
    "path": "/tmp/test_content.txt",
    "size": 9,
    "state": "file"
}

# 在目标主机上发现换行符没有生效，这算不算 Ansible 1.1 的一个 Bug？
[root@ceph-1 ~]# cat /tmp/test_content.txt
xxxxxxx\n[root@ceph-1 ~]#
```

从上面的例子中可以看到，当 content 参数值中包含转义符如\n 时，Ansible 的 copy 模块无法对其进行转义，而是原封不动地输出。那么如何才能得到想要的结果呢？这个问题会在 1.6 节介绍 copy 模块执行流程时进行解答，在 1.6 节里将通过改造相关代码来实现这里想要的结果。接下来将分析 copy 模块的代码，厘清该模块的执行逻辑。copy 模块的代码位于源码目录下的 library/copy 文件中，具体代码如下：

```
# 代码位置：library/copy
import os
import shutil
import time

def main():

    # 获取通用的 Ansible 模块对象，简化写法
    module = AnsibleModule(...)

    # 提取相关参数
    src    = os.path.expanduser(module.params['src'])
    dest   = os.path.expanduser(module.params['dest'])
    backup = module.params['backup']
    force  = module.params['force']
    original_basename = module.params.get('original_basename',None)

    # 如果 src 文件不存在或者不可读，则直接返回错误信息
    if not os.path.exists(src):
        module.fail_json(msg="Source %s failed to transfer" % (src))
    if not os.access(src, os.R_OK):
        module.fail_json(msg="Source %s not readable" % (src))

    # 计算源文件的 MD5 值
    md5sum_src = module.md5(src)
    md5sum_dest = None

    if os.path.exists(dest):
        # 如果 dest 存在，则开始考虑 force 参数，如果 force 设置为 no，则直接退出，
          同时提示文件存在
        if not force:
            module.exit_json(msg="file already exists", src=src, dest=dest,
changed=False)
        # 对于目录情况，可以配合 original_basename 参数，生成最后的目标文件路径
        if (os.path.isdir(dest)):
            basename = os.path.basename(src)
            if original_basename:
                basename = original_basename
            dest = os.path.join(dest, basename)
        # 如果可读，计算原目标文件的 MD5 值
        if os.access(dest, os.R_OK):
            md5sum_dest = module.md5(dest)
    else:
        # 对于目标文件所在目录也不存在的情况，直接返回错误信息
        if not os.path.exists(os.path.dirname(dest)):
```

```
            module.fail_json(msg="Destination directory %s does not exist"
% (os.path.dirname(dest)))
    # 对于目标文件所在目录不可写的情况，直接返回错误信息
    if not os.access(os.path.dirname(dest), os.W_OK):
        module.fail_json(msg="Destination %s not writable" % (os.path.
dirname(dest)))

    backup_file = None
    if md5sum_src != md5sum_dest or os.path.islink(dest):
        # 如果源文件和目标文件的 MD5 值不同，才执行后续的动作
        try:
            if backup:
                # 如果设置了备份项并且目标文件存在，则先备份目标文件
                if os.path.exists(dest):
                    backup_file = module.backup_local(dest)
            # allow for conversion from symlink.
            if os.path.islink(dest):
                os.unlink(dest)
                open(dest, 'w').close()
            # TODO:pid + epoch should avoid most collisions, hostname/mac for
              those using nfs?
            # might be an issue with exceeding path length
            dest_tmp = "%s.%s.%s.tmp" % (dest,os.getpid(),time.time())
            # 最后的核心操作就是使用 shutil.copyfile()方法将源文件复制一份，得到临
              时文件
            shutil.copyfile(src, dest_tmp)
            # 将生成的临时文件替换成最终的目标文件，原子替换方式
            module.atomic_replace(dest_tmp, dest)
        except shutil.Error:
            module.fail_json(msg="failed to copy: %s and %s are the same" %
(src, dest))
        except IOError:
            module.fail_json(msg="failed to copy: %s to %s" % (src, dest))
        changed = True
    else:
        # 如果源文件和目标文件是同一个文件，设置 changed=False 后直接返回
        changed = False

    # 最后的返回结果
    res_args = dict(
        dest = dest, src = src, md5sum = md5sum_src, changed = changed
    )
    # 如果有备份文件，则写入备份的文件名
    if backup_file:
        res_args['backup_file'] = backup_file

    module.params['dest'] = dest
    # 获取文件信息
    file_args = module.load_file_common_arguments(module.params)
    res_args['changed'] = module.set_file_attributes_if_different(file_
args, res_args['changed'])

    module.exit_json(**res_args)
```

```
# this is magic, see lib/ansible/module_common.py
#<<INCLUDE_ANSIBLE_MODULE_COMMON>>
main()
```

上面的代码中给出了详细的注释，整个 copy 模块的代码清晰、明了，并且实现过程对新手非常友好，没有任何复杂的写法。copy 模块中的代码主要是将源文件移动到目的路径上，即在 copy 模块代码执行之前，本地的文件就已经被上传到了目标主机的某个临时目录下。因此，如果 content 参数中出现无法正确转义的情况，便不能在 copy 模块中修改了，只能在生成该源文件的地方进行修改。为此，可以在获取 src 值的代码后面加上调试信息，打印源文件位置及源文件内容，方便进行查看与分析，具体如下：

```
# 代码位置: library/copy
# ...

def main():

    # ...

    # 提取相关参数
    src    = os.path.expanduser(module.params['src'])

    # 打印调试信息
    debug_data = {
      'src': src,
      'content': open(src, 'rb').read()
    }
    # 直接返回，不进行后面的操作
    module.exit_json(**debug_data)

    # ...
```

修改测试环境的路径为/root/.pyenv/versions/2.7.18/envs/ansible1.1/share/ansible/copy，然后再来看 content 参数的示例，具体如下：

```
# 先删除 ceph-1 上的目标文件，再执行改动的 copy 模块
[root@ceph-1 ~]# rm -f /tmp/test_content.txt

(ansible1.1) [root@master ~]# ansible ceph-1 -i hosts -m copy -a 'dest=
/tmp/test_content.txt content="xxxxxxx\n"'
ceph-1 | success >> {
    "changed": false,
    "content": "xxxxxxx\\n",
    "src": "/root/.ansible/tmp/ansible-1602686551.95-120685118068937/source"
}
```

🔔 **说明：** 在执行上述指令之前，应先将原先生成的/tmp/test_content.txt 文件删掉，否则 copy 模块代码不会执行。

从上面的结果中可以看到，在执行 copy 模块之前，内容为 content 参数值的源文件已经生成且内容中的转义字符前被添加了一个斜线。

4．file模块

file 模块也是一个常用的模块，它能帮助使用者设置远端文件、系统链接、目录的属性，也可以创建和删除系统链接或目录，但是该模块无法创建 Linux 上的普通文件。file 模块支持的参数如下：

- path：操作的文件对象，可以是文件或者目录。如果对象是系统链接（state=link），还需要设置 src 参数后才能创建该系统链接。
- state：有 4 个选项值，分别为 file、link、directory 和 absent，默认为 file。当该值为 absent 时，表示删除 path 对应的文件。
- mode：设置文件或者目录的权限值，如 0644。
- owner/group：设置操作文件的所有者和所属组。
- src：只有在 state=link 时才有效，用于设置系统链接。
- seuser/serole/setype/selevel/context：和 selinux 设置相关，一般较少用到。
- recurse：递归操作，主要在 state=directory 时有效。

下面是关于 file 模块的一个简单示例，结果表明 file 模块能创建目录，但是无法创建普通文件。

```
# 操作 ceph-1 节点，清空目录下的文件
[root@ceph-1 ~]# ls
[root@ceph-1 ~]#

# 在 master 节点上开始操作，创建普通文件失败
(ansible1.1) [root@master ~]# ansible ceph-1 -i hosts -m file -a "path=
~/test_file.txt state=file mode=600"
ceph-1 | FAILED >> {
    "failed": true,
    "msg": "file (/root/test_file.txt) does not exist, use copy or template
module to create",
    "path": "/root/test_file.txt",
    "state": "absent"
}
# 创建目录成功
(ansible1.1) [root@master ~]# ansible ceph-1 -i hosts -m file -a "path=
~/test_file state=directory mode=600"
ceph-1 | success >> {
    "changed": true,
    "group": "root",
    "mode": "0600",
    "owner": "root",
    "path": "/root/test_file",
    "size": 6,
    "state": "directory"
}

# 查看 ceph-1 节点，mode 参数也生效了
[root@ceph-1 ~]# ls -ld test_file/
```

```
drw------- 2 root root 6 Oct 21 14:11 test_file/
```

在基本了解了 file 模块的使用后，可以继续阅读 file 模块中的代码，具体如下（笔者注释掉了与 selinux 相关的代码以方便阅读）：

```python
# 源码位置：library/file

import shutil
import stat
import grp
import pwd
try:
    # 判断是否有 selinux 模块
    import selinux
    HAVE_SELINUX=True
except ImportError:
    HAVE_SELINUX=False

# 模块说明，使用该模块前请仔细阅读
# ...

def main():

    # 不应该是全局变量
    global module

    # 模块的参数信息，其中 state 只有 4 个选项值，分别为 file、directory、link 和
      absent，默认为 file
    module = AnsibleModule(...)

    params = module.params
    state  = params['state']
    # 这里就是可以写~/xxx.txt 的原因
    params['path'] = path = os.path.expanduser(params['path'])

    # ...

    src = params.get('src', None)
    if src:
        src = os.path.expanduser(src)

    # src 不为空的另一种用法是：src 有值，path 为目录且 state 不等于 link 时，就可以得
      到完整的 path 值
    if src is not None and os.path.isdir(path) and state != "link":
        params['path'] = path = os.path.join(path, os.path.basename(src))

    file_args = module.load_file_common_arguments(params)

    # state 的值为 link 时，必须要有 src 和 path，否则直接返回错误信息
    if state == 'link' and (src is None or path is None):
        module.fail_json(msg='src and dest are required for "link" state')
    # 其他情况必须要有 path
    elif path is None:
        module.fail_json(msg='path is required')
```

```
changed = False

prev_state = 'absent'

# 检查 path 文件是否存在，使用的是 lexists()方法，这样即使软链接失效了，该方法也
  会返回 True；而 exists()方法对于失效的软链接返回 False
if os.path.lexists(path):
    if os.path.islink(path):
        prev_state = 'link'
    elif os.path.isdir(path):
        prev_state = 'directory'
    else:
        prev_state = 'file'

if prev_state != 'absent' and state == 'absent':
    try:  # 删除 link、directory 和 file 三种对象
        if prev_state == 'directory':
            if os.path.islink(path):
                os.unlink(path)                # 链接文件，直接删除
            else:
                try:
                    # shutil.rmtree()方法将递归删除文件夹下的所有子文件夹和子文件
                    shutil.rmtree(path, ignore_errors=False)
                except:
                    module.exit_json(msg="rmtree failed")
        else:
            # 对于普通文件和链接文件，直接使用 unlink()方法删除即可
            os.unlink(path)
    except Exception, e:
        module.fail_json(path=path, msg=str(e))
    module.exit_json(path=path, changed=True)

if prev_state != 'absent' and prev_state != state:
    # 主动探测的文件类型和模块设置的 state 值不一致
    module.fail_json(path=path, msg='refusing to convert between %s and
%s for %s' % (prev_state, state, src))

if prev_state == 'absent' and state == 'absent':
    # 其他类型的文件不处理，直接返回
    module.exit_json(path=path, changed=False)

if state == 'file':

    if prev_state != 'file':
        # 对于文件不存在的情况，将跳转到这里，输出相应的错误信息并返回
        module.fail_json(path=path, msg='file (%s) does not exist, use
copy or template module to create' % path)

    # 设置文件信息，如所属用户、所属组及 mode 值等
    changed = module.set_file_attributes_if_different(file_args, changed)
    module.exit_json(path=path, changed=changed)

elif state == 'directory':
```

```
        # 对于目录
        if prev_state == 'absent':
            # 如果目录不存在，则调用 os.makedirs() 方法创建，从这里的调用方法可知 file
              模块是支持多级目录创建的
            os.makedirs(path)
            changed = True

        changed = module.set_directory_attributes_if_different(file_args,
changed)
        recurse = params['recurse']
        if recurse:
            # 如果设置了递归参数，则用常规的方法来遍历目录
            for root,dirs,files in os.walk( file_args['path'] ):
                # root 为本次遍历的目录，dirs 为本次遍历目录下的所有目录，files 为本
                  次遍历目录下的所有文件
                for dir in dirs:
                    dirname=os.path.join(root,dir)
                    tmp_file_args = file_args.copy()
                    tmp_file_args['path']=dirname
                    # 统一修改子目录的信息，如所属用户、所属组及 mode 值等
                    changed = module.set_directory_attributes_if_different
(tmp_file_args, changed)
                for file in files:
                    filename=os.path.join(root,file)
                    tmp_file_args = file_args.copy()
                    tmp_file_args['path']=filename
                    # 统一修改目录下的文件信息，如所属用户、所属组及 mode 值等
                    changed = module.set_file_attributes_if_different(tmp_
file_args, changed)
        module.exit_json(path=path, changed=changed)

    elif state == 'link':

        if os.path.isabs(src):
            abs_src = src
        else:
            module.fail_json(msg="absolute paths are required")
        if not os.path.exists(abs_src):
            module.fail_json(path=path, src=src, msg='src file does not exist')

        if prev_state == 'absent':
            # 如果没有该链接则创建链接
            os.symlink(src, path)
            changed = True
        elif prev_state == 'link':
            old_src = os.readlink(path)
            if not os.path.isabs(old_src):
                # 获取 path 路径下的链接地址
                old_src = os.path.join(os.path.dirname(path), old_src)
            if old_src != src:
                # 如果需要链接的地址和新的地址不一致，需要先删除原链接，然后再链接到新
                  的路径上
                os.unlink(path)
```

```
                os.symlink(src, path)
                changed = True
        else:
            module.fail_json(dest=path, src=src, msg='unexpected position
reached')

        # 设置权限、属主及上下文数据
        # ...

        module.exit_json(dest=path, src=src, changed=changed)

    module.fail_json(path=path, msg='unexpected position reached')

# this is magic, see lib/ansible/module_common.py
#<<INCLUDE_ANSIBLE_MODULE_COMMON>>
main()
```

file 模块的代码并不多，整个模块的执行流程也非常清晰。同 copy 模块一样，笔者也为该模块做了详细的注释，有 Python 基础的读者基本都能理解这些代码的含义。从 file 模块的代码中能看到，file 模块并不支持文件创建。如果想让 file 模块支持这个功能，需要怎么调整代码呢？笔者对其进行了如下调整：

```
# 源码位置: library/file
# ...

def main():
    # ...

    if state == 'file':

        if prev_state != 'file':
            # 注释掉没有 path 文件时就退出的代码
            # module.fail_json(path=path, msg='file (%s) does not exist, use
copy or template module to create' % path)
            # 新增没有 path 文件时就创建一个空文件的代码。如果上一级目录不存在，则直接
              创建上级目录
            base_dir = os.path.dirname(path) if os.path.dirname(path) else
os.getcwd()
            if not os.path.exists(base_dir):
                os.makedirs(os.path.dirname(path))
            with open(path, 'a'):
                os.utime(path, None)
        # ...
    elif state == 'directory':
        # ...
    elif state == 'link':
        # ...

    # ...
```

同步上述改动后的代码到虚拟环境中，修改文件地址如下：

```
/root/.pyenv/version/2.7.18/envs/ansible1.1/share/ansible/file
```

如何查看修改后的效果呢？笔者给出了如下操作示例：

```
# 先到 ceph-1 节点确认 /root/test_file 目录下没有相应文件
[root@ceph-1 ~]# ls test_file/
[root@ceph-1 ~]#

# 回到 master 节点的虚拟环境中执行 file 模块创建文件
(ansible1.1) [root@master ~]# ansible ceph-1 -i hosts -m file -a "path=
~/test_file/test_create_file.txt mode=660"
ceph-1 | success >> {
    "changed": true,
    "group": "root",
    "mode": "0660",
    "owner": "root",
    "path": "/root/test_file/test_create_file.txt",
    "size": 0,
    "state": "file"
}

(ansible1.1) [root@master ~]# ansible ceph-1 -i hosts -m file -a "path=
~/test_file/1/2/test_create_file.txt mode=600"
ceph-1 | success >> {
    "changed": true,
    "group": "root",
    "mode": "0600",
    "owner": "root",
    "path": "/root/test_file/1/2/test_create_file.txt",
    "size": 0,
    "state": "file"
}

# 在 ceph-1 节点上可以清楚地看到几个测试的文件已经创建了
[root@ceph-1 ~]# yum install tree
[root@ceph-1 ~]# tree .
.
└── test_file
    ├── 1
    │   └── 2
    │       └── test_create_file.txt
    └── test_create_file.txt

3 directories, 2 files
[root@ceph-1 ~]# ll test_file/test_create_file.txt
-rw-rw---- 1 root root 0 Oct 21 16:32 test_file/test_create_file.txt
[root@ceph-1 ~]# ll test_file/1/2/test_create_file.txt
-rw------- 1 root root 0 Oct 21 16:34 test_file/1/2/test_create_file.txt
```

可以看到，改造后的 file 模块实现了想要的功能。对于一些特殊的需求，都可以按上面的方式对 Ansible 进行改造。另外一种途径就是编写自己的 Ansible 模块实现特定的需求，这样可以不用修改 Ansible 的内置模块。

1.2.3　编写 Ansible 1.1 的 Playbook

学习完 1.2.2 节介绍的 Ansible 的一些常用模块后，就可以试着编写一些简单的 Playbook 文件了。这里首先介绍 Playbook 的一些基本语法，然后通过实例进一步加深读者对 Playbook 的理解。

```
(ansible1.1) [root@master ~]# cat test_playbook1.yml
# 注意，这里的 hosts 正常应该与下面的 gather_facts 对齐，在代码编辑器中显示正常，
# 但是在这里略显不齐，后续笔者不再说明相关问题
- hosts: nodes
  gather_facts: no
  tasks:
    - name: run first shell
      shell: "cat /etc/hosts | grep `hostname` | awk '{print $1}'"
      register: shell_out

    - name: debug local ip
      debug:
        msg: "hostname=${inventory_hostname}, ip=${shell_out.stdout}"

    - name: create directory
      file:
        path: /tmp/test_dir
        state: directory
```

上面的 Playbook 文件是按照 YAML 格式进行书写的，称之为剧本集，这是 Ansible 解析 Playbook 的标准格式。该 Playbook 中只有一个 Play（剧本）且第一层属性有 hosts、gather_facts 和 tasks。其中，第一个 hosts 指定该 Play 中所有任务运行的主机组或者主机，tasks 值代表该 Play 的任务集，每一个任务只能执行 Ansible 的一个模块。例如上面有 3 个任务，分别调用了 Ansible 的 shell、debug 和 file 模块，这些任务的书写规则和模块息息相关。因此，想要学好 Ansible Playbook 的编写方法，首先要熟练使用 Ansible 模块，其次是掌握 Playbook 书写中的一些特殊语法。例如第一个任务中的 register，表示将 shell 模块执行的结果放到一个字典结果中并赋给 shell_out 变量，而后续的任务中可以通过 ${shell_out.stdout}得到上一个 shell 模块执行的结果。在 Ansible 1.1 中，变量的使用是${变量名}形式，而在后续的版本中已经全部改为{{ 变量名 }}形式了。上述 Playbook 的执行结果如下：

```
(ansible1.1) [root@master ~]# ansible-playbook -i hosts test_playbook1.yml

PLAY [nodes] *********************

TASK: [run first shell] *********************
changed: [ceph-3]
changed: [ceph-2]
changed: [ceph-1]
```

```
TASK: [debug local ip] *********************
ok: [ceph-1]
ok: [ceph-2]
ok: [ceph-3]

TASK: [create directory] *********************
ok: [ceph-2]
ok: [ceph-3]
ok: [ceph-1]

PLAY RECAP *********************
ceph-1          : ok=3   changed=1   unreachable=0   failed=0
ceph-2          : ok=3   changed=1   unreachable=0   failed=0
ceph-3          : ok=3   changed=1   unreachable=0   failed=0
```

由于 Ansible 1.1 的调试模块（debug）存在一些问题，这里本是想将第一个 shell 命令的结果打印出来，结果却没有显示。这一问题在 Ansible 1.2 中已经修复。为了在 Ansible 1.1 中也能看到结果，可以加上-v 选项（也可以是-vv、-vvv 选项）打印出详细的执行信息，具体如下：

```
TASK: [run first shell] *********************
changed: [ceph-3] => {"changed": true, "cmd": "cat /etc/hosts | grep
`hostname` | awk '{print $1}' ", "delta": "0:00:00.009521", "end": "2020-
10-21 19:43:32.705166", "rc": 0, "start": "2020-10-21 19:43:32.695645",
"stderr": "", "stdout": "192.168.26.122\nshell=True, args=cat /etc/hosts
| grep `hostname` | awk '{print $1}' "}
changed: [ceph-2] => {"changed": true, "cmd": "cat /etc/hosts | grep
`hostname` | awk '{print $1}' ", "delta": "0:00:00.007132", "end": "2020-
10-21 19:43:32.723753", "rc": 0, "start": "2020-10-21 19:43:32.716621",
"stderr": "", "stdout": "192.168.26.121\nshell=True, args=cat /etc/hosts
| grep `hostname` | awk '{print $1}' "}
changed: [ceph-1] => {"changed": true, "cmd": "cat /etc/hosts | grep
`hostname` | awk '{print $1}' ", "delta": "0:00:00.016025", "end": "2020-
10-21 19:43:32.775012", "rc": 0, "start": "2020-10-21 19:43:32.758987",
"stderr": "", "stdout": "192.168.26.120\nshell=True, args=cat /etc/hosts
| grep `hostname` | awk '{print $1}' "}

TASK: [debug local ip] *********************
ok: [ceph-3] => {"msg": "hostname=ceph-3, ip=192.168.26.122\nshell=True,
args=cat /etc/hosts | grep `hostname` | awk '{print $1}' "}
ok: [ceph-2] => {"msg": "hostname=ceph-2, ip=192.168.26.121\nshell=True,
args=cat /etc/hosts | grep `hostname` | awk '{print $1}' "}
ok: [ceph-1] => {"msg": "hostname=ceph-1, ip=192.168.26.120\nshell=True,
args=cat /etc/hosts | grep `hostname` | awk '{print $1}' "}

TASK: [create directory] *********************
ok: [ceph-3] => {"changed": false, "group": "root", "mode": "0755", "owner":
"root", "path": "/tmp/test_dir", "size": 6, "state": "directory"}
ok: [ceph-2] => {"changed": false, "group": "root", "mode": "0755", "owner":
"root", "path": "/tmp/test_dir", "size": 6, "state": "directory"}
ok: [ceph-1] => {"changed": false, "group": "root", "mode": "0755", "owner":
"root", "path": "/tmp/test_dir", "size": 6, "state": "directory"}

PLAY RECAP *********************
```

```
ceph-1        : ok=3    changed=1    unreachable=0    failed=0
ceph-2        : ok=3    changed=1    unreachable=0    failed=0
ceph-3        : ok=3    changed=1    unreachable=0    failed=0
```

可以看到，加上-v 选项，可以打印出更多的模块执行日志，方便查看和分析模块的运行过程。

1.2.4　小结

本节介绍了 Ansible 1.1 中的几个常用模块，首先从模块的功能开始分析，其次对模块的源码进行分析，最后面对一些不完善或者可以改进的地方，通过改造模块代码实现了相应需求。这种介绍方式可以让读者更好地掌握 Ansible 模块。Ansible 1.1 的其他常用模块如 user、lineinfile 和 service 等，由于参数和代码较多，不适合在这里展开，有兴趣的读者可以自行学习和研究。此外，本节还介绍了 Ansible 中的 Playbook、Play 和 Task 等概念，最后用一个简单的案例进行了演示。

1.3　Ansible 1.1 源码概览

Ansible 是一个非常典型的开源项目，其项目结构、代码风格及注释的书写等十分规范。Ansible 的创始人 Michael DeHanan 更是在 IT 圈内家喻户晓，以至于 Ansible 一经发布就能在各国的运维人员之间迅速流行。本节将介绍 Ansible 1.1 的大部分源码文件，让读者对 Ansible 的源码有一个初步的了解，为后续的源码分析做好准备。

1.3.1　非源码文件与目录

如图 1-5 所示为笔者在 VSCode 中导入的 Ansible 1.1 源码，其中，lib/ansible 下的所有文件为 Ansible 1.1 的源代码内容，其他则是一个完整 Python 项目所需要包含的文件，例如项目的说明文件 README.md、项目的文档手册（全部位于 docs 目录下）、项目提供的命令（ansible 和 ansible-playbook 等命令）及 Ansible 1.1 项目包含的所有模块文件（library 目录）等。

在源码包的 examples 目录下提供了一些简单的 Playbook 示例，可以帮助开发者迅速掌握 Playbook 文件的编写方法。此外，setup.py 是 Ansible 项目的构建工具，通过在该源码包目录下执行 python setup.py install 命令，就可以在当前 Linux 系统中成功安装 Ansible 1.1 工具。

图 1-5　Ansible 1.1 的完整源码

1.3.2 源码文件

Ansible 项目的代码均位于 lib/ansible 目录下，下面大致浏览一下 Ansible 1.1 的源码目录结构，如图 1-6 所示。

下面对一些非常重要的文件或者目录进行说明。

1. module_common.py

module_common.py 文件在 1.2.2 节中提到过，其中包含 Ansible 模块的公共代码部分，这些内容会在后续和 Ansible 中的模块代码合并，组成一个完整可运行的 Python 文件，最后被传送到远端主机上执行。

2. errors.py

errors.py 文件中定义了 Ansible 的全局异常，这些异常继承自 Exception 类，并没有其他额外操作。

3. constants.py

图 1-6　lib/ansible 目录下的源码结构

constants.py 文件比较关键，下面会针对该文件的代码内容进行说明。请看如下的代码片段：

```
# 源码位置：lib/ansible/constants.py

# 忽略一些模块导入
# ...

def get_config(p, section, key, env_var, default):
    if env_var is not None:
        value = os.environ.get(env_var, None)    # 获取环境变量
        if value is not None:
            return value
    if p is not None:
        try:
            return p.get(section, key)    # 从配置文件中获取 section 对
                                          # 应的 key 的值

        except:
            return default
    # 异常或者其他情况返回默认值
return default
```

get_config()方法比较简单，即从配置文件（ansible.cfg）中获取相应 section 中的 key 对应的值。如果环境变量中存在该 key 变量，则优先取环境变量的值；如果环境变量和配

置文件中都取不到，则返回默认值（default）。

再来看如下代码片段：

```
# 源码位置: lib/ansible/constants.py
# ...

def load_config_file():
    # Python 2 中的写法，在 Python 3 中会有所变化
    p = ConfigParser.ConfigParser()
    path1 = os.path.expanduser(os.environ.get('ANSIBLE_CONFIG', "~/.ansible.cfg"))
    path2 = os.getcwd() + "/ansible.cfg"
    path3 = "/etc/ansible/ansible.cfg"

    if os.path.exists(path1):
        p.read(path1)
    elif os.path.exists(path2):
        p.read(path2)
    elif os.path.exists(path3):
        p.read(path3)
    else:
        return None
    return p

def shell_expand_path(path):
# 如果 os.path.expanduser() 方法无效，则需要使用 shell_expand_path() 方法
# 如果传入的 path 为 None，则默认为 ANSIBLE_PRIVATE_KEY_FILE 常量值
    if path:
        path = os.path.expanduser(path)
return path
```

从上面的代码逻辑中可以看到，Ansible 导入配置文件（ansible.cfg）的优先顺序为：~/.ansible.cfg、当前命令执行目录下的 ansible.cfg 和/etc/ansible/ansible.cfg。在这个配置文件中可以定义很多用于控制 Ansible 运行方式的参数，如控制 ansible 命令执行时启动的进程数（forks）、SSH 远程连接的方式（是基于 paramiko 模块、SSH 命令，还是本地操作）等。此外，shell_expand_path()方法的作用只是根据用户展开 path 值。

再来看如下代码片段：

```
# 源码位置: lib/ansible/constants.py
# ...

# 导入 ansible.cfg 配置文件
p = load_config_file()

# 获取当前用户
active_user   = pwd.getpwuid(os.geteuid())[0]

# 设置 Ansible 模块的位置
if getattr(sys, "real_prefix", None):
    DIST_MODULE_PATH = os.path.join(sys.prefix, 'share/ansible/')
else:
    DIST_MODULE_PATH = '/usr/share/ansible/'
```

上述代码主要是导入 ansible.cfg 配置文件并获取当前活跃用户及 Ansible 模块路径。
下面在 Python 交互模式下演示上述代码的执行结果，具体操作如下：

```
(ansible1.1) [root@master ~]# python
Python 2.7.18 (default, Oct 13 2020, 23:55:15)
[GCC 4.8.5 20150623 (Red Hat 4.8.5-39)] on linux2
Type "help", "copyright", "credits" or "license" for more information.
>>> import pwd
>>> import os
>>> pwd.getpwuid(os.geteuid())[0]
'root'
>>> import sys
>>> getattr(sys, "real_prefix", None)
'/root/.pyenv/versions/2.7.18'
>>> os.path.join(sys.prefix, 'share/ansible/')
'/root/.pyenv/versions/ansible1.1/share/ansible/'
```

通过上述操作可以很清楚地看到之前源码中每一步的执行结果。下面继续看
constants.py 中的后续代码：

```
# 源码位置: lib/ansible/constants.py
# ...

# 配置文件中的默认 section
DEFAULTS='defaults'

# 定义的常量值
DEFAULT_HOST_LIST        = shell_expand_path(get_config(p, DEFAULTS,
'hostfile',         'ANSIBLE_HOSTS',         '/etc/ansible/hosts'))
DEFAULT_MODULE_PATH      = shell_expand_path(get_config(p, DEFAULTS,
'library',          'ANSIBLE_LIBRARY',       DIST_MODULE_PATH))
DEFAULT_REMOTE_TMP       = shell_expand_path(get_config(p, DEFAULTS,
'remote_tmp',       'ANSIBLE_REMOTE_TEMP',    '$HOME/.ansible/tmp'))
DEFAULT_MODULE_NAME      = get_config(p, DEFAULTS, 'module_name',
    None,                           'command')
DEFAULT_PATTERN          = get_config(p, DEFAULTS, 'pattern',
      None,

# ...
```

上述代码定义了较多的常量值，这些常量值将从 ansible.cfg 中获取，如果取不到则会
有一个默认值，而这些常量值会直接影响 ansible 命令的执行。以下是一些重要的常量值
及其说明。

- DEFAULT_HOST_LIST：指定 Ansible 文件中 inventory 文件（即 host 文件）的路径，
  该值可在 Ansible 的配置文件（ansible.cfg）中由 hostfile 变量指定，默认值为/etc/
  ansible/hosts。
- DEFAULT_MODULE_PATH：Ansible 自定义模块的代码目录。
- DEFAULT_REMOTE_TMP：Ansible 远程执行时会新建一个临时目录，用于存放一
  些代码文件和上传的文件等，默认值为$HOME/.ansible/tmp。
- DEFAULT_MODULE_NAME：当执行 ansible 命令时，如果不指定执行模块，则执

行默认模块 command。

```
(ansible1.1) [root@master ~]# ansible ceph-1 -i hosts -a "pwd"
ceph-1 | success | rc=0 >>
/root
shell=False, args=['pwd']
```

说明：最后一行输出是前面介绍 shell 模块时做的一些修改。如果是干净无修改的环境，则不会出现最后一行内容。如果 shell=False，说明使用的是 command 模块；如果 shell=True，则说明是 shell 模块。

- DEFAULT_FORKS：Ansible 运行时启动的进程数，默认最多启动 5 个进程任务。
- DEFAULT_TIMEOUT：SSH 连接超时时间。
- DEFAULT_TRANSPORT：SSH 连接方式。在 1.4 节中将看到 Ansible 提供的 5 种连接插件，其中比较重要的两个插件为 paramiko 插件和 ssh 插件。前者基于 paramiko 模块封装了远程连接和操作的接口，而后者使用 SSH 命令本身完成远程连接和操作的接口。Ansible 1.1 中默认使用 paramiko 插件远程操作节点，而之后的版本则默认使用 ssh 插件（即 ssh.py 文件中的代码）。

其余的配置这里就不再一一介绍了，后面如果用到，会单独说明。

4．color.py

color.py 文件提供了一个 stringc()方法，可以实现在控制台上打印不同颜色的文本信息。请看下面的示例代码：

```
(ansible1.1) [root@master ~]# python
Python 2.7.18 (default, Oct 13 2020, 23:55:15)
[GCC 4.8.5 20150623 (Red Hat 4.8.5-39)] on linux2
Type "help", "copyright", "credits" or "license" for more information.
>>> from ansible.color import stringc
>>> print(stringc('这是红色', 'red'))
这是红色
>>> print(stringc('这是蓝色', 'blue'))
这是蓝色
```

读者可以在自己的 Linux 系统上测试这几行语句，看是否有颜色变化。支持的颜色如下：

```
# 源码位置: lib/ansible/color.py
# ...

codeCodes = {
    'black':      '0;30', 'bright gray':   '0;37',
    'blue':       '0;34', 'white':         '1;37',
    'green':      '0;32', 'bright blue':   '1;34',
    'cyan':       '0;36', 'bright green':  '1;32',
    'red':        '0;31', 'bright cyan':   '1;36',
    'purple':     '0;35', 'bright red':    '1;31',
    'yellow':     '0;33', 'bright purple': '1;35',
```

```
        'dark gray': '1;30', 'bright yellow': '1;33',
        'normal':    '0'
    }

    # ...
```

其实现原理也非常简单，只需要在文本前后分别加上"\033[颜色编号 m"和"\033[0m"（参考 stringc()方法的源码），就可以在 Linux 终端打印不同的颜色。其实将\033 替换成\e 也是没有问题的，可以手动在 Linux 的终端执行命令"echo -e "\e[0;34mhello,world\e[0m""，看看最终能否打印出蓝色的"hello, world"。以下是在 Ansible 1.1 中用于打印各种颜色的 stringc()方法的源码：

```
# 源码位置：lib/ansible/color.py
# ...

def stringc(text, color):
    """打印带颜色的字符串"""

    if ANSIBLE_COLOR:
        return "\033["+codeCodes[color]+"m"+text+"\033[0m"
    else:
        return text
```

5. callbacks.py

callbacks.py 文件用于将返回的结果以统一风格进行显示。例如下面两个 ansible 命令：

```
(ansible1.1) [root@master ~]# ansible ceph-1 -i hosts -m shell -a "hostname"
ceph-1 | success | rc=0 >>
ceph-1

(ansible1.1) [root@master ~]# ansible ceph-1 -i hosts -m shell -a "hostname-
no-exist"
ceph-1 | FAILED | rc=127 >>
/bin/sh: hostname-no-exist: command not found

(ansible1.1) [root@master ~]# ansible ceph-1 -i hosts -m file -a "path=~/test
/create_dir state=directory"
ceph-1 | success >> {
    "changed": true,
    "group": "root",
    "mode": "0755",
    "owner": "root",
    "path": "/root/test/create_dir",
    "size": 6,
    "state": "directory"
}
```

在以上命令的执行结果中，第一行表示执行 Ansible 模块是否成功，第二行之后才是 ansible 命令指定模块的执行结果。由此可以猜到 Ansible 对每一台主机 shell 模块返回结果的封装格式为"目标主机 | success"或者"FAILED | rc=命令返回码 >>"。在 callbacks.py 文件中实现格式化字符串的代码如下：

```
# 源码位置：lib/ansible/callbacks.py
# ...

def regular_generic_msg(hostname, result, oneline, caption):
'''非 command 模块的输出格式'''

    if not oneline:
        return "%s | %s >> %s\n" % (hostname, caption, utils.jsonify(result,
format=True))
    else:
        return "%s | %s >> %s\n" % (hostname, caption, utils.jsonify(result))


# ...

def command_generic_msg(hostname, result, oneline, caption):
    ''' command 模块的输出格式'''

    rc     = result.get('rc', '0')              # 获取命令执行返回码
    stdout = result.get('stdout','')            # 获取命令执行的正确输出
    stderr = result.get('stderr', '')           # 获取命令执行的错误输出
    msg    = result.get('msg', '')              # 获取模块执行的返回信息

    hostname = hostname.encode('utf-8')         # 执行的主机名
    caption  = caption.encode('utf-8')

    if not oneline:                             # 整理多行的输出格式
        buf = "%s | %s | rc=%s >>\n" % (hostname, caption, result.get('rc',0))
        if stdout:
            buf += stdout
        if stderr:
            buf += stderr
        if msg:
            buf += msg
        return buf + "\n"
    else:                                       # 整理单行的输出格式
        if stderr:
            return "%s | %s | rc=%s | (stdout) %s (stderr) %s" % (hostname,
caption, rc, stdout, stderr)
        else:
            return "%s | %s | rc=%s | (stdout) %s" % (hostname, caption, rc,
stdout)

def host_report_msg(hostname, module_name, result, oneline):
'''对单个主机整理 json 格式的结果'''

    failed = utils.is_failed(result)            # 判断是否失败
    msg = ''
    if module_name in [ 'command', 'shell', 'raw' ] and 'ansible_job_id' not
in result and result.get('parsed',True) != False:
    # cmmand、shell 和 raw 模块的判断，调用 command_generic_msg() 方法得到返回信息
        if not failed:
            msg = command_generic_msg(hostname, result, oneline, 'success')
        else:
```

```
            msg = command_generic_msg(hostname, result, oneline, 'FAILED')
        else:
            # 其余模块，成功或失败，返回相应的封装信息
            if not failed:
                msg = regular_generic_msg(hostname, result, oneline, 'success')
            else:
                msg = regular_generic_msg(hostname, result, oneline, 'FAILED')
        return msg
```

regular_generic_msg()和 command_generic_msg()两个方法的返回内容是不是和前面 Ansible 的输出非常类似？下面在 Python 的交互模式下调用这两个方法，即可实现和前面一样的输出，具体代码如下：

```
(ansible1.1) [root@master ~]# python
Python 2.7.18 (default, Oct 13 2020, 23:55:15)
[GCC 4.8.5 20150623 (Red Hat 4.8.5-39)] on linux2
Type "help", "copyright", "credits" or "license" for more information.
>>> from ansible.callbacks import regular_generic_msg, command_generic_
msg, host_report_msg
>>> print(command_generic_msg('ceph-1', {'rc': '127', 'stdout': 'ceph-1'},
True, 'FAILED'))
ceph-1 | FAILED | rc=127 | (stdout) ceph-1
>>> print(command_generic_msg('ceph-1', {'rc': '127', 'stdout': 'ceph-1'},
False, 'FAILED'))
ceph-1 | FAILED | rc=127 >>
ceph-1
>>> print(host_report_msg('ceph-1', 'shell', {'rc': '127', 'stdout':
'ceph-1'}, False))
ceph-1 | FAILED | rc=127 >>
ceph-1
>>> from ansible.callbacks import regular_generic_msg
>>> result = { "changed": True,  "group": "root", "mode": "0755", "owner":
"root", "path": "/root/test/create_dir", "size": 6, "state": "directory"}
>>> print(regular_generic_msg('ceph-1', result, False, 'success'))
ceph-1 | success >> {
    "changed": true,
    "group": "root",
    "mode": "0755",
    "owner": "root",
    "path": "/root/test/create_dir",
    "size": 6,
    "state": "directory"
}

>>> print(host_report_msg('ceph-1', 'file', result, False))
ceph-1 | success >> {
    "changed": true,
    "group": "root",
    "mode": "0755",
    "owner": "root",
    "path": "/root/test/create_dir",
    "size": 6,
    "state": "directory"
}
```

通过上面的示例操作就可以很清楚地了解这些函数的作用了。至于具体的实现细节，直接看代码即可，每个函数的内容都不多，十分易懂。callbacks.py 文件中其他几个重要类的代码如下：

```python
# 源码位置：lib/ansible/callbacks.py
# ...


class AggregateStats(object):

    def __init__(self):

        self.processed  = {}
        self.failures   = {}
        self.ok         = {}
        self.dark       = {}
        self.changed    = {}
        self.skipped    = {}

    def _increment(self, what, host):

        self.processed[host] = 1
        prev = (getattr(self, what)).get(host, 0)
        getattr(self, what)[host] = prev+1

    def compute(self, runner_results, setup=False, poll=False, ignore_
errors=False):
        # 统计 ok、failures、changed、skipped 的次数，方便后续 Playbook 最后显示结果
        # ...


    def summarize(self, host):
        ''' 返回单台主机的信息，比如成功数和失败数等'''

        return dict(
            ok          = self.ok.get(host, 0),
            failures    = self.failures.get(host, 0),
            unreachable = self.dark.get(host,0),
            changed     = self.changed.get(host, 0),
            skipped     = self.skipped.get(host, 0)
        )

# ...
```

AggregateStats 类在执行 Ansible 的 Playbook 时用于统计每台主机的任务执行情况，该类中定义的 compute()方法会在每次任务执行完毕后进行一次统计。最后，Playbook 执行完毕后会统计各个节点的执行情况，即调用 summarize()方法对各主机的执行结果进行统一展示，类似于下面这样：

```
PLAY RECAP *********************
ceph-1                     : ok=3    changed=1    unreachable=0    failed=0
ceph-2                     : ok=3    changed=1    unreachable=0    failed=0
ceph-3                     : ok=3    changed=1    unreachable=0    failed=0
```

6. utils目录

utils 目录下定义了一些辅助代码，如进行模板文件渲染的辅助代码文件（template.py）、插件加载器文件（plugins.py）等。

7. runner目录

runner 目录下的代码可以说是整个 Ansible 的核心代码，后续会在 1.4 节和 1.5 节中详细分析该目录下的核心源码。

8. playbook目录

playbook 目录下只有三个代码文件，分别为__init__.py、play.py 和 task.py，用来解析 Playbook 中的 task 任务及 Ansible 自定义的任务语法，如 when、delegate_to 和 ignore_errors 等。在第 2 章 Ansible1.9.6 的源码分析中会重点追踪 Playbook 的解析执行过程，这里不再过多描述。

9. inventory目录

inventory 目录下的代码文件主要用于解析 hosts 文件内容，同时也包含获取变量的插件代码。在 inventory 模块中需要了解以下几个重要概念：

- 组（group）：每个组中有多个主机，每个主机必定要属于一个组。
- 主机（host）：表示单个主机节点。
- 变量（variable）：其实就是一个字典，保存 key-value 值。

inventory 中的变量可分为 host 变量和 group 变量。group 变量最终会添加到该组的所有节点的 host 变量中。这些变量用于 Playbook 中的任务控制，或者渲染下发到该节点的模板文件等。

还记得前面测试案例中定义的 hosts 文件吗？其中定义的 nodes 组的节点形式为 ceph-[1:3]，该节点将被解析成 ceph-1、ceph-2 和 ceph-3，而这部分工作正是 inventory 目录下的 expand_hosts.py 文件完成的。实现代码如下：

```
# 源码文件: lib/ansible/inventory/expand_hosts.py
# ...

def detect_range(line = None):
    '''
    辅助函数，检查是否包含 Ansible 返回的表达式，类似[1:9]这样的格式

    包含返回则返回 True，否则返回 False
    '''
    if (not line.startswith("[") and
        line.find("[") != -1 and
        line.find(":") != -1 and
        line.find("]") != -1 and
```

```
            line.index("[") < line.index(":") < line.index("]")):
        return True
    else:
        return False

def expand_hostname_range(line = None):
    '''
    辅助函数，扩展类似 xx[1:9].cn 这样的字符串，最终得到完整的主机列表
    '''
    all_hosts = []
    if line:
        # 例如表达式 db[1:6]-node，会考虑将其分成三部分处理
        # 头部: 'db'
        # 返回: [1:6]
        # 尾部: '-node'

        (head, nrange, tail) = line.replace('[','|').replace(']','|').
split('|')
        bounds = nrange.split(":")
        # 检查范围表达式
        if len(bounds) != 2:
            raise errors.AnsibleError("host range incorrectly specified")
        beg = bounds[0]                  # 起始范围
        end = bounds[1]                  # 结束范围
        if not beg:
            beg = "0"                    # 如果没有设置开始值，则从 0 开始
        if not end:
            # 如果没有设置结束值，则直接抛错
            raise errors.AnsibleError("host range end value missing")
        if beg[0] == '0' and len(beg) > 1:
            rlen = len(beg) # range length formatting hint
            if rlen != len(end):
                raise errors.AnsibleError("host range format incorrectly
specified!")
            # 这里会将 01:12 的形式扩展成 01、02、03 等
            fill = lambda _ : str(_).zfill(rlen) # range sequence
        else:
            fill = str

        try:
            # 如果是[a:d]这样的形式，也支持扩展，得到 a、b、c、d
            i_beg = string.ascii_letters.index(beg)
            i_end = string.ascii_letters.index(end)
            if i_beg > i_end:
                raise errors.AnsibleError("host range format incorrectly
specified!")
            # 得到完整的字母序列
            seq = string.ascii_letters[i_beg:i_end+1]
        # 不是字母的范围，则使用 range() 得到数字范围内的完整列表
        except ValueError:
            seq = range(int(beg), int(end)+1)

        for rseq in seq:
```

```
        hname = ''.join((head, fill(rseq), tail))    # 组成完整的主机名
        all_hosts.append(hname)                        # 加入主机列表中

    return all_hosts
```

可以看到，上述代码文件中定义了两个方法：

- detect_range()：检测输入的文本中是否包含 "xxx[小数:数字]" 这样的形式；
- expand_hostname_range()：将输入的节点文本进行扩展，得到节点列表。

这里在 Python 的交互模式下对这两个方法进行调用，具体如下：

```
>>> from ansible.inventory.expand_hosts import detect_range, expand_
hostname_range
>>> detect_range('ceph-[1:3]')
True
>>> expand_hostname_range('ceph-[1:3]')
['ceph-1', 'ceph-2', 'ceph-3']
>>> expand_hostname_range('ceph-[x:z]')
['ceph-x', 'ceph-y', 'ceph-z']
```

接下来是 Ansible 中定义的 Host 类，其定义位于 inventory 目录下的 host.py 文件中，具体代码如下：

```python
# 源码位置: lib/ansible/inventory/host.py
# ...

class Host(object):
    ''' Ansible 单个主机 '''

    __slots__ = [ 'name', 'vars', 'groups' ]

    def __init__(self, name=None, port=None):
        # 主机名
        self.name = name
        # 主机变量
        self.vars = {}
        # 所属组，一个 host 可以属于多个组
        self.groups = []
        if port and port != C.DEFAULT_REMOTE_PORT:
            self.set_variable('ansible_ssh_port', int(port))

        if self.name is None:
            # 必须要有主机名
            raise Exception("host name is required")

    def add_group(self, group):
        # 给 host 添加一个组
        self.groups.append(group)

    def set_variable(self, key, value):
        # 设置该 host 拥有的变量
        self.vars[key]=value

    def get_groups(self):
```

```
    # 获取节点所属的全部组，如果节点属于组 1，并且组 1 是组 2 的一个子组，那么节点也
      属于组 2
    groups = {}
    for g in self.groups:
        groups[g.name] = g
        # 获取该组的全部子组
        ancestors = g.get_ancestors()
        for a in ancestors:
            groups[a.name] = a
    return groups.values()

def get_variables(self):

    results = {}
    groups = self.get_groups()
    for group in sorted(groups, key=lambda g: g.depth):
        results.update(group.get_variables())
    # 合并更新主机变量
    results.update(self.vars)
    # 默认自带的一些变量，如 inventory_hostname 等
    results['inventory_hostname'] = self.name
    results['inventory_hostname_short'] = self.name.split('.')[0]
    results['group_names'] = sorted([ g.name for g in groups if g.name !=
'all'])
    return results
```

　　上面的代码比较简单且已经添加了详细的注释。从 get_variables()方法中可以看到，
每个 host 节点都有一个固定的 inventory_hostname 变量，该变量表示的正是 host 文件中指
定的目标主机名。接着是 group.py 文件中定义的 Group 类，具体代码如下：

```
# 源码位置：lib/ansible/inventory/group.py
# ...

class Group(object):
    ''' Ansible 中主机的组 '''

    __slots__ = [ 'name', 'hosts', 'vars', 'child_groups', 'parent_groups', 'depth' ]

    def __init__(self, name=None):

        self.depth = 0
        self.name = name
        self.hosts = []
        self.vars = {}
        self.child_groups = []
        self.parent_groups = []
        if self.name is None:
            raise Exception("group name is required")

    def add_child_group(self, group):

        if self == group:
            raise Exception("can't add group to itself")
        self.child_groups.append(group)
```

```
        group.depth = max([self.depth+1, group.depth])
        group.parent_groups.append(self)

    def add_host(self, host):
        # 添加主机
        self.hosts.append(host)
        host.add_group(self)

    def set_variable(self, key, value):

        self.vars[key] = value

    def get_hosts(self):
        # 获取组内所有节点，包括子组；节点使用集合形式去重
        hosts = set()
        for kid in self.child_groups:
            hosts.update(kid.get_hosts())
        hosts.update(self.hosts)
        return list(hosts)

    def get_variables(self):
        return self.vars.copy()

    def _get_ancestors(self):
        # 获取父辈组
        results = {}
        for g in self.parent_groups:
            # 返回结果：组名-组对象
            results[g.name] = g
            # 递归
            results.update(g._get_ancestors())
        return results

    def get_ancestors(self):

        return self._get_ancestors().values()
```

上述代码中有一个地方需要注意，相比 Host 类，Group 类要考虑父组和子组的情况，其中包含一些常用的方法，如获取组内所有 host 节点、获取组变量，以及添加 host 和子组等方法。

接着来看 inventory 目录下提供的最重要的一个类——Inventory。该类的作用就是解析 hosts 文件，获取 hosts 文件中定义的组、Ansible 主机及相应的变量，默认的 hosts 文件路径为/etc/ansible/hosts。这个类中的代码并不复杂，主要会结合 Host 和 Group 对象进行一些操作。下面简单演示一下 Inventory 类的用法。

```
(ansible1.1) [root@master ~]# cat hosts
[master]
master

[nodes]
ceph-[1:3]
```

```
[nodes:vars]
ansible_ssh_user=root
ansible_ssh_pass=@SHENcong19920522

[master:vars]
ansible_connection=local
(ansible1.1) [root@master ~]# python
Python 2.7.18 (default, Oct 13 2020, 23:55:15)
[GCC 4.8.5 20150623 (Red Hat 4.8.5-39)] on linux2
Type "help", "copyright", "credits" or "license" for more information.
>>> from ansible import inventory
>>> inventory_manager = inventory.Inventory('/root/hosts')
>>> inventory_manager.list_hosts('all')
['ceph-1', 'ceph-2', 'ceph-3', 'master']
>>> inventory_manager.list_hosts('ceph*')
['ceph-1', 'ceph-2', 'ceph-3']

# 可以看到, 通过 groups_for_host() 方法可以得到 host 所属的 group
>>> [group.name for group in inventory_manager.groups_for_host('ceph-1')]
['all', 'nodes']
# 通过 groups_list() 方法可以获得所有组及组的所有节点信息
>>> inventory_manager.groups_list()
{'ungrouped': [], 'all': ['ceph-2', 'master', 'ceph-3', 'ceph-1'], 'master'
: ['master'], 'nodes': ['ceph-2', 'ceph-3', 'ceph-1']}
>>> inventory_manager.list_groups()
['all', 'master', 'nodes', 'ungrouped']
>>> [group.name for group in inventory_manager.get_groups()]
['ungrouped', 'all', 'master', 'nodes']
# 获取组变量信息
>>> inventory_manager.get_group_variables('nodes')
{'ansible_ssh_user': 'root', 'ansible_ssh_pass': '@SHENcong19920522'}
>>> inventory_manager.get_group_variables('all')
{}
>>> inventory_manager.get_group_variables('master')
{'ansible_connection': 'local'}
# 获取节点变量信息
>>> inventory_manager.get_variables('master')
{'inventory_hostname': 'master', 'inventory_hostname_short': 'master',
'ansible_connection': 'local', 'group_names': ['master']}
```

通过上面的操作示例再结合 Inventory 类的代码, 可以很容易地理解该类的作用。

10. callback_plugins目录

callback_plugins 目录下定义了一个回调模块类 CallbackModule, 这个类中只定义了一些方法, 并没有任何操作。

1.3.3　小结

本节主要对 Ansible 项目的源码结构进行了初步介绍。一个完整的项目除了源码本身外，还包括使用样例、详细的文档说明和依赖的模块等，有的项目甚至提供了一键部署和安装脚本功能，这些都是非常值得学习的地方。此外，本节还对 Ansible 的部分源码目录及源码文件进行了介绍，为后续完整分析 Ansible 工具源码打下基础。

1.4 Ansible 1.1 底层通信代码解析

1.3 节简单介绍了 Ansible 1.1 的整个源码目录和文件，并对部分重点文件中的代码进行了解析，然而笔者对 runner 目录中的代码并未做过多介绍。本节将详细剖析该目录下的一个核心模块——节点的通信模块。Ansible 提供了基于 paramiko 模块、SSH 命令和 Zeromq 的 3 种远程节点的通信方式，以及两种本地通信方式。每种通信方式都对应一个 Python 文件，每个文件中都定义了一个 Connection 类且在该类中有如下几个方法：

- connect()：本地通信不用执行任何操作，其他方式需要与远端节点建立连接。
- exec_command()：最核心的方法，在本地或者远程连接的主机上执行 shell 命令。
- put_file()：将文件上传到本地或者远端主机的某个位置。
- fetch_file()：从本地或者远端主机的某个位置下载文件。
- close()：断开连接。对于本地通信而言，什么也不需要做。

下面笔者通过几种通信模式对这些方法进行具体讲解，同时附上改造和测试的代码。

1.4.1 本地连接插件

本地连接方式有两种，即本地模式（local）和 chroot 模式。这两种模式分别对应 Ansible 的 lib/ansible/runner/connection_plugins 目录下的 local.py 和 chroot.py 文件。后者基于 chroot 命令，主要是把根目录换成指定的目录，这样可以提高系统的安全性，防止登录用户的权利过大，其实现的本地通信代码和普通模式并无太大区别。本节主要分析 local.py 文件，以下是该文件中 Connection 类定义的核心方法。

1. connect()方法

由于是本地通信，所以 connect()不需要进行任何连接操作，具体代码如下：

```
# 源码位置：lib/ansible/runner/connection_plugins/local.py
# ...

class Connection(object):
    # ...

    def connect(self, port=None):
        ''' 本地连接，无须任何操作 '''

        return self
```

```
# ...
```

2. exec_command()方法

exec_command()方法直接使用subprocess模块实现在本地执行shell命令并获取返回结果。当对该 shell 命令使用 sudo 进行提权时，需要能自动处理 Linux 系统要求输入提权密码的问题，而该 sudo 密码会事先写在配置文件（hosts）中。这段代码非常有用，后续将会基于该代码进行改造，形成一段可以直接使用的代码段。具体代码如下：

```
# 源码位置: lib/ansible/runner/connection_plugins/local.py
# ...
class Connection(object):
    # ...

    def exec_command(self, cmd, tmp_path, sudo_user, sudoable=False,
executable='/bin/sh'):
        '''本地执行命令'''

        if not self.runner.sudo or not sudoable:
            # 对于非 sudo 情况
            if executable:
                local_cmd = [executable, '-c', cmd]
            else:
                local_cmd = cmd
        else:
            # 构建 sudo 的命令，对于 centos 而言，加上 sudo 命令即可，不过生成的命令较
              复杂
            local_cmd, prompt = utils.make_sudo_cmd(sudo_user, executable,
cmd)

        vvv("EXEC %s" % (local_cmd), host=self.host)
        # 调用 subprocess 模块
        p = subprocess.Popen(local_cmd, shell=isinstance(local_cmd, basestring),
                        cwd=self.runner.basedir, executable=executable
or None,
                        stdin=subprocess.PIPE,
                        stdout=subprocess.PIPE, stderr=subprocess.PIPE)

        # 需要有 sudo_pass
        if self.runner.sudo and sudoable and self.runner.sudo_pass:
            # 文件加锁
            fcntl.fcntl(p.stdout, fcntl.F_SETFL,
                    fcntl.fcntl(p.stdout, fcntl.F_GETFL) | os.O_NONBLOCK)
            fcntl.fcntl(p.stderr, fcntl.F_SETFL,
                    fcntl.fcntl(p.stderr, fcntl.F_GETFL) | os.O_NONBLOCK)
            sudo_output = ''
            # 如果 sudo 结果的输出是以前面的输入语句结尾，则输入相应的 sudo 密码
            # 否则读取相应的输出
            while not sudo_output.endswith(prompt):
                # 读描述符、写描述符及错误描述符，设置读取的超时时间
```

```
                rfd, wfd, efd = select.select([p.stdout, p.stderr], [],
                                    [p.stdout, p.stderr], self.runner.timeout)
                # 有输出
                if p.stdout in rfd:
                    chunk = p.stdout.read()
                # 有错误输出
                elif p.stderr in rfd:
                    chunk = p.stderr.read()
                else:
                    # 无输出，则直接抛出异常
                    stdout, stderr = p.communicate()
                    raise errors.AnsibleError('timeout waiting for sudo
password prompt:\n' + sudo_output)
                if not chunk:
                    # 也是无输出，直接抛出异常，sudo 命令的输出结束
                    stdout, stderr = p.communicate()
                    raise errors.AnsibleError('sudo output closed while
waiting for password prompt:\n' + sudo_output)
                sudo_output += chunk
            # 输入密码
            p.stdin.write(self.runner.sudo_pass + '\n')
            # 文件解锁
            fcntl.fcntl(p.stdout, fcntl.F_SETFL, fcntl.fcntl(p.stdout,
fcntl.F_GETFL) & ~os.O_NONBLOCK)
            fcntl.fcntl(p.stderr, fcntl.F_SETFL, fcntl.fcntl(p.stderr,
fcntl.F_GETFL) & ~os.O_NONBLOCK)

        # 非 sudo 情况，直接得到正确和错误的输出
        stdout, stderr = p.communicate()
        return (p.returncode, '', stdout, stderr)
```

上面这段代码比较复杂，用到了一些 Python 的高级库，阅读起来比较吃力。可以将上述代码分解成两部分进行阅读。

- 当没有使用 sudo 命令提权时，直接使用 subprocess 模块获取 shell 命令输出，可参考示例 1。
- 当使用 sudo 命令获取 root 权限时，先保证该用户加入了 sudoers 列表，然后使用 sudo 提权，此时需要输入相应的密码才能保证命令被正确提权并执行。简化版的操作参考示例 2。

示例 1：非 sudo 形式

```
(ansible1.1) [root@master ~]# python
Python 2.7.18 (default, Oct 13 2020, 23:55:15)
[GCC 4.8.5 20150623 (Red Hat 4.8.5-39)] on linux2
Type "help", "copyright", "credits" or "license" for more information.
>>> import subprocess
>>> p = subprocess.Popen('/bin/sh -c "pwd"', shell='/bin/sh', cwd='/tmp',
stdin=subprocess.PIPE, stdout=subprocess.PIPE, stderr=subprocess.PIPE)
```

```
>>> p.communicate()
('/tmp\n', '')
>>>
```

示例 2：sudo 形式（CentOS 7）

先在 master 节点上新建一个用户 halo 并设置其密码为 halo.1234!。

```
(ansible1.1) [root@master ~]# useradd -m halo
# ...
(ansible1.1) [root@master ~]# passwd halo
Changing password for user halo.
New password:
BAD PASSWORD: The password contains the user name in some form
Retype new password:
passwd: all authentication tokens updated successfully.
```

然后将 halo 用户写入/etc/sudoers 文件中（位置任意），这样 halo 用户便能使用 sudo 命令进行提权。具体写入语句如下：

```
# 代码位置：/etc/sudoers
# ...

halo    ALL=(ALL)       ALL
```

改造前面的 exec_command()方法，先去掉 tmp_path 和 sudo_user 参数，前者只是为了兼容但并未使用，后者在现有的 Python 2.7.5 中已经无效。改造后的代码如下：

```python
# -*- coding: utf-8 -*-

"""
改造 Ansible 1.1 源码中 local.py 文件中的 exec_command()方法
"""
import os
import random
import pipes
import subprocess
import fcntl
import select


def make_sudo_cmd(executable, cmd):
    """
    Ansible 1.1 中的 make_sudo_cmd()方法，去掉了 sudo_user
    """
    randbits = ''.join(chr(random.randint(ord('a'), ord('z'))) for x in
xrange(32))
    prompt = '[sudo via ansible, key=%s] password: ' % randbits
    sudocmd = '%s -k && %s %s -S -p "%s" %s -c %s' % (
        'sudo', 'sudo', '-H',
        prompt, executable or '$SHELL', pipes.quote(cmd))
    return ('/bin/sh -c ' + pipes.quote(sudocmd), prompt)


def exec_command(cmd, sudoable=False, executable='/bin/sh', sudo_pass=
None):
```

```
        '''本地执行命令'''

    if not sudoable:
        if executable:
            local_cmd = [executable, '-c', cmd]
        else:
            local_cmd = cmd
    else:
        local_cmd, prompt = make_sudo_cmd(executable, cmd)

    print("EXEC %s" % (local_cmd))
    p = subprocess.Popen(local_cmd, shell=isinstance(local_cmd, basestring),
                    executable=executable or None,
                    stdin=subprocess.PIPE,
                    stdout=subprocess.PIPE, stderr=subprocess.PIPE)

    if sudoable and sudo_pass:
        # 对于使用 sudo 且存在密码的情况，需要在后续输入 sudo_pass 的值
        fcntl.fcntl(p.stdout, fcntl.F_SETFL,
                fcntl.fcntl(p.stdout, fcntl.F_GETFL) | os.O_NONBLOCK)
        fcntl.fcntl(p.stderr, fcntl.F_SETFL,
                fcntl.fcntl(p.stderr, fcntl.F_GETFL) | os.O_NONBLOCK)
        sudo_output = ''
        while not sudo_output.endswith(prompt):
            rfd, wfd, efd = select.select([p.stdout, p.stderr], [],
                                    [p.stdout, p.stderr], 10)
            if p.stdout in rfd:
                chunk = p.stdout.read()
            elif p.stderr in rfd:
                chunk = p.stderr.read()
            else:
                stdout, stderr = p.communicate()
                raise Exception('timeout waiting for sudo password prompt:\n'
+ sudo_output)
            if not chunk:
                stdout, stderr = p.communicate()
                raise Exception('sudo output closed while waiting for password
prompt:\n' + sudo_output)
            sudo_output += chunk
        p.stdin.write(sudo_pass + '\n')
        fcntl.fcntl(p.stdout, fcntl.F_SETFL, fcntl.fcntl(p.stdout, fcntl.
F_GETFL) & ~os.O_NONBLOCK)
        fcntl.fcntl(p.stderr, fcntl.F_SETFL, fcntl.fcntl(p.stderr, fcntl.
F_GETFL) & ~os.O_NONBLOCK)

    stdout, stderr = p.communicate()
    return (p.returncode, '', stdout, stderr)
```

将上面这段代码命名为 test_local_exec_command.py 并保存到/home/halo 目录下，接着使用 halo 用户执行如下操作：

```
[halo@master ~]$ python
Python 2.7.5 (default, Apr 2 2020, 13:16:51)
[GCC 4.8.5 20150623 (Red Hat 4.8.5-39)] on linux2
Type "help", "copyright", "credits" or "license" for more information.
```

```
>>> from test_local_exec_command import exec_command
>>> print(exec_command('whoami'))
EXEC ['/bin/sh', '-c', 'whoami']
(0, '', 'halo\n', '')
# 输入错误的密码
>>> print(exec_command('whoami', sudoable=True, sudo_pass='halo.1233!'))
EXEC /bin/sh -c 'sudo -k && sudo -H -S -p "[sudo via ansible, key=yumd
fxjgsileqavkrggwvqtmkpashvhj] password: " /bin/sh -c whoami'
(1, '', '', 'Sorry, try again.\n[sudo via ansible, key=yumdfxjgsileqavkrggwv
qtmkpashvhj] password: \nsudo: no password was provided\nsudo: 1 incorrect
password attempt\n')
# 输入正确的密码
>>> print(exec_command('whoami', sudoable=True, sudo_pass='halo.1234!'))
EXEC /bin/sh -c 'sudo -k && sudo -H -S -p "[sudo via ansible, key=yvydp
iymnsnsxrcllzadwkhqkzsexmux] password: " /bin/sh -c whoami'
(0, '', 'root\n', '')
>>>
```

上述操作先进入 Python 的交互式环境，然后导入编写的 exec_command()方法。接着
调用该方法执行 whoami 命令并打印输出，此时应该输出 halo 用户本身。然后设置
sudoable=True，分别输入错误的 sudo_pass 密码和正确的 sudo_pass 密码，可以看到，最后
一次输出了正确的提权结果。

3．put_file()方法

put_file()方法是本地连接插件的上传方法，即本地文件的移动动作，直接使用 shutil
模块的 copyfile()方法即可。具体代码如下：

```
# 源码位置: lib/ansible/runner/connection_plugins/local.py
# ...

class Connection(object):
    # ...

    def put_file(self, in_path, out_path):
        ''' 在本地传输文件 '''

        vvv("PUT %s TO %s" % (in_path, out_path), host=self.host)
        if not os.path.exists(in_path):
            raise errors.AnsibleFileNotFound("file or module does not exist:
%s" % in_path)
        try:
            # 直接使用 shutil 模块的 copyfile()方法实现将本地文件上传到本地
            shutil.copyfile(in_path, out_path)
        except shutil.Error:
            # 打印并抛出异常
            # ...
        except IOError:
            # 打印并抛出异常
            # ...

    # ...
```

4．fetch_file()方法

fetch_file()方法用于将本地文件下载到本地，其实就是 put_file()方法的一个逆过程，交换源地址和目的地址即可。具体代码如下：

```
# 源码位置: lib/ansible/runner/connection_plugins/local.py
# ...

class Connection(object):
# ...

    def fetch_file(self, in_path, out_path):
        vvv("FETCH %s TO %s" % (in_path, out_path), host=self.host)
        ''' 对于本地连接插件的下载，直接使用 copy 方式即可 '''
        self.put_file(in_path, out_path)

    # ...
```

5．close()方法

对于本地通信，不用建立连接也不用关闭连接。

那么该如何使用本地通信模式呢？只需要将本地的主机连接方式（ansible_connection）设置为 local 即可。具体操作如下：

```
(ansible1.1) [root@master ~]# cat hosts
[master]
master

[nodes]
ceph-[1:3]

[nodes:vars]
ansible_ssh_user=root
ansible_ssh_pass=@SHENcong19920522

[master:vars]
ansible_connection=local

(ansible1.1) [root@master ~]# ansible master -i hosts -m ping
master | success >> {
    "changed": false,
    "ping": "pong"
}
```

Ansible 工具安装的节点为 master，如果要操控自身，就不用像其他节点那样输入认证（SSH 端口、SSH 登录密码等）信息，直接设置连接方式为本地连接即可。通过-vvv选项可以看到前面代码中调用 vvv()函数的打印信息，这表明当设置连接方式为本地连接后，在进行相关操作时，会调用 local.py 文件中的相关类与方法。

1.4.2　基于 paramiko 模块连接插件

在 Ansible 1.1 中，远程主机的 SSH 通信方式默认是基于 paramiko 模块的，这个默认配置在 constants.py 文件中可见，具体如下：

```
# 源码位置: lib/ansible/constants.py
# ...

DEFAULT_TRANSPORT         = get_config(p, DEFAULTS, 'transport',
'ANSIBLE_TRANSPORT',        'paramiko')

# ...
```

paramiko 模块是 Python 中一个非常流行的第三方模块，它实现了 SSH2 远程安全连接，支持认证及密钥方式，可以实现远程命令执行、文件传输和中间 SSH 代理等功能。从 Ansible 1.1 之后就不再默认使用 paramiko 模块进行远程通信了，而是转为使用 SSH 命令方式。在之后的 Ansible 2 中亦是如此，但 Ansible 仍旧保留并不断升级了 paramiko 等在内的众多用于远程通信的插件，并且同样支持使用 ansible_connection 变量进行指定。paramiko 插件的源码路径为 lib/ansible/runner/connection_plugins/paramiko_ssh.py。下面依旧以前面列出的 5 个方法展开分析。

1．connect()方法

基于 paramiko 模块进行远程连接时，Ansible 会使用内存作为缓存，将每次的连接信息存到字典中，对应的 key 值为"%s__%s__" % (self.host, self.user)。而真正实现连接操作的是_connect_uncached()方法。connect()方法的具体代码如下：

```
# 源码位置: lib/ansible/runner/connection_plugins/paramiko_ssh.py
# ...

SSH_CONNECTION_CACHE = {}
SFTP_CONNECTION_CACHE = {}

class Connection(object):
    ''' 基于 Paramiko 模块进行 SSH 远程连接 '''
    # ...

    def _cache_key(self):
        # 缓存的 key 字段
        return "%s__%s__" % (self.host, self.user)

    def connect(self):
        cache_key = self._cache_key()
        if cache_key in SSH_CONNECTION_CACHE:
            # 使用连接缓存信息
            self.ssh = SSH_CONNECTION_CACHE[cache_key]
        else:
```

```
            # 连接并缓存连接结果
            self.ssh = SSH_CONNECTION_CACHE[cache_key] = self._connect_
uncached()
        return self

    def _connect_uncached(self):
        ''' 激活连接对象 '''

        if not HAVE_PARAMIKO:
            # 没有安装 paramiko 模块，直接抛出异常
            raise errors.AnsibleError("paramiko is not installed")

        vvv("ESTABLISH CONNECTION FOR USER: %s on PORT %s TO %s" % (self.user,
self.port, self.host), host=self.host)

        ssh = paramiko.SSHClient()
        ssh.set_missing_host_key_policy(paramiko.AutoAddPolicy())

        allow_agent = True
        if self.password is not None:
            # 是否允许连接到 SSH 代理，如果有密码就不用
            allow_agent = False
        try:
            # 密钥文件
            if self.runner.private_key_file:
                key_filename = os.path.expanduser(self.runner.private_key_
file)
            else:
                key_filename = None
            # 连接的核心方法
            ssh.connect(self.host, username=self.user, allow_agent=allow_
agent, look_for_keys=True,
                key_filename=key_filename, password=self.password,
                timeout=self.runner.timeout, port=self.port)
        except Exception, e:
            # 打印异常，同时区分异常类型
            # ...

        return ssh

    # ...
```

这里对 paramiko 模块的连接函数进行了封装，添加了连接缓存，避免重复的连接请求。

2．exec_command()方法

远程执行命令是 paramiko 模块的核心操作。Ansible 同样对 paramiko 模块的代码进行了进一步封装和完善，大部分代码用于处理 sudo 模式下的密码输入情况。exec_command()方法的具体代码如下：

```
# 源码位置: lib/ansible/runner/connection_plugins/paramiko_ssh.py
# ...

class Connection(object):
    # ...

    def exec_command(self, cmd, tmp_path, sudo_user, sudoable=False,
executable='/bin/sh'):
        ''' 在远端主机上执行命令'''

        bufsize = 4096
        try:
            # 打开通道
            chan = self.ssh.get_transport().open_session()
        except Exception, e:
            # 打开通道失败, 抛出异常
            # ...

        if not self.runner.sudo or not sudoable:
            # 非 sudo 模式远程执行 shell 命令
            if executable:
                quoted_command = executable + ' -c ' + pipes.quote(cmd)
            else:
                quoted_command = cmd
            vvv("EXEC %s" % quoted_command, host=self.host)
            # 调用通道的 exec_command() 方法执行命令
            chan.exec_command(quoted_command)
        else:
            # sudo usually requires a PTY (cf. requiretty option), therefore
            # we give it one, and we try to initialise from the calling
            # environment
            chan.get_pty(term=os.getenv('TERM', 'vt100'),
                        width=int(os.getenv('COLUMNS', 0)),
                        height=int(os.getenv('LINES', 0)))
            # 生成 sudo 模式执行的命令, prompt 是生成的 sudo 指令中-p 选项指定的需要输
            #   入密码时的提示语句
            shcmd, prompt = utils.make_sudo_cmd(sudo_user, executable, cmd)
            vvv("EXEC %s" % shcmd, host=self.host)
            sudo_output = ''
            try:
                # 同样也是调用通道的 exec_command() 方法远程执行命令
                chan.exec_command(shcmd)
                # 对于需要输入 sudo 密码的情况, 需要额外处理
                if self.runner.sudo_pass:
                    # 这个 prompt 和 local.py 中的一样, 是 sudo 命令-p 选项指定的提示
                    #   语句
                    while not sudo_output.endswith(prompt):
                        chunk = chan.recv(bufsize)
                        if not chunk:
                            # 异常情况, 抛出异常
                            # ...
                        # 将通道传回的文本合成, while 会判断输出是否包含输入密码的关键字
                        sudo_output += chunk
```

```
                # 如果出现需要输入密码的情况，那么就在通道中输入密码并加上回车
                chan.sendall(self.runner.sudo_pass + '\n')
        except socket.timeout:
            raise errors.AnsibleError('ssh timed out waiting for sudo.\n'
+ sudo_output)
        # 获取正确输出通道结果
        stdout = ''.join(chan.makefile('rb', bufsize))
        # 获取错误输出通道结果
        stderr = ''.join(chan.makefile_stderr('rb', bufsize))
        return (chan.recv_exit_status(), '', stdout, stderr)

    # ...
```

上述代码中已经添加了详细的注释，理解起来并不困难，因为这些是 paramiko 模块的基本使用语句，另外，paramiko 的官方文档也非常详细。下面将这段代码改造一下，实现一个没有其他依赖的 Connection 类（去掉 self.runner），然后进行测试并解析这段代码的含义。本次主要测试如下几个案例：

- 在 master 节点上使用 paramiko 模块远程执行一些普通命令，如 hostname -s、ps -ef | grep sshd 等。
- 在 master 节点上使用 paramiko 模块远程执行 sudo 命令，这次使用具有 sudo 权限的用户（sudo 不免密用户）。
- 在上一个案例基础上，使用 sudo 免密用户在远程执行 sudo 命令。

下面直接对 paramiko_ssh.py 文件中的 Connection 类进行二次改造，将 exec_command() 方法依赖的 make_sudo_cmd() 方法直接复制过来作为对象方法调用。此外，去掉 sudo_user 参数，同时也去掉 exec_command() 方法中的 tmp_path 参数（并未用到）。具体代码如下：

```
#   -*- coding: utf-8   -*-

"""
改造 Ansible 1.1 中 paramiko_ssh.py 文件中的代码
"""
import os
import pipes
import random

import paramiko

class Connection:
    def __init__(self, host, port, user, password, timeout=10):

        self.ssh = None
        self.sftp = None
        self.host = host
        self.port = port
        self.user = user
        self.password = password
        self.timeout = timeout

    def make_sudo_cmd(self, executable, cmd):
```

```
        """
        连接插件的辅助函数，用于生成 sudo 命令
        """
        randbits = ''.join(chr(random.randint(ord('a'), ord('z'))) for x in
xrange(32))
        prompt = '[sudo via ansible, key=%s] password: ' % randbits
        # 使用 sudo 提权的命令
        sudocmd = '%s -k && %s %s -S -p "%s" %s -c %s' % (
            'sudo', 'sudo', '-H',
            prompt, executable or '$SHELL', pipes.quote(cmd))
        return ('/bin/sh -c ' + pipes.quote(sudocmd), prompt)

    def connect(self):
        ''' 激活连接对象'''
        print("ESTABLISH CONNECTION FOR USER: %s on PORT %s TO %s" %
(self.user, self.port, self.host))

        self.ssh = paramiko.SSHClient()
        self.ssh.set_missing_host_key_policy(paramiko.AutoAddPolicy())

        allow_agent = True
        if self.password is not None:
            allow_agent = False
        try:
            # 核心连接方法
            self.ssh.connect(self.host, username=self.user, allow_agent=
allow_agent,
                        look_for_keys=True, key_filename=None, password=
self.password,
                        timeout=self.timeout, port=self.port)
        except Exception as e:
            # 连接异常处理
            msg = str(e)
            if "PID check failed" in msg:
                raise Exception("paramiko version issue, please upgrade
paramiko on the machine
                        running ansible")
            elif "Private key file is encrypted" in msg:
                msg = 'ssh %s@%s:%s : %s\nTo connect as a different user, use
-u <username>.' % (
                    self.user, self.host, self.port, msg)
                raise Exception(msg)
            else:
                raise Exception(msg)


    def exec_command(self, cmd, sudoable=False, sudo_pass=None, executable=
'/bin/sh'):

        bufsize = 4096
        try:
            # 打开连接通道
            chan = self.ssh.get_transport().open_session()
        except Exception as e:
```

```
        msg = "Failed to open session"
        if len(str(e)) > 0:
            msg += ": %s" % str(e)
        raise Exception(msg)

    if not sudoable:
        # 非 sudo 形式比较简单，直接调用通道对象的 exec_command()方法即可
        if executable:
            quoted_command = executable + ' -c ' + pipes.quote(cmd)
        else:
            quoted_command = cmd
        print("[%s] EXEC %s" % (self.host, quoted_command))
        chan.exec_command(quoted_command)
    else:
        # 对于远程执行 sudo 命令，较为复杂
        chan.get_pty(term=os.getenv('TERM', 'vt100'),
                width=int(os.getenv('COLUMNS', 0)),
                height=int(os.getenv('LINES', 0)))
        # 得到完整的 sudo 命令 shcmd 及输入密码时的提示字符串 prompt
        shcmd, prompt = self.make_sudo_cmd(executable, cmd)
        print("[%s] EXEC %s" % (self.host, shcmd))
        sudo_output = ''
        try:
            # 执行命令
            chan.exec_command(shcmd)
            if sudo_pass:
                # 如果输入了 sudo 密码，则远端一定会传回需要输入密码的文本
                while not sudo_output.endswith(prompt):
                    chunk = chan.recv(bufsize)
                    if not chunk:
                        raise Exception('ssh connection closed waiting for
password prompt')
                    sudo_output += chunk
                # 收到需要输入密码的信息后，向通道输入 sudo 密码
                chan.sendall(sudo_pass + '\n')
        except socket.timeout:
            raise Exception('ssh timed out waiting for sudo.\n' +
sudo_output)

    stdout = ''.join(chan.makefile('rb', bufsize))
    stderr = ''.join(chan.makefile_stderr('rb', bufsize))
    return (chan.recv_exit_status(), '', stdout, stderr)
```

首先在 ceph-1 节点中新建 halo 和 store 两个用户，其中，将 halo 用户配置成 sudo 不免密，即使用 sudo 时需要输入 halo 账号的密码进行确认，将 store 用户配置成 sudo 免密，即使用 sudo 直接提权，不用输入密码。具体在/etc/sudoers 中的配置如下：

```
store ALL=(ALL) NOPASSWD: ALL
halo  ALL=(ALL) ALL
```

接着就可以在 Python 的交互模式下调用新改造的 Connection 类，具体操作如下：

```
(ansible1.1) [root@master ~]# ls test_paramiko.py
test_paramiko.py
```

```
(ansible1.1) [root@master ~]# python
Python 2.7.18 (default, Oct 13 2020, 23:55:15)
[GCC 4.8.5 20150623 (Red Hat 4.8.5-39)] on linux2
Type "help", "copyright", "credits" or "license" for more information.
>>> from test_paramiko import Connection
/root/.pyenv/versions/ansible1.1/lib/python2.7/site-packages/paramiko/
transport.py:33: CryptographyDeprecationWarning: Python 2 is no longer
supported by the Python core team. Support for it is now deprecated in
cryptography, and will be removed in a future release.
  from cryptography.hazmat.backends import default_backend
>>> conn = Connection('ceph-1', 22, 'halo', 'halo.1234!')
>>> conn.connect()
ESTABLISH CONNECTION FOR USER: halo on PORT 22 TO ceph-1
>>> conn.exec_command('hostname')
[ceph-1] EXEC /bin/sh -c hostname
(0, '', 'ceph-1\n', '')
>>> conn.exec_command('cat /etc/sudoers')
[ceph-1] EXEC /bin/sh -c 'cat /etc/sudoers'
(1, '', '', 'cat: /etc/sudoers: Permission denied\n')
>>> conn.exec_command('cat /etc/sudoers', sudoable=True, sudo_pass=
'halo.1234')
[ceph-1] EXEC /bin/sh -c 'sudo -k && sudo -H -S -p "[sudo via ansible,
key=bgtwpyugvkmuabjchzhfhuoikfrndfwt] password: " /bin/sh -c '"'"'cat
/etc/sudoers'"'"''
(0, '', '省略输出内容', '')
```

上述操作演示了使用 halo/halo.1234！创建 Connection 对象并远程连接 ceph-1 节点，然后依次调用 conn.exec_command()方法远程执行 hostname 和 cat /etc/sudoers 命令的过程，并且后一条指令分别使用了非 sudo 模式和 sudo 模式。可以看到，在使用 sudoable=True 及指定 sudo_pass 后，成功获取了命令执行结果而没有权限问题。

🔲注意：如果设置使用 sudo 提权但没有自动输入密码进行处理，程序会一直卡住无法继续执行，最后只能强行终止（按 Ctrl+C 组合键）。

接下来测试 sudo 免密用户，操作基本一致，具体如下：

```
>>> conn = Connection('ceph-1', 22, 'store', 'store.1234')
>>> conn.connect()
ESTABLISH CONNECTION FOR USER: store on PORT 22 TO ceph-1
>>> conn.exec_command('whoami')
[ceph-1] EXEC /bin/sh -c whoami
(0, '', 'store\n', '')
>>> conn.exec_command('cat /etc/sudoers')
[ceph-1] EXEC /bin/sh -c 'cat /etc/sudoers'
(1, '', '', 'cat: /etc/sudoers: Permission denied\n')
>>> conn.exec_command('cat /etc/sudoers', sudoable=True)
[ceph-1] EXEC /bin/sh -c 'sudo -k && sudo -H -S -p "[sudo via ansible,
key=xdqzfpjtluimmmbkclexuduwfbfczkom] password: " /bin/sh -c '"'"'cat
/etc/sudoers'"'"''
(0, '', '...', '')
```

由于是 sudo 免密，因此不用输入 sudo_pass，便可以直接得到其执行结果。对 Ansible

1.1 代码改造和调用后，就可以清楚地理解源代码的作用与含义了。

3．put_file()方法

put_file()方法用于将本地文件上传到远端主机的某个位置。参数 in_path 为本地待上传文件地址，out_path 为远端的目的地址。具体代码如下：

```python
# 源码位置: lib/ansible/runner/connection_plugins/paramiko_ssh.py
# ...

class Connection(object):

    # ...

    def put_file(self, in_path, out_path):
        '''从本地上传文件到远端服务器'''
        vvv("PUT %s TO %s" % (in_path, out_path), host=self.host)
        if not os.path.exists(in_path):
            raise errors.AnsibleFileNotFound("file or module does not exist: %s" % in_path)
        try:
            # 打开 sftp 通道
            self.sftp = self.ssh.open_sftp()
        except Exception, e:
            raise errors.AnsibleError("failed to open a SFTP connection (%s)" % e)
        try:
            # 上传文件
            self.sftp.put(in_path, out_path)
        except IOError:
            raise errors.AnsibleError("failed to transfer file to %s" % out_path)

    # ...
```

4．fetch_file()方法

fetch_file()方法用于从远端服务器上下载文件，这里相比 put_file()方法多了一个先从缓存取 sftp 信息的过程。具体代码如下：

```python
# 源码位置: lib/ansible/runner/connection_plugins/paramiko_ssh.py
# ...

class Connection(object):

    # ...

    def _connect_sftp(self):
        cache_key = "%s__%s__" % (self.host, self.user)
        if cache_key in SFTP_CONNECTION_CACHE:
            return SFTP_CONNECTION_CACHE[cache_key]
        else:
            result = SFTP_CONNECTION_CACHE[cache_key] = self.connect().
```

```
ssh.open_sftp()
        return result

    def fetch_file(self, in_path, out_path):
        ''' 保存远端文件到本地的一个具体路径上 '''
        vvv("FETCH %s TO %s" % (in_path, out_path), host=self.host)
        try:
            self.sftp = self._connect_sftp()
        except Exception, e:
            raise errors.AnsibleError("failed to open a SFTP connection
(%s)", e)
        try:
            # 核心就是调用 get()方法
            self.sftp.get(in_path, out_path)
        except IOError:
            raise errors.AnsibleError("failed to transfer file from %s" %
in_path)

    # ...
```

5. close()方法

close()方法用于清除缓存，同时关闭打开的 SFTP 通道及 SSH 通道。具体代码如下：

```
# 源码位置: lib/ansible/runner/connection_plugins/paramiko_ssh.py
# ...

class Connection(object):

    # ...

    def close(self):
        ''' 终止连接 '''
        cache_key = self._cache_key()
        SSH_CONNECTION_CACHE.pop(cache_key, None)
        SFTP_CONNECTION_CACHE.pop(cache_key, None)
        if self.sftp is not None:
            self.sftp.close()
        self.ssh.close()

    # ...
```

以上就是 Ansible 1.1 默认的远程通信的代码及提供的操作接口，其实就是在 paramiko 模块上进行了进一步封装。在第 4 章中将看到，即使是在 Ansible 2.8.16 版本中，基于 paramiko 模块的连接插件整体上并没有发生很大变化，提供的方法仍然存在，但是在实现细节上考虑了更多可能出现的异常情况。

1.4.3　基于 SSH 命令连接插件

该部分代码的完整路径为 lib/ansible/runner/connection_plugins/ssh.py，和前面一样，

本节将重点分析 Connection 类的 5 个方法，并一一进行实验，以达到彻底理解的目的。

1. connect()方法

connect()方法中主要包括根据对象信息生成 SSH 命令的通用选项参数，这里用到的是 SSH 命令的-o 选项。具体代码如下：

```
# 源码位置：lib/ansible/runner/connection_plugins/ssh.p
# ...

class Connection(object):
    '''基于 SSH 命令的连接类 '''
    # ...

    def connect(self):
        '''连接远端主机 '''

        vvv("ESTABLISH CONNECTION FOR USER: %s" % self.user, host=self.host)

        # 生成 SSH 远程连接命令，带上众多优化参数
        self.common_args = []
        extra_args = C.ANSIBLE_SSH_ARGS
        if extra_args is not None:
            self.common_args += shlex.split(extra_args)
        else:
            self.common_args += ["-o", "ControlMaster=auto",
                                 "-o", "ControlPersist=60s",
                                 "-o", "ControlPath=/tmp/ansible-ssh-%h-%p-%r"]
        self.common_args += ["-o", "StrictHostKeyChecking=no"]
        # 设置 SSH 连接端口
        if self.port is not None:
            self.common_args += ["-o", "Port=%d" % (self.port)]
        # 带密钥
        if self.runner.private_key_file is not None:
            self.common_args +=
                ["-o", "IdentityFile="+os.path.expanduser(self.runner.
private_key_file)]
        # 带密码认证
        if self.password:
            self.common_args += ["-o", "GSSAPIAuthentication=no",
                                 "-o", "PubkeyAuthentication=no"]
        else:
            self.common_args += ["-o", "KbdInteractiveAuthentication=no",
                                 "-o", "PasswordAuthentication=no"]
        self.common_args += ["-o", "User="+self.user]
        self.common_args += ["-o", "ConnectTimeout=%d" % self.runner.timeout]

        # 可以添加 vvv()方法，查看 self.comm_args 的最终结果

        return self

    # ...
```

注意，这里的 vvv()方法用于打印调试信息，后面会在该 connect()方法最后添加一个 vvv()方法打印最后的 self.common_args 结果，然后在 ansible 命令中加上-vvv 选项查看生成 SSH 命令的选项参数。

2. exec_command()方法

基于 SSHPASS 和 SSH 命令实现远程执行 shell 命令。这里先对 SSHPASS 命令的参数进行简单介绍，可以直接通过 man sshpass 查看命令的使用手册。SSHPASS 命令的常用参数如下：

- -p：指定 SSH 远程登录密码。
- -f：指定文件，文件的第一行就是密码。
- -d：指定从运行程序继承的文件描述符，密码从打开的文件描述符中读取。

下面先看几个关于 SSHPASS 和 SSH 命令的组合操作，分别实现远程执行命令、上传文件及下载文件，这些也是 exec_command()方法实现的核心。具体操作如下：

```
(ansible1.1) [root@master ~]# sshpass -p @SHENcong19920522 ssh ceph-1
"hostname -s"
ceph-1
(ansible1.1) [root@master ~]# echo @SHENcong19920522 | sshpass -d0 ssh
ceph-1 "hostname -s"
ceph-1
# 上传文件——基于 SCP 命令
(ansible1.1) [root@master ~]# sshpass -d0 scp hosts root@ceph-1:/tmp <
password.txt
# 下载文件——基于 SCP 命令
(ansible1.1) [root@master ~]# sshpass -d0 scp root@ceph-1:/tmp/hosts
/root/shencong/ < password.txt
(ansible1.1) [root@master ~]# cat put.txt
put /root/hosts /tmp/hosts
# 上传文件——基于 SFTP 命令
(ansible1.1) [root@master ~]# sshpass -p@SHENcong19920522 sftp ceph-1 put
< put.txt
Connected to ceph-1.
sftp> put /root/hosts /tmp/hosts
Uploading /root/hosts to /tmp/hosts
# 下载文件——基于 SFTP 命令
(ansible1.1) [root@master ~]# sshpass -p@SHENcong19920522 sftp ceph-1 <
get.txt
Connected to ceph-1.
sftp> get /tmp/hosts /root/shencong/hosts
Fetching /tmp/hosts to /root/shencong/hosts
```

有了上面的基础之后，再来看一下 exec_command()的实现代码，具体如下：

```
# 源码位置: lib/ansible/runner/connection_plugins/ssh.py
# ...

class Connection(object):
    # ...
```

```
    def _password_cmd(self):
        if self.password:
            try:
                # 本地执行命令，调用 subprocess 模块
                p = subprocess.Popen(["sshpass"], stdin=subprocess.PIPE,
                    stdout=subprocess.PIPE, stderr=subprocess.PIPE)
                p.communicate()
            except OSError:
                raise errors.AnsibleError("to use -c ssh with passwords, you
must install the sshpass
                            program")
            (self.rfd, self.wfd) = os.pipe()
            return ["sshpass", "-d%d" % self.rfd]
        return []

    def _send_password(self):
        # 发送密码
        if self.password:
            os.close(self.rfd)
            os.write(self.wfd, "%s\n" % self.password)
            os.close(self.wfd)

    def exec_command(self, cmd, tmp_path, sudo_user,sudoable=False,
executable='/bin/sh'):
        ''' 远程执行命令 '''

        ssh_cmd = self._password_cmd()
        # 合成 SSH 命令，第一部分是 SSHPASS 命令，接着是 SSH 及通用参数，最后是主机名。
            待执行的命令在后面合成
        ssh_cmd += ["ssh", "-tt", "-q"] + self.common_args + [self.host]

        if not self.runner.sudo or not sudoable:
            # 非 sudo 直接合成完整的远程执行命令
            if executable:
                ssh_cmd.append(executable + ' -c ' + pipes.quote(cmd))
            else:
                ssh_cmd.append(cmd)
        else:
            # 需要先生成 sudo 形式的命令，然后再与前面的命令合成完整的命令
            sudocmd, prompt = utils.make_sudo_cmd(sudo_user, executable, cmd)
            ssh_cmd.append(sudocmd)

        vvv("EXEC %s" % ssh_cmd, host=self.host)
        try:
            # Make sure stdin is a proper (pseudo) pty to avoid: tcgetattr
errors
            import pty
            master, slave = pty.openpty()
            p = subprocess.Popen(ssh_cmd, stdin=slave,
                        stdout=subprocess.PIPE, stderr=subprocess.PIPE)
            stdin = os.fdopen(master, 'w', 0)
        except:
            p = subprocess.Popen(ssh_cmd, stdin=subprocess.PIPE,
                        stdout=subprocess.PIPE, stderr=subprocess.PIPE)
```

```
        stdin = p.stdin

    # 如果需要输入密码，会在这里进行处理
    self._send_password()

    if self.runner.sudo and sudoable and self.runner.sudo_pass:
        fcntl.fcntl(p.stdout, fcntl.F_SETFL,
                fcntl.fcntl(p.stdout, fcntl.F_GETFL) | os.O_NONBLOCK)
        sudo_output = ''
        # 循环检查是否要求输入 sudo 密码
        while not sudo_output.endswith(prompt):
            rfd, wfd, efd = select.select([p.stdout], [], [p.stdout],
self.runner.timeout)
            if p.stdout in rfd:
                chunk = p.stdout.read()
                if not chunk:
                    raise errors.AnsibleError('ssh connection closed
waiting for sudo password
                                                    prompt')
                sudo_output += chunk
            else:
                stdout = p.communicate()
                raise errors.AnsibleError('ssh connection error waiting
for sudo password prompt')
        stdin.write(self.runner.sudo_pass + '\n')
        fcntl.fcntl(p.stdout, fcntl.F_SETFL, fcntl.fcntl(p.stdout,
fcntl.F_GETFL) & ~os.O_NONBLOCK)


    stdout = ''
    stderr = ''
    while True:
        rfd, wfd, efd = select.select([p.stdout, p.stderr], [], [p.stdout,
p.stderr], 1)
        if p.stdout in rfd:
            dat = os.read(p.stdout.fileno(), 9000)
            stdout += dat
            if dat == '':
                p.wait()
                break
        elif p.stderr in rfd:
            dat = os.read(p.stderr.fileno(), 9000)
            stderr += dat
            if dat == '':
                p.wait()
                break
        elif p.poll() is not None:
            break
    stdin.close() # 从 stdout 读出数据后就要关闭 stdin (see also issue #848)

    if p.returncode != 0 and stderr.find('Bad configuration option:
ControlPersist') != -1:
        # 抛出异常
        # ...
```

```
        return (p.returncode, '', stdout, stderr)
    # ...
```

上面的代码使用 subprocess 模块执行最后生成的 SSHPASS 和 SSH 组合命令，如果设置了 sudo_pass 值，会输入提权的密码，确保程序正常执行，其执行逻辑和 paramiko 模块基本一致。那么如何像前面一样对这里的代码进行改造，形成自己的 Python 代码库呢？这个就当作课后作业留给读者，这里不再演示。

3. put_file()方法

put_file()方法就是使用前面的 SSHPASS+SCP/SFTP 命令实现文件的远程上传功能。具体代码如下：

```python
# 源码位置: lib/ansible/runner/connection_plugins/ssh.p
# ...

class Connection(object):

    # ...

    def put_file(self, in_path, out_path):
        '''从本地向远端上传文件'''
        vvv("PUT %s TO %s" % (in_path, out_path), host=self.host)
        if not os.path.exists(in_path):
            # 本地文件不存在，抛出异常
            raise errors.AnsibleFileNotFound("file or module does not exist:
%s" % in_path)
        # 生成 sshpass 命令部分
        cmd = self._password_cmd()

        # 使用 scp 或者 sftp 命令组成上传命令
        if C.DEFAULT_SCP_IF_SSH:
            cmd += ["scp"] + self.common_args
            cmd += [in_path,self.host + ":" + out_path]
            indata = None
        else:
            cmd += ["sftp"] + self.common_args + [self.host]
            indata = "put %s %s\n" % (in_path, out_path)
        # 使用 subprocess 模块执行 cmd 命令
        p = subprocess.Popen(cmd, stdin=subprocess.PIPE,
                         stdout=subprocess.PIPE, stderr=subprocess.PIPE)
        # 对于传入了 sudo_pass 的情况，需要输入 sudo 密码
        self._send_password()
        # 获取执行结果
        stdout, stderr = p.communicate(indata)

        if p.returncode != 0:
```

```
        raise errors.AnsibleError("failed to transfer file to %s:\n%s\
n%s" % (out_path, stdout, stderr))
    # ...
```

4．fetch_file ()方法

fetch_file()方法同样是使用 SSHPASS+SCP/SFTP 命令实现文件的远程下载功能。具体代码如下：

```
# 源码位置：lib/ansible/runner/connection_plugins/ssh.py
# ...

class Connection(object):
    ''' 基于 SSH 命令的 SSH 连接 '''
    # ...

    def fetch_file(self, in_path, out_path):
        ''' 从远端下载文件到本地 '''
        vvv("FETCH %s TO %s" % (in_path, out_path), host=self.host)
        cmd = self._password_cmd()

        # 使用 SCP 或者 SFTP 命令生成下载文件的命令
        if C.DEFAULT_SCP_IF_SSH:
            cmd += ["scp"] + self.common_args
            cmd += [self.host + ":" + in_path, out_path]
            indata = None
        else:
            cmd += ["sftp"] + self.common_args + [self.host]
            indata = "get %s %s\n" % (in_path, out_path)
        # 使用 subprocess 模块执行 cmd 命令
        p = subprocess.Popen(cmd, stdin=subprocess.PIPE,
                        stdout=subprocess.PIPE, stderr=subprocess.PIPE)
        # 处理可能输入密码的情况
        self._send_password()
        stdout, stderr = p.communicate(indata)

        if p.returncode != 0:
            raise errors.AnsibleError("failed to transfer file from %s:\n%s\
n%s" % (in_path, stdout, stderr))

    # ...
```

5．close()方法

close()方法在此处并没有任何操作。

至此，Ansible 1.1 的最底层的连接过程讲解完毕。在大多数情况下使用的是 local.py、paramiko_ssh.py 和 ssh.py 这 3 个插件的 Connection 类，而 chroot.py 和 fireball.py 并不常用，有兴趣的读者可以自行研究其源码，这里不再过多介绍。

1.4.4 小结

本节分析了 Ansible 1.1 中的三个连接插件的源码，即本地插件（local.py）、paramiko 插件（paramiko_ssh.py）及 SSH 插件（ssh.py）。通过详细解读源码并测试其核心类与方法，不仅可以理解插件的连接过程，还可以将插件源码改造后用在其他项目中。

1.5 ping 模块的执行流程

本节将分析 Ansible 1.1 工具背后的运行逻辑，同时也会深入学习 Ansible 1.1 中核心模块的源码，这些源码全部位于 lib/anisble/runner 目录下。

1.5.1 ansible 命令

本节将继续使用 1.1 节中搭建的调试环境。为了能更好地看到效果，将 1.1 节中调试的 hosts 文件修改如下：

```
[shen@localhost Desktop]$ cat /etc/hosts
127.0.0.1        localhost.localdomain localhost
::1      localhost6.localdomain6 localhost6


192.168.26.120  ceph-1
192.168.26.121  ceph-2
192.168.26.122  ceph-3

[shen@localhost Desktop]$ cat hosts
[nodes]
ceph-[1:3]

[nodes:vars]
ansible_ssh_user=root
ansible_ssh_pass=@SHENcong19920522
```

调试的参数和 1.1 节中保持不变，接下来我们将分析 ansible all -i ~/Desktop/hosts -m ping -vvv 这行命令背后的执行逻辑。我们直接从源码中 bin 目录下的 ansible 命令文件开始追踪，其代码非常简单，具体如下：

```
# 源码位置: bin/ansible
# ...

if __name__ == '__main__':
    cli = Cli()
    # 解析选项和参数
    (options, args) = cli.parse()
```

```
try:
    # 解析参数并运行 ansible 模块
    (runner, results) = cli.run(options, args)
    # 得到结果并解析结果
    for result in results['contacted'].values():
        if 'failed' in result or result.get('rc', 0) != 0:
            sys.exit(2)
    if results['dark']:
        sys.exit(2)
except errors.AnsibleError, e:
    # Generic handler for ansible specific errors
    print "ERROR: %s" % str(e)
    sys.exit(1)
```

从上面的代码中可以看到，ansible 命令中最核心的就是 cli.run()方法。首先来看 Cli 类的定义，其代码如下：

```
# 源码位置: bin/ansible
# ...

class Cli(object):
    ''' ansible 命令背后的代码 '''

    def __init__(self):
        self.stats = callbacks.AggregateStats()
        self.callbacks = callbacks.CliRunnerCallbacks()

    def parse(self):
        '''为 bin/ansible 创建选项解析器 '''

        parser = utils.base_parser(
            constants=C,
            runas_opts=True,
            subset_opts=True,
            async_opts=True,
            output_opts=True,
            connect_opts=True,
            check_opts=True,
            diff_opts=False,
            usage='%prog <host-pattern> [options]'
        )

        # ansible 命令支持的选项，比较常见的有-a、-m、--list-hosts
        parser.add_option('-a', '--args', dest='module_args',
            help="module arguments", default=C.DEFAULT_MODULE_ARGS)
        parser.add_option('-m', '--module-name', dest='module_name',
            help="module name to execute (default=%s)" % C.DEFAULT_MODULE_
NAME,
            default=C.DEFAULT_MODULE_NAME)
        parser.add_option('--list-hosts', dest='listhosts', action='store_
true',
            help="dump out a list of hosts matching input pattern, does not
execute any modules!")
```

```
        options, args = parser.parse_args()
        # 得到选项解析结果
        self.callbacks.options = options

        if len(args) == 0 or len(args) > 1:
            parser.print_help()
            sys.exit(1)
        return (options, args)

    def run(self, options, args):
        ''' 使用 Runner 库执行 SSH 操作 '''

        pattern = args[0]

        # 传入参数中指定的 hosts 路径，如果没有，则默认为/etc/ansible/hosts
        inventory_manager = inventory.Inventory(options.inventory)
        if options.subset:
            inventory_manager.subset(options.subset)
        # 根据传入的 pattern 参数找出操作的目标主机
        hosts = inventory_manager.list_hosts(pattern)
        if len(hosts) == 0:
            print >>sys.stderr, "No hosts matched"
            sys.exit(1)

        # 如果是--list-hosts 选项，则只是列出 ansible 主机列表，然后退出
        if options.listhosts:
            for host in hosts:
                print '    %s' % host
            sys.exit(0)

        # command 模块必须要带参数
        if options.module_name == 'command' and not options.module_args:
            print >>sys.stderr, "No argument passed to command module"
            sys.exit(1)

        # 一些选项处理
        # ...

        # 最核心的类
        runner = Runner(
            module_name=options.module_name, module_path=options.module_path,
            module_args=options.module_args,
            remote_user=options.remote_user, remote_pass=sshpass,
            inventory=inventory_manager, timeout=options.timeout,
            private_key_file=options.private_key_file,
            forks=options.forks,
            pattern=pattern,
            callbacks=self.callbacks, sudo=options.sudo,
            sudo_pass=sudopass, sudo_user=options.sudo_user,
            transport=options.connection, subset=options.subset,
            check=options.check,
            diff=options.check
        )
```

```
        if options.seconds:
            # 后台异步执行
            # ...
        else:
            # 一般模块调用会执行这里的语句
            results = runner.run()

        return (runner, results)

    # ...
# ...
```

上面的代码逻辑非常清晰，(options, args) = cli.parse()就是解析命令行的参数，包括-a、-m 及--list-hosts 选项；在 cli.parse()方法中通过 utils.base_parser()创建一个基本的 parser，而这个 parser 则包含众多的通用选项，如-i、-v 和-f 等。接下来就是调用 cli.run()方法执行 ansible 模块并获取相应的执行结果。而 run()方法的功能如下：

（1）根据传入的 hosts 文件或者默认的 hosts 文件生成 inventory_manager 对象。

（2）通过 inventory_manager 对象及输入的目标主机格式可以得到匹配的主机列表。如果匹配的主机为 0，则打印错误信息并退出。

（3）如果是--list-hosts 选项，只需要列出匹配的 Ansible 主机列表，然后直接退出。

（4）单独处理 command 模块不带参数的情况，遇到了直接打印提示信息并退出。

（5）处理一些其他选项，如是否要求运行中输入密码、是否为 sudo 形式等。

（6）根据前面得到的参数，如模块、模块参数和 inventory_manager 等来构建运行对象（runner），然后调用 runner.run()方法运行模块任务。

（7）返回 runner 对象及模块执行结果。

可以看到，run()方法中的核心执行流程转向了 Runner 类，整个 run()方法的核心又是调用 runner 对象的 run()方法。接下来的 1.5.2 节中将介绍 runner 目录下的一些核心代码，为后续追踪 ping 模块的执行过程做好准备。

1.5.2　runner 目录下的核心源码

在真正分析模块的执行代码之前，需要先介绍一下 runner 目录下的一些核心代码，只有理解了这些核心代码，才能更好地理解 ansible 模块运行的过程。

1. connection_plugins连接插件

connection_plugins 是 Ansible 中的连接插件模块，该部分插件代码位于 lib/ansible/connection-plugins 目录下，而该目录下的不同代码文件（除了__init__.py 文件）对应不同的连接方式，这里选择底层连接通信的方式。如果设置 ansible_connection 为 local，则连接方式为本地连接，将调用 local.py 文件中的 Connection 类；如果设置 ansible_connection

为 paramiko，则与远端主机的通信方式将使用 paramiko 模块，将调用 paramiko_ssh.py 文件中的 Connection 类。那么这一步该如何选择呢？请看下面的代码：

```
# 源码位置：lib/ansible/runner/__init__.py
# ...
import connection

class Runner(object):
    ''' ansible 的核心 API 接口 '''

    def __init__(self,
        ...
    ):
        # ...
        # 重要，实例化 Connection 类
        self.connector        = connection.Connection(self)
        # ...

    def _executor_internal_inner(self, host, module_name, module_args,
inject, port, is_chained=False,
        complex_args=None):
        # ...

        # 非常重要的参数，贯穿后面的大部分方法
        conn = self.connector.connect(actual_host, actual_port, actual_
user, actual_pass, actual_transport)
        # ...

    # ...
```

继续追踪导入的 connection 模块代码，具体如下：

```
# 源码位置：lib/ansible/runner/connection.py
# ...

class Connection(object):
    ''' 远端主机统一的连接类 '''

    def __init__(self, runner):
        self.runner = runner

    def connect(self, host, port, user, password, transport):
        conn = None
        # 获取对应的连接插件中的 Connection 对象
        conn = utils.plugins.connection_loader.get(transport, self.runner,
host, port, user=user, password=password)
        if conn is None:
            raise AnsibleError("unsupported connection type: %s" % transport)
        # 注意，后面所有的 connect() 方法返回的是连接插件的 Connection 对象本身
        # 所以这个 self.active 表示的就是连接对象
        self.active = conn.connect()
        return self.active
```

　　上述代码中定义的 Connection 类用于对接 Ansible 的底层通信模式，如果在 ansible.cfg 中指定 ansible_connection 参数的值为 ssh，此时 utils.plugins.connection_loader.get()方法将加载 connection_plugins/ssh.py 文件中的 Connection 类并依据后面的参数初始化该类。注意，每种通信方式中 Connection 对象的 connect()方法最后返回的都是 self，也就是对象本身。因此 self.active 保存的就是 ssh.py 文件中的 Connection 对象。模块加载的代码也非常简单，使用了 imp 模块，这里不再详细分析。为了能让读者更容易理解上述代码，笔者准备了一个简单的操作案例，具体如下：

```
>>> from ansible.runner import Runner
>>> from ansible import inventory
>>> from ansible.runner import connection
>>> inventory_manager = inventory.Inventory('/root/hosts')
>>> runner = Runner(inventory=inventory_manager)
>>> conn = connection.Connection(runner)
>>> conn.connect('ceph-1', 22, 'root', '@SHENcong19920522', 'ssh')
<ansible.runner.connection_plugins.ssh.Connection object at 0x7f9b77501b10>
>>> conn_paramiko = conn.connect('ceph-1', 22, 'root', '@SHENcong19920522',
 'paramiko')
<ansible.runner.connection_plugins.paramiko_ssh.Connection object at 0x7
f9b7748e1d0>
>>> conn_paramiko.exec_command('hostname', None, None)
(0, '', 'ceph-1\n', '')
```

　　上述案例首先创建了一个 Runner 对象，接着用该对象初始化 connection.py 文件中的 Connection 类，并在 connect()方法中传入了相应的 SSH 通信方式。可以看到返回的结果是对应连接插件中定义的 Connection 类对象，这里甚至可以使用刚刚创建的 conn_paramiko 对象去执行远程操作（即调用 exec_command()方法）。

2. action_plugins动作插件

　　这里姑且称 action_plugins 为模块的动作插件，其代码均位于 lib/ansible/runner/action_plugins 目录下。每个动作插件对应一个代码文件，每个动作插件中都定义了一个 ActionModule 类且都定义了 run()方法，而这个 run()便是该动作插件的核心调用方法。后面在分析模块的执行过程时我们会看到：每个模块会执行各自的动作插件中的 run()方法。例如 copy 模块，会调用 action_plugins 下 copy.py 插件中的 run()方法；而对于 ping 模块，没有对应的动作插件（ping.py），则会调用通用动作插件（normal.py）中的 run()方法。

　　在 Runner 对象的_executor_internal_inner()方法中，加载动作插件并执行的代码如下：

```
# 源码位置: lib/ansible/runner/__init__.py
# ...

class Runner(object):
    # ...

    def _executor_internal_inner(self, host, module_name, module_args,
inject, port, is_chained=False, complex_args=None):
```

```
        # ...

        # 根据模块的名称，获取对应动作插件的 ActionModule 对象
        if module_name in utils.plugins.action_loader:
            if self.background != 0:
                # 抛出异常，省略异常信息
                raise errors.AnsibleError(...)
            handler = utils.plugins.action_loader.get(module_name, self)
        elif self.background == 0:
            handler = utils.plugins.action_loader.get('normal', self)
        else:
            handler = utils.plugins.action_loader.get('async', self)

        # ...

        # 调用动作插件中的 run() 方法
        result = handler.run(conn, tmp, module_name, module_args, inject,
complex_args)

        # ...

    # ...
```

上面的代码首先判断执行的模块是否有对应名称的动作插件，如果有则直接加载该动作插件中的 ActionModule 类；而对于非异步执行且没有对应插件的模块，会默认加载 normal.py 插件中的 ActionModule 类；如果需要异步执行，则加载 async.py 插件中的 ActionModule 类。为了能更好地理解上述语句，继续拆解上面的代码如下：

```
(ansible1.1) [root@master ~]# ls ~/.pyenv/versions/ansible1.1/lib/python2.7/
site-packages/ansible/runner/action_plugins/
add_host.py  async.py  copy.py  debug.py  fail.py  fetch.py  group_by.py
__init__.py  normal.py  pause.py  raw.py  script.py  template.py
```

以上是 Ansible 1.1 所支持的动作插件，接下来执行如下操作：

```
>>> from ansible import utils
>>> 'fetch' in utils.plugins.action_loader
True
>>> 'copy' in utils.plugins.action_loader
True
>>> 'ping' in utils.plugins.action_loader
False
>>> from ansible.runner import Runner
>>> from ansible import inventory
>>> inventory_manager = inventory.Inventory('/root/hosts')
>>> runner = Runner(inventory=inventory_manager)
>>> utils.plugins.action_loader.get('copy', runner)
<ansible.runner.action_plugins.copy.ActionModule object at 0x7fa3f65d4550>
```

从上面的操作中可以看到，通过模块名字就能判断是否有对应的插件。如果有，则可以通过 utils.plugins.action_loader.get()方法获取对应插件中的 ActionModule 对象。这里可以思考一下，上面的代码中为何可以用 module_name in utils.plugins.action_loader 这样的判断语句呢？这涉及 Python 中一个非常重要的概念：魔法函数。in 判断操作涉及的魔法函

数为__contains__()，而这个魔法函数正好在 plugins.py 文件中有赋值定义，具体如下：

```python
# 源码位置: lib/ansible/utils/plugins.py
# ...

class PluginLoader(object):
    # ...

    def find_plugin(self, name):
        """通过名称找到对应的插件"""
        suffix = ".py"
        if not self.class_name:
            suffix = ""
        for i in self._get_paths():
            path = os.path.join(i, "%s%s" % (name, suffix))
            if os.path.exists(path):
                return path
        return None

    def has_plugin(self, name):
        """检查对应名称的插件是否存在"""
        return self.find_plugin(name) is not None

    # 魔法属性
    __contains__ = has_plugin

    # ...

action_loader      = PluginLoader('ActionModule',   'ansible.runner.
action_plugins',      C.DEFAULT_ACTION_PLUGIN_PATH,      'action_plugins')
```

　　到这里就很清楚了，对于 module_name in utils.plugins.action_loader 这样的操作，Ansible 会调用 PluginLoader 对象下的 has_plugin()方法，该方法的第一个参数正是待判断的模块名（module_name）。此外，has_plugin()方法还会调用对象内部的 find_plugin()方法继续进行判断，而 find_plugin()方法会在当前插件目录下寻找 module_name.py（这里 module_name 表示模块名称，如 module_name=copy，则对应为 copy.py）文件是否存在。最后，对应的动作插件如果存在则返回 True，否则返回 False，这便是前面调用时看到的现象。

3. 其他核心方法

　　了解了前面介绍的两个插件后，继续来看 Runner 类中定义的其他方法，具体代码如下：

```python
# 源码位置: lib/ansible/runner/__init__.py
# ...

class Runner(object):
    ''' Ansible 的核心 API 接口'''

    # ...
```

```
def _transfer_str(self, conn, tmp, name, data):
    ''' 传输字符串到远端文件中 '''

    # data 就是要传输的内容
    if type(data) == dict:
        data = utils.jsonify(data)

    # 在本地创建一个临时文件
    afd, afile = tempfile.mkstemp()
    afo = os.fdopen(afd, 'w')
    try:
        if not isinstance(data, unicode):
            #ensure the data is valid UTF-8
            data.decode('utf-8')
        else:
            data = data.encode('utf-8')
        # 将需要上传的内容保存到临时文件中
        afo.write(data)
    except:
        raise errors.AnsibleError("failure encoding into utf-8")
    afo.flush()
    afo.close()

    # tmp 为远端的临时目录，name 为文件名，都由外部参数传入
    remote = os.path.join(tmp, name)
    try:
        # 调用 put_file() 方法将文件上传到远端服务器上，其中，afile 为本地文件地
        址，remote 为远端文件地址
        conn.put_file(afile, remote)
    finally:
        # 删除本地的临时文件
        os.unlink(afile)
    # 返回上传的远端文件地址
    return remote

# ...
```

上面的代码也容易理解，其中比较重要的就是 conn 参数，其正是 1.5.2 节中介绍的 conn。因此，这里上传文本的逻辑便非常清晰了：首先将文本内容保存到本地的临时文件中，然后通过 conn.put_file() 方法上传到远端的服务器上，最后删除本地的临时文件。

```
# 源码位置: lib/ansible/runner/__init__.py
# ...

class Runner(object):

    # ...

    def _low_level_exec_command(self, conn, cmd, tmp, sudoable=False,
executable=None):
        ''' 通过 SSH 远程执行命令并返回结果'''

        if executable is None:
```

```
        executable = '/bin/sh'

    sudo_user = self.sudo_user
    # 获取远程执行命令的输出
    rc, stdin, stdout, stderr = conn.exec_command(cmd, tmp, sudo_user,
sudoable=sudoable, executable=executable)

    # 正确的输出结果
    if type(stdout) not in [ str, unicode ]:
        out = ''.join(stdout.readlines())
    else:
        out = stdout

    # 错误输出
    if type(stderr) not in [ str, unicode ]:
        err = ''.join(stderr.readlines())
    else:
        err = stderr

    # 返回不同情况的结果
    if rc is not None:
        return dict(rc=rc, stdout=out, stderr=err)
    else:
        return dict(stdout=out, stderr=err)

    # ...
```

　　了解了 conn 参数的含义后，再看上面的代码会不会觉得非常简单呢？这里最核心的动作就是调用 conn 的 exec_command()方法远程执行 shell 命令并返回相应的结果，下面通过几行代码展示其功能，操作示例如下：

```
>>> from ansible.runner import Runner, connection
>>> from ansible import inventory
>>> inventory_manager = inventory.Inventory('/root/hosts')
>>> runner = Runner(inventory=inventory_manager)
>>> conn = connection.Connection(runner).connect('ceph-1', 22, 'root',
'@SHENcong19920522', 'paramiko')
>>> runner._low_level_exec_command(conn, 'hostname', None)
{'stdout': 'ceph-1\n', 'stderr': '', 'rc': 0}
>>> runner._low_level_exec_command(conn, 'cat /etc/xxx', None)
{'stdout': '', 'stderr': 'cat: /etc/xxx: No such file or directory\n', 'rc': 1}
```

　　笔者先构建了一个 Inventory 对象，由此得到 Runner 对象。接着和前面一样得到远程连接对象 conn，最后将其作为参数传入 runner._low_level_exec_command()方法中去执行远程操作。上面分别测试了执行正确命令和错误命令的两种情况。清楚了核心方法的作用后，就可以继续追踪 Runner 类中的另外两个方法——_remote_md5()和_make_tmp_path()了，具体代码如下：

```
# 源码位置: lib/ansible/runner/__init__.py
# ...
```

```python
class Runner(object):

    # ...

    def _remote_md5(self, conn, tmp, path):
        '''远程执行 md5sum 命令，获取文件的 MD5 值，如果没有文件则返回 0 '''

        path = pipes.quote(path)
        test = "rc=0; [ -r \"%s\" ] || rc=2; [ -f \"%s\" ] || rc=1; [ -d \"%s\" ] && rc=3" % (path, path, path)
        # 兼容各种系统，计算文件 MD5 值的命令
        md5s = [
            "(/usr/bin/md5sum %s 2>/dev/null)" % path,   # Linux
            "(/sbin/md5sum -q %s 2>/dev/null)" % path,   # ?
            "(/usr/bin/digest -a md5 %s 2>/dev/null)" % path,   # Solaris 10+
            "(/sbin/md5 -q %s 2>/dev/null)" % path,      # Freebsd
            "(/usr/bin/md5 -n %s 2>/dev/null)" % path,   # Netbsd
            "(/bin/md5 -q %s 2>/dev/null)" % path        # Openbsd
        ]

        # 利用 shell 命令中的 || 特性，前一个命令执行失败，才会执行 || 后面的命令
        cmd = " || ".join(md5s)
        cmd = "%s; %s || (echo \"${rc}  %s\")" % (test, cmd, path)
        data = self._low_level_exec_command(conn, cmd, tmp, sudoable=False)
        data2 = utils.last_non_blank_line(data['stdout'])
        try:
            # 只取 MD5 的结果并返回。输出结果有两列，取第一列，即 MD5 的结果
            return data2.split()[0]
        except IndexError:
            # 输出错误信息
            # ...
            return "INVALIDMD5SUM"

    # ************************************************************

    def _make_tmp_path(self, conn):
        '''在远端主机上创建临时目录并返回这个临时目录的路径'''

        # 唯一的目录名
        basefile = 'ansible-%s-%s' % (time.time(), random.randint(0, 2**48))
        # 生成完整的临时目录路径
        basetmp = os.path.join(C.DEFAULT_REMOTE_TMP, basefile)
        if self.sudo and self.sudo_user != 'root' and basetmp.startswith('$HOME'):
            basetmp = os.path.join('/tmp', basefile)

        # 创建临时目录的命令
        cmd = 'mkdir -p %s' % basetmp
        if self.remote_user != 'root':
            cmd += ' && chmod a+rx %s' % basetmp
        cmd += ' && echo %s' % basetmp

        # 远程执行创建临时目录的命令
        result = self._low_level_exec_command(conn, cmd, None, sudoable=
```

```
False)
        rc = utils.last_non_blank_line(result['stdout']).strip() + '/'
        return rc
```

上面的代码中添加了详细的注释，逻辑也非常清晰。下面依旧通过调用_remote_md5()
和_make_tmp_path()方法来进一步理解它们的作用，具体操作如下：

```
[root@ceph-1 ~]# cat test_md5
33
测试计算文件的 MD5 值
[root@ceph-1 ~]# md5sum /root/test_md5
733e2d6cf1cb88f7f6cebc27b6cc1b95  /root/test_md5
[root@ceph-1 ~]# ls ~/.ansible/
[root@ceph-1 ~]#
```

接着继续在虚拟环境中的 Python 交互模式下执行如下操作：

```
# 这里 conn 及 runner 的获取和上一个操作案例一致，这里就不再书写重复的代码了
>>> runner._remote_md5(conn, None, '/root/test_md5')
'733e2d6cf1cb88f7f6cebc27b6cc1b95'
>>> runner._make_tmp_path(conn)
'/root/.ansible/tmp/ansible-1602943126.8-228804349893657/'
```

上述代码执行后，ceph-1 节点上 Ansible 默认的临时目录下又生成了一个新的目录，
具体如下：

```
[root@ceph-1 ~]# ls ~/.ansible/tmp/
ansible-1602943126.8-228804349893657
```

接下来是一个非常核心的方法——_copy_module()，其代码如下：

```
# 源码位置: lib/ansible/runner/__init__.py
# ...

class Runner(object):

    # ...

    def _copy_module(self, conn, tmp, module_name, module_args, inject,
complex_args=None):
        '''通过 SFTP 传输 module，但是不执行'''

        # 如果 complex args 是 None，直接设置为{}

        if module_name.startswith("/"):
            raise errors.AnsibleFileNotFound("%s is not a module" % module_
name)

        # 找到模块的本地路径，下面将演示这个代码的输出
        in_path = utils.plugins.module_finder.find_plugin(module_name)
        if in_path is None:
            # 模块不存在，则抛出异常
            raise errors.AnsibleFileNotFound("module %s not found in %s" %
                (module_name, utils.plugins.module_finder.print_paths()))
```

```
        # 远端临时路径
        out_path = os.path.join(tmp, module_name)

        module_data = ""
        is_new_style=False

        with open(in_path) as f:
            # 模块中的代码内容
            module_data = f.read()
            # 每个模块中都有这个module_common.REPLACER字符串内容
            if module_common.REPLACER in module_data:
                is_new_style=True

            complex_args_json = utils.jsonify(complex_args)
            encoded_args = "\"\"\"%s\"\"\"" % module_args.replace("\"","\\\"")
            encoded_lang = "\"\"\"%s\"\"\"" % C.DEFAULT_MODULE_LANG
            encoded_complex = "\"\"\"%s\"\"\"" % complex_args_json.replace
("\\", "\\\\")

            # 将模块代码中的一些内容替换成module_common中指定的内容
            module_data = module_data.replace(module_common.REPLACER,
                              module_common.MODULE_COMMON)

            """
            module_data中添加了module_common.MODULE_COMMON内容后，就会有
            module_common.REPLACER_ARGS、module_common.REPLACER_LANG 和
            module_common.REPLACER_COMPLEX,
            具体可以去看module_common.MODULE_COMMON的内容
            """
            module_data = module_data.replace(module_common.REPLACER_ARGS,
encoded_args)
            module_data = module_data.replace(module_common.REPLACER_LANG,
encoded_lang)
            module_data= module_data.replace(module_common.REPLACER_COMPLEX,
                              encoded_complex)

            if is_new_style:
                facility = C.DEFAULT_SYSLOG_FACILITY
                if 'ansible_syslog_facility' in inject:
                    facility = inject['ansible_syslog_facility']
                module_data = module_data.replace('syslog.LOG_USER', "syslog.%s"
% facility)

        lines = module_data.split("\n")
        shebang = None
        if lines[0].startswith("#!"):
            shebang = lines[0]
            # 获取#!后面的解释器
            args = shlex.split(str(shebang[2:]))
            interpreter = args[0]
            interpreter_config = 'ansible_%s_interpreter' % os.path.basename
(interpreter)

            if interpreter_config in inject:
```

```
            lines[0] = shebang = "#!%s %s" % (inject[interpreter_config],
" ".join(args[1:]))
            module_data = "\n".join(lines)

    # 最后将完整的模块代码上传到远端的临时目录上
    self._transfer_str(conn, tmp, module_name, module_data)

    return (out_path, is_new_style, shebang)

# ...
```

为了帮助读者理解上面的代码，先完成一个简单的演示，具体如下：

```
(ansible1.1) [root@master ~]# python
Python 2.7.18 (default, Oct 13 2020, 23:55:15)
[GCC 4.8.5 20150623 (Red Hat 4.8.5-39)] on linux2
Type "help", "copyright", "credits" or "license" for more information.
>>> from ansible import utils
>>> utils.plugins.module_finder.find_plugin('ping')
'/root/.pyenv/versions/ansible1.1/share/ansible/ping'
>>> utils.plugins.module_finder.find_plugin('copy')
'/root/.pyenv/versions/ansible1.1/share/ansible/copy'
```

上面通过 find_plugin()方法得到的地址正是模块文件的全路径，即_copy_module()方法中的 in_path 值就是对应输入模块的完整路径，而 out_path 是一个在临时目录下的完整路径。此外，前面在介绍 ansible 模块代码时发现每个模块中都有如下一行文本：

```
#<<INCLUDE_ANSIBLE_MODULE_COMMON>>
```

从 module_common.py 中定义的 REPLACER 变量可知，该变量的值就是 Ansible 的所有模块中出现的那行文本（见上面的那行代码）。该变量的定义语句如下：

```
# 源码位置: lib/ansible/module_common.py
# ...

REPLACER = "#<<INCLUDE_ANSIBLE_MODULE_COMMON>>"
REPLACER_ARGS = "<<INCLUDE_ANSIBLE_MODULE_ARGS>>"
REPLACER_LANG = "<<INCLUDE_ANSIBLE_MODULE_LANG>>"
REPLACER_COMPLEX = "<<INCLUDE_ANSIBLE_MODULE_COMPLEX_ARGS>>"

MODULE_COMMON = """

# == BEGIN DYNAMICALLY INSERTED CODE ==

MODULE_ARGS = <<INCLUDE_ANSIBLE_MODULE_ARGS>>
MODULE_LANG = <<INCLUDE_ANSIBLE_MODULE_LANG>>
MODULE_COMPLEX_ARGS = <<INCLUDE_ANSIBLE_MODULE_COMPLEX_ARGS>>

# ...

"""
```

_copy_module()方法的整体逻辑并不复杂，首先读取模块的全部内容并保存到 module_data 中。接着用 module_common.MODULE_COMMON 替换出现的 module_common. REPLACER，此时 module_data 便有了 module_common.MODULE_ARGS、module_common.

MODULE_LANG 及 module_common.MODULE_COMPLEX_ARGS 这 3 个字符串，接着 Ansible 会将模块的参数及相关信息写入这 3 个变量中，得到一个完整的可运行的模块代码。最后调用前面介绍的_transfer_str()方法将该模块上传到远端主机的临时路径上（out_path）并返回相应的信息。

这里同样调用_copy_module()方法进行测试，以便让读者更直观地理解该方法的含义。下面会测试两个模块：ping 模块和 copy 模块。首先查看操作节点 ceph-1 的/tmp 目录，具体操作如下：

```
[root@ceph-1 tmp]# ls
[root@ceph-1 tmp]# pwd
/tmp
```

接着在虚拟环境中的 Python 交互模式下执行如下操作：

```
>>> from ansible import inventory
>>> from ansible.runner import Runner, connection
>>> inventory_manager = inventory.Inventory('/root/hosts')
>>> runner = Runner(inventory=inventory_manager)
>>> conn = connection.Connection(runner).connect('ceph-1', 22, 'root',
'@SHENcong19920522', 'paramiko')
>>> runner._copy_module(conn, '/tmp/', 'ping', 'data=hello', {})
('/tmp/ping', True, '#!/usr/bin/python')
>>> runner._copy_module(conn, '/tmp/', 'copy', 'src=/etc/hosts dest=/tmp', {})
('/tmp/copy', True, '#!/usr/bin/python')
```

此时在目标节点的/tmp 目录下就多了两个模块文件，即 ping 和 copy，具体如下：

```
[root@ceph-1 tmp]# ls
copy  ping
```

查看 ping 和 copy 模块中的代码可知，参数 MODULE_ARGS、MODULE_LANG 及 MODULE_COMPLEX_ARGS 的值分别如下：

```
# ping 模块中的部分代码
MODULE_ARGS = """data=hello"""
MODULE_LANG = """C"""
MODULE_COMPLEX_ARGS = """{}"""

# copy 模块中的部分代码
MODULE_ARGS = """src=/etc/hosts dest=/tmp"""
MODULE_LANG = """C"""
MODULE_COMPLEX_ARGS = """{}"""
```

直接运行 ping 模块的代码，结果如下（copy 模块的代码还缺少一步，需要将原文件从 master 节点复制到 ceph-1 节点上，因此暂时不能直接运行）：

```
[root@ceph-1 tmp]# python ping
{"changed": false, "ping": "hello"}
```

以上输出结果就是 Ansible 执行 ping 模块的核心输出结果，只不过这个结果最终会被封装一层用于控制台展示。而在 Runner 中还有一个核心方法用于实现上传完整的模块文件并远程执行的功能，它就是_execute_module()方法，具体代码如下：

```
# 源码位置：lib/ansible/runner/__init__.py
# ...

class Runner(object):
    # ...

    def _execute_module(self, conn, tmp, module_name, args, async_jid=None,
async_module=None, async_limit=None, inject=None, persist_files=False,
complex_args=None):

        '''运行已经传过去的模块'''

        # 处理 fireball 模式
        # ...

        # 调用_copy_module()上传完整的模块代码
        (remote_module_path, is_new_style, shebang) = self._copy_module
(conn, tmp, module_name, args, inject, complex_args)

        # 获取一些环节变量
        environment_string = self._compute_environment_string(inject)

        cmd_mod = ""
        if self.sudo and self.sudo_user != 'root':
            # deal with possible umask issues once sudo'ed to other user
            cmd_chmod = "chmod a+r %s" % remote_module_path
            self._low_level_exec_command(conn, cmd_chmod, tmp, sudoable=False)

        cmd = ""
        if not is_new_style:
            # 一般模块的执行不会到这里
            # ...
        else:
            if async_jid is None:
                cmd = "%s" % (remote_module_path)
            else:
                cmd = " ".join([str(x) for x in [remote_module_path, async_
jid, async_limit, async_module]])

        if not shebang:
            # 执行的模块脚本一定要以#!开头，后面便是指定该脚本的解释器
            raise errors.AnsibleError("module is missing interpreter line")

        cmd = " ".join([environment_string, shebang.replace("#!",""), cmd])
        if tmp.find("tmp") != -1 and C.DEFAULT_KEEP_REMOTE_FILES != '1' and
not persist_files:
            # 这里除了执行已经上传的模块文件的代码之外，还会在执行完成后将其删除
            cmd = cmd + "; rm -rf %s >/dev/null 2>&1" % tmp
        res = self._low_level_exec_command(conn, cmd, tmp, sudoable=True)
        # 将结果转换成 JSON 格式
        data = utils.parse_json(res['stdout'])
```

```
        if 'parsed' in data and data['parsed'] == False:
            data['msg'] += res['stderr']
    # 返回 ReturnData 对象
    return ReturnData(conn=conn, result=data)
```

从上面的执行逻辑中可以看到，_execute_module()方法先调用_copy_module()方法将完整的模块代码上传到远端主机的临时路径上，接着调用_low_level_exec_command()方法远程执行该模块代码（类似前面在目标主机上手动输入 python /tmp/ping 命令）。此外，在该方法中还添加了 rm -rf 命令，用于执行完模块代码后删除该文件。这里同样可以在 Python 交互模式下手工调用该方法，执行效果如下：

```
>>> from ansible import inventory
>>> from ansible.runner import Runner, connection
>>> inventory_manager = inventory.Inventory('/root/hosts')
>>> runner = Runner(inventory=inventory_manager)
>>> conn = connection.Connection(runner).connect('ceph-1', 22, 'root',
'@SHENcong19920522', 'paramiko')
>>> return_data = runner._execute_module(conn, '/tmp', 'ping', 'data=
hello', inject={})
>>> return_data.result
{u'changed': False, u'ping': u'hello'}
>>> return_data.is_successful()
True
>>> return_data.communicated_ok()
True
>>> return_data.comm_ok
True
>>> return_data.host
'ceph-1'
```

🔔说明：由于 Ansible 中会使用 rm -rf 命令来清除临时文件，这会导致在前一个案例中执行了_execute_module()方法后会删除/tmp 目录。因此，当再次执行 runner._execute_module(conn, '/tmp/', 'ping', 'data=hello', {})时就会失败，原因是/tmp 目录已经被删除，导致模块代码无法上传。此时需要手动创建/tmp 目录后才能成功执行该语句。

通过本节的学习，读者是否对 Ansible 工具有了更加深入的理解呢？接下来我们将通过 VSCode 跟踪 ping 模块的执行流程，了解 Ansible 模块的运行原理。

1.5.3　ping 模块的运行流程

本节将简单修改调试的 ansible 命令（修改 launch.json 内容），为 ping 模块加上一个 data=hello 参数。修改后的命令如下：

```
{
    "version": "0.2.0",
    "configurations": [
        {
```

```
            "name": "Python: Current File",
            "type": "python",
            "request": "launch",
            "program": "${file}",
            "env": {"PYTHONPATH":"${workspaceRoot}/lib"},
            "args": [
                "all",
                "-i", "~/Desktop/hosts",
                "-m", "ping",
                "-a", "data=hello",
                "-vvv"
            ],
            "console": "integratedTerminal"
        }
    ]
}
```

选中 bin/ansible 文件，然后选择 Run 菜单下的 Star Debugging 选项后开启调试模式。根据前面的经验，最终调用的是 lib/ansible/runner 目录下的__init__.py 中 Runner 对象的 run()方法。run()方法的具体内容如下：

```python
# 源码位置：lib/ansible/runner/__init__.py
# ...

class Runner(object):

    # ...

    def run(self):
        '''对匹配的主机执行 Ansible 模块'''

        # 找到所有匹配的 Ansible 主机
        hosts = self.inventory.list_hosts(self.pattern)
        if len(hosts) == 0:
            # 没有匹配的主机，返回空的结果
            self.callbacks.on_no_hosts()
            return dict(contacted={}, dark={})

        global multiprocessing_runner
        # 这行代码很关键，multiprocessing_runner 就是相应运行的 Runner 实例
        multiprocessing_runner = self
        results = None

        # 获取动作插件，ping 模块没有想要的动作插件，所以 p 得到的为 None
        p = utils.plugins.action_loader.get(self.module_name, self)

        if p and getattr(p, 'BYPASS_HOST_LOOP', None):
            # ...
        elif self.forks > 1:
            # 多进程执行
            try:
                results = self._parallel_exec(hosts)
            except IOError, ie:
                print ie.errno
```

```
        if ie.errno == 32:
            # broken pipe from Ctrl+C
            raise errors.AnsibleError("interupted")
        raise
    else:
        # 单进程执行
        results = [ self._executor(h) for h in hosts ]
    return self._partition_results(results)

# ...
```

在调试 ping 模块时，由于没有设置-f 选项指定启动的进程数（默认 self.forks = 5），因此程序会进入 if 的中间分支并调用 self._parallel_exec(hosts)方法去执行 Ansible 模块。如果设置了单进程（self.forks=1）执行，则会调用 self._executor()方法依次执行匹配的 host。继续往后追踪 self._parallel_exec()方法可知，其最终调用的仍然是 self._executor()方法：

```
# 源码位置：lib/ansible/runner/__init__.py
# ...

def _executor_hook(job_queue, result_queue):

    if HAS_ATFORK:
        atfork()

    # 注册终止信号
    signal.signal(signal.SIGINT, signal.SIG_IGN)
    while not job_queue.empty():
        try:
            host = job_queue.get(block=False)
            # 多进程任务队列中依旧是执行 Runner 对象的_executor()方法
            # 注意，全局变量 multiprocessing_runner 等于运行的 Runner 实例
            # 由 Runner 类中 run()方法的 multiprocessing_runner=self 可知
            result_queue.put(multiprocessing_runner._executor(host))
        except Queue.Empty:
            pass
        except:
            traceback.print_exc()

# ...

class Runner(object):

    # ...

    def _parallel_exec(self, hosts):

        # Python 多进程间共享数据，Queue 是多进程安全的队列，可以使用 Queue 实现多进
        #   程之间的数据传递
        manager = multiprocessing.Manager()
        # 任务队列，队列中主要是目标主机
        job_queue = manager.Queue()
        for host in hosts:
            job_queue.put(host)
```

```
# 结果队列
result_queue = manager.Queue()

workers = []
for i in range(self.forks):
    # 启动多进程执行 Ansible 模块，多进程执行的方法为 _executor_hook()
    prc = multiprocessing.Process(target=_executor_hook,
        args=(job_queue, result_queue))
    prc.start()
    workers.append(prc)

try:
    # 同步等待所有的任务执行完毕
    for worker in workers:
        worker.join()
except KeyboardInterrupt:
    # 如果外界输入 Ctrl+C 等终止信号，则逐个停止多进程
    for worker in workers:
        worker.terminate()
        worker.join()

# 获取 Ansible 模块执行结果 result_queue
results = []
try:
    while not result_queue.empty():
        # 如果队列不为空，则从结果队列中取回模块执行结果
        results.append(result_queue.get(block=False))
except socket.error:
    raise errors.AnsibleError("<interrupted>")
return results
```

　　这里的多进程实现非常简单，稍微了解过 Python 中多进程模块（multiprocessing）的读者应该都能看懂上面的代码。下面简要分析一下这些代码，帮助读者理解 Ansible 的多进程实现原理。上面的代码通过 multiprocessing 模块中的 SyncManager 对象创建了两个用于多进程共享数据的队列。

- job_queue：执行的任务队列。一开始会将匹配的 Ansible 主机压入队列，对本次调试的程序而言将会有 3 个元素，即 ceph-1、ceph-2 和 ceph-3。
- result_queue：模块执行的结果队列。

　　启动多进程时，会将上述两个队列作为参数传入启动进程的目标方法中。在目标方法 _executor_hook() 中会从共享的任务列表中取出一个待执行的目标主机，然后调用 Runner 对象的 _executor() 方法获取相应的结果，同时将模块的执行结果放入多进程的共享队列 result_queue 中。综合上面的分析，可以得到一个重要信息：不管是多进程还是单进程执行，任务执行的粒度都是 host。对于本次调试命令而言，因为 self.fork=5，但实际只有 3 个目标主机，当有 3 个启动进程抢占了任务队列中的主机元素后，剩余启动的两个进程会因为没有任务而很快结束并跳转至同步等待状态（worker.join()）。

　　至此，最核心的调用方法已经找到了，正是 Runner 对象中的 _executor() 方法。接下来

对 ping 模块的调试可以直接从这里断点，由于执行的命令是多进程的，所以_executor()方法会被多次调用。首先来看_executor()方法的具体实现，代码如下：

```
# 源码位置: lib/ansible/runner/__init__.py
# ...

class Runner(object):

    # ...

    def _executor(self, host):
        try:
            exec_rc = self._executor_internal(host)
            # 返回的结果一定要是 ReturnData 类型
            if type(exec_rc) != ReturnData:
                raise Exception("unexpected return type: %s" % type(exec_rc))
            # 作者的注释
            # redundant, right?
            if not exec_rc.comm_ok:
                self.callbacks.on_unreachable(host, exec_rc.result)
            return exec_rc
        except errors.AnsibleError, ae:
            msg = str(ae)
            # 通知不可达的回调
            self.callbacks.on_unreachable(host, msg)
            return ReturnData(host=host, comm_ok=False, result=dict(failed=
True, msg=msg))
        except Exception:
            msg = traceback.format_exc()
            self.callbacks.on_unreachable(host, msg)
            return ReturnData(host=host, comm_ok=False, result=dict(failed=
True, msg=msg))

    # ...
```

通过上述代码可知，Ansible 模块执行的核心方法从_executor()方法再次转到了_executor_internal()方法。该方法的输入为目标主机，输出为执行结果，并且结果类型必须是 ReturnData。以下是 ReturnData 类的实现源码：

```
# 源码位置: lib/ansible/runner/return_data.py

from ansible import utils

class ReturnData(object):

    __slots__ = [ 'result', 'comm_ok', 'host', 'diff' ]

    def __init__(self, conn=None, host=None, result=None,
        comm_ok=True, diff=dict()):

        # conn 中包含连接信息，主要是连接的主机
        if conn is not None:
            self.host = conn.host
```

```
            delegate = getattr(conn, 'delegate', None)
            if delegate is not None:
                self.host = delegate
        else:
            self.host = host

        # result 就是直接的结果
        self.result = result
        # 执行是否正常
        self.comm_ok = comm_ok

        # if these values are set and used with --diff we can show
        # changes made to particular files
        self.diff = diff

        # 如果 self.result 为 str 或者 unicode 格式，需要转换成 JSON 格式
        if type(self.result) in [ str, unicode ]:
            self.result = utils.parse_json(self.result)

        if self.host is None:
            raise Exception("host not set")
        if type(self.result) != dict:
            raise Exception("dictionary result expected")

    def communicated_ok(self):
        # 判断通信是否成功
        return self.comm_ok

    def is_successful(self):
        # 判断模块执行是否成功，提供一个简单的方法判断
        return self.comm_ok and (self.result.get('failed', False) == False)
and (self.result.get('rc',0) == 0)
```

ReturnData 类的代码也比较简单，即用连接信息（conn）、模块执行的结果（result）等参数来实例化该类。它对模块执行的结果进行二次封装，同时添加一些方法简化操作，如判断执行结果是否成功，对返回结果为 str 或者 unicode 的情况进行处理等。下面继续学习_executor_internal()方法，具体代码如下：

```
# 源码位置: lib/ansible/runner/__init__.py
# ...

class Runner(object):

    # ...

    def _executor_internal(self, host):
        '''执行任意模块，一次或者多次'''

        # 获取主机变量
        host_variables = self.inventory.get_variables(host)
        # 获取主机的连接方式
        host_connection = host_variables.get('ansible_connection', self.
transport)
```

```
        if host_connection in [ 'paramiko', 'ssh' ]:
            port = host_variables.get('ansible_ssh_port', self.remote_port)
            if port is None:
                port = C.DEFAULT_REMOTE_PORT
        else:
            # fireball, local, etc
            port = self.remote_port


        # 一些信息保存到 inject 中
        # ...


        # 主要是处理 Playbook 中编写的 with_items 语法，用于 for 循环
        # ...


        if items is None:
            return self._executor_internal_inner(host, self.module_name,
self.module_args, inject, port, complex_args=self.complex_args)
        elif len(items) > 0:
            aggregrate = {}
            all_comm_ok = True
            all_changed = False
            all_failed = False
            results = []
            for x in items:
                inject['item'] = x
                result = self._executor_internal_inner(host, self.module_
name, self.module_args, inject, port, complex_args=self.complex_args)
                results.append(result.result)
                # 处理结果
                # ...

            # 处理整体结果，方便实例化 ReturnData
            # ...
            return ReturnData(host=host, comm_ok=all_comm_ok, result=rd_
result)
        else:
            # 跳过该任务，直接返回并记录跳过信息
            self.callbacks.on_skipped(host, None)
            return ReturnData(host=host, comm_ok=True, result=dict(skipped=
True))
```

笔者对源代码进行了简化，去掉了一些简单的变量赋值和处理 with_items 语法的代码。可以看到，代码中最核心的部分就是调用 self._executor_internal_inner() 方法获取相应的结果，而这个方法正是 1.5.2 节中 connection_plugins 和 action_plugins 两个插件加载和调用的地方。下面同样对 self._executor_internal_inner() 方法进行简化，只给出最核心的代码部分。

```
# 源码位置：lib/ansible/runner/__init__.py
# ...

class Runner(object):

    # ...
```

```python
    def _executor_internal_inner(self, host, module_name, module_args,
inject, port, is_chained=False, complex_args=None):
        ''' 决定如何调用模块 '''
        new_args = ""
        if type(module_args) == dict:
            for (k,v) in module_args.iteritems():
                new_args = new_args + "%s='%s' " % (k,v)
            module_args = new_args

        # 后面用于合成完整模块代码的 3 个参数，1.5.2 节中已经介绍过
        module_name = utils.template(self.basedir, module_name, inject)
        module_args = utils.template(self.basedir, module_args, inject)
        complex_args = utils.template(self.basedir, complex_args, inject)

        # 获取动作插件中定义的核心类
        if module_name in utils.plugins.action_loader:
            if self.background != 0:
                raise errors.AnsibleError("async mode is not supported with
the %s module" % module_name)
            handler = utils.plugins.action_loader.get(module_name, self)
        elif self.background == 0:
            handler = utils.plugins.action_loader.get('normal', self)
        else:
            handler = utils.plugins.action_loader.get('async', self)

        # 针对 Playbook 中模块执行条件的具体处理
        # ...

        conn = None
        actual_host = inject.get('ansible_ssh_host', host)
        actual_port = port
        actual_user = inject.get('ansible_ssh_user', self.remote_user)
        actual_pass = inject.get('ansible_ssh_pass', self.remote_pass)
        actual_transport = inject.get('ansible_connection', self.transport)
        if actual_transport in [ 'paramiko', 'ssh' ]:
            actual_port = inject.get('ansible_ssh_port', port)

        # 处理 Playbook 中的 delegate_to 语法，将执行的目标指定为相应的主机
        # ...

        # 处理远程连接中端口的一些异常情况
        # ...

        try:
            # 加载连接插件，连接远程主机，得到相应的通信底层文件的 Connection 对象
            conn = self.connector.connect(
                    actual_host, actual_port, actual_user, actual_pass,
actual_transport)
            if delegate_to or host != actual_host:
                conn.delegate = host

        except errors.AnsibleConnectionFailed, e:
            result = dict(failed=True, msg="FAILED: %s" % str(e))
```

```
              return ReturnData(host=host, comm_ok=False, result=result)

        tmp = ''
        if getattr(handler, 'NEEDS_TMPPATH', True):
            # 创建远端的临时目录
            tmp = self._make_tmp_path(conn)

        # 执行动作插件模块中的 run() 方法
        result = handler.run(conn, tmp, module_name, module_args, inject,
complex_args)

        # 关闭连接
        conn.close()

        # 整理返回结果
        # ...

        return result
```

分析了_executor_internal_inner()方法的代码后可知，该方法主要做了 3 件事情：

（1）加载底层通信的连接插件，得到对应插件的 Connection 对象：conn。前面也演示过这样的操作，并且通过得到的 conn 可以对连接的主机进行远程执行命令（exec_command()）、上传文件（put_file()）及下载文件（get_file()）等操作。

（2）加载模块对应的动作插件（模块名.py），如果没有则默认加载 normal.py 插件。如果是异步操作，则加载 async.py 插件，然后执行动作插件的 run()方法。

（3）整理 run()方法最后返回的结果。

此时可知，每个模块执行的核心步骤其实就在对应的插件代码中。那么对于这里调试的 ping 模块而言，其执行动作是什么样的呢？直接查看 normal.py 文件中的 ActionModule 类即可，其实现源码如下：

```
# 源码位置: lib/ansible/runner/action_plugins/normal.py
# ...
class ActionModule(object):

    def __init__(self, runner):
        self.runner = runner

    def run(self, conn, tmp, module_name, module_args, inject, complex_
args=None, **kwargs):
        ''' 传送和执行非 copy 或者非 template 模块 '''

        # 获取模块参数
        module_args = self.runner._complex_args_hack(complex_args, module_
args)
        if self.runner.check:
            # 只是检查
            if module_name in [ 'shell', 'command' ]:
                # 对于 shell 和 command 模块不支持 check 模式
                return ReturnData(conn=conn, comm_ok=True, result=dict
(skipped=True,
```

```
                    msg='check mode not supported for %s' % module_name))
            # 其他模块的 check 模式，只需在模块参数字符串后加上 CHECKMODE=True 即可
            module_args += " CHECKMODE=True"

        # shell 和 command 为相同的模块源码，只不过在模块参数中使用了一个区分标识
        if module_name == 'shell':
            module_name = 'command'
            module_args += " #USE_SHELL"

        return self.runner._execute_module(
        conn, tmp, module_name, module_args, inject=inject, complex_args=
complex_args
    )
```

　　看到这里，很多问题都已经豁然开朗了。对于 ping 模块或者其他加载 normal.py 插件的模块而言，动作插件最后调用的正是前面演示的 self.runner._execute_module()方法。这里还有一个能解释之前疑惑的地方，即 shell 模块和 command 模块如何实现共用一个模块代码呢？上述代码中针对这个问题给出了清晰的说明：对于 shell 模块，会将模块名称重新改为 command 并在模块参数的最后加上#USE_SHELL 标识，剩下的会在 command 模块源码中去识别和区分。

　　至此，Ansible 1.1 中关于 ping 模块的执行流程已经全部讲解完毕，下面用一个流程图来展示上述模块的执行过程，如图 1-7 所示。

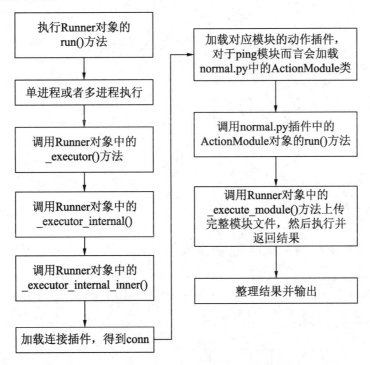

图 1-7　Ansible 1.1 中的模块执行流程

1.6　copy 模块的执行流程

1.5 节中笔者分析了整个 Ansible 模块的执行流程，也剖析了 runner 目录下的大部分核心代码。本节将以另一个典型的模块（copy 模块）为例，进一步分析其核心的执行逻辑，帮助读者更好地学习 Ansible 模块的分析方法。

1.6.1　copy 模块的动作插件

通过 1.5 节的分析可以发现，模块执行期间最大的区别就是加载不同的动作插件，ping、shell 和 command 模块的动作插件均为 normal.py 文件，而本节要分析的 copy 模块的动作插件则为 copy.py 文件，其代码如下：

```
# 源码位置：lib/ansible/runner/action_plugins/copy.py
# ...

class ActionModule(object):

    def __init__(self, runner):
        self.runner = runner

    def run(self, conn, tmp, module_name, module_args, inject, complex_args=
None, **kwargs):
        #从 options 中获取相关的参数信息
        options = {}
        if complex_args:
            options.update(complex_args)
        options.update(utils.parse_kv(module_args))
        source  = options.get('src', None)
        content = options.get('content', None)
        dest    = options.get('dest', None)

        # 几个异常情况处理，直接返回错误信息
        # ...

        if 'first_available_file' in inject:
            # 处理一种单独的配置情况
            # ...
        elif content is not None:
            # copy 模块中有 content 参数且有值，则将 content 的内容保存到临时文件中，
              tmp_content 是生成的临时文件地址
            fd, tmp_content = tempfile.mkstemp()
            f = os.fdopen(fd, 'w')
            try:
                f.write(content)
            except Exception, err:
                # 异常情况下，删除临时创建的文件并返回错误信息
```

```
            os.remove(tmp_content)
            result = dict(failed=True, msg="-错误信息-")
            return ReturnData(conn=conn, result=result)
        f.close()
        # 本地待上传文件的地址
        source = tmp_content
    else:
        # 直接得到完整的源文件路径地址
        source = utils.template(self.runner.basedir, source, inject)
        source = utils.path_dwim(self.runner.basedir, source)

    # 计算待上传文件的 MD5 值
    local_md5 = utils.md5(source)
    if local_md5 is None:
        result=dict(failed=True, msg="could not find src=%s" % source)
        return ReturnData(conn=conn, result=result)

    # 如果目标地址有/, 则表示的是目录, 需重新获取 dest 值, 这是最终上传的目的地址
        的值
    if dest.endswith("/"):
        base = os.path.basename(source)
        dest = os.path.join(dest, base)

    # 计算远端对应路径下的文件的 MD5 值
    remote_md5 = self.runner._remote_md5(conn, tmp, dest)
    # 以上方法输出中, 对于目录的情况会输出 3
    if remote_md5 == '3':
        # Destination is a directory
        if content is not None:
            os.remove(tmp_content)
            # 对于 dest 为目录的情况, 不能使用 content 参数
            result = dict(failed=True, msg="can not use content with a dir
as dest")
            return ReturnData(conn=conn, result=result)
        # 处理 dest 在远端服务器上为目录的情况, 生成新的 dest 值并计算目标文件的 MD5 值
        dest = os.path.join(dest, os.path.basename(source))
        remote_md5 = self.runner._remote_md5(conn, tmp, dest)

    exec_rc = None
    if local_md5 != remote_md5:
        # 对于本地文件和远端文件不一致或者远端文件不存在的情况
        if self.runner.diff:
            # 如果设置了相关参数, 会通过该方法获取本地文件和远端文件的区别
            diff = self._get_diff_data(conn, tmp, inject, dest, source)
        else:
            diff = {}

        # 处理只检查不执行的情况
        # ...

        # 将本地文件先上传到远端临时文件中, tmp_src 为文件上传到远端服务器的地址
        tmp_src = tmp + 'source'
        conn.put_file(source, tmp_src)
```

```
        if content is not None:
            # 删除 content 参数创建的临时文件
            os.remove(tmp_content)
        # 修正文件的权限问题，只需要给所有用户加上可读权限
        if self.runner.sudo and self.runner.sudo_user != 'root':
            self.runner._low_level_exec_command(conn, "chmod a+r %s" %
tmp_src, tmp)

            # 执行 copy 模块代码，同时将模块的 src 参数修改成远端的地址
        module_args = "%s src=%s original_basename=%s" %
            (module_args, pipes.quote(tmp_src), pipes.quote(os.path.basename
(source)))
            # 调用 Runner 对象的 _execute_module() 方法上传完整的 copy 模块代码并执行
        return self.runner._execute_module(
            conn, tmp, 'copy', module_args, inject=inject, complex_args=
complex_args)
        else:
            # 对于本地文件和远端文件相同的情况，则不用上传本地文件
            if content is not None:
                os.remove(tmp_content)
        tmp_src = tmp + os.path.basename(source)
        # 修改模块参数中的 src 值，替换成远端文件的地址；上传的目标地址不变
        module_args = "%s src=%s" % (module_args, pipes.quote(tmp_src))
        if self.runner.check:
            module_args = "%s CHECKMODE=True" % module_args
        return self.runner._execute_module(
            conn, tmp, 'file', module_args, inject=inject, complex_args=
complex_args)
```

这里也可以看到 copy 模块和 ping 模块执行的一个不同点：ping 模块的代码生成并上传之后可以直接运行并得到想要的结果（很多模块也是这样的，如 user、shell、command 及 file 等模块），而 copy 模块需要先将本地文件上传到远端主机上，然后修改相应的 src 参数，最后才能生成 copy 模块代码并上传到远端主机上去执行。完整的 copy 模块代码只是将已经上传到临时目录下的源文件复制到远端主机的目的地址上，然后返回相应的成功信息。

1.6.2 copy 模块遗留问题的解决

在 1.2 节中遗留了一个关于 copy 模块的小问题：无法对 content 参数中出现的转义字符进行翻译。来看之前的问题示例：

```
(ansible1.1) [root@master ~]# ansible ceph-1 -i hosts  -m copy -a "content=
'xxxxxx\nyyyyyy' dest=/root/test_copy.txt"
ceph-1 | success >> {
    "changed": false,
    "group": "root",
    "mode": "0644",
    "owner": "root",
    "path": "/root/test_copy.txt",
```

```
    "size": 14,
    "state": "file"
}

# 在 ceph-1 节点上查看传输的文件
[root@ceph-1 ~]# cat test_copy.txt
xxxxxx\nyyyyyy[root@ceph-1 ~]#
```

通过分析 copy 模块的插件代码，可以很快定位到问题的源头，具体如下：

```
options.update(utils.parse_kv(module_args))
```

上面这一行语句在对传入的模块参数 module_args 进行解析时造成转义符无法转义。可以通过下面的示例代码对此现象进行测试。

```
import shlex
from optparse import OptionParser

optParser = OptionParser()
optParser.add_option('-a', '--args', dest='module_args', help="module
arguments")
option, args = optParser.parse_args()

def parse_kv(args):
    '''将 key=value 这样形式的字符串转换成字典'''

    options = {}
    if args is not None:
        # 切割字符串
        vargs = shlex.split(str(args), posix=True)
        for x in vargs:
            # 通过=来得到 key 和 value，形成对应关系
            if x.find("=") != -1:
                k, v = x.split("=",1)
                options[k]=v
    return options
# 得到命令行参数
args = optParser.parse_args()
# 得到模拟的模块参数字符串
module_args = args[0].module_args
# 解析结果
print(parse_kv(module_args))
```

运行上述代码，可以得到和之前相同的现象，具体如下：

```
[root@master ~]# python test.py -a "content='xxxxx\nyyyyy' dest=/root/
test_copy"
{'content': 'xxxxx\\nyyyyy', 'dest': '/root/test_copy'}
```

然而转义符处理起来比较棘手，如果处理不当，还会引起新的 Bug，如果确实有这方面的需求，可以简单地按照自己的需求修改这里的 parse_kv()方法。假设 content 内容比较正常，不会包含一些特殊字符，则可以进行如下改造：

```
import shlex
from optparse import OptionParser
```

```
optParser = OptionParser()
optParser.add_option('-a', '--args', dest='module_args', help="module
arguments")
option, args = optParser.parse_args()

def parse_kv_copy(args):

    options = {}
    if args is not None:
        # 将\\替换成S#
        args = args.replace('\\', '$#')
        vargs = shlex.split(str(args), posix=True)
        for x in vargs:
            if x.find("=") != -1:
                k, v = x.split("=",1)
                # 解析完后再将$#还原成\\
                options[k] = eval(repr(v).replace('$#', '\\'))
    return options

args = optParser.parse_args()
module_args = args[0].module_args
data = parse_kv_copy(module_args)
print(data['content'])
```

先将反斜杠替换成$#，然后在生成 key=value 时再将反斜杠替换回来，这样就可以避免出现前面的问题。当然这是一种不太合适的策略，如果 content 中出现$#符号就会导致转换出错，因此改造 Ansible 1.1 的源码时要慎重考虑。下面运行改造后的代码，效果如下：

```
[root@master ~]# python test.py -a "content='xxxxx\nyyyyy\tkkkk\txxxxx\
tmmmm' dest=/root/test_copy"
xxxxx
yyyyy   kkkk    xxxxx    mmmm
```

按照这样的方式改造 copy 模块动作插件中的代码，具体代码如下：

```
# 源码位置: lib/ansible/runner/action_plugins/copy.py
# ...

import shlex

class ActionModule(object):
    # ...

    def parse_kv_copy(self, args):

        options = {}
        if args is not None:
            args = args.replace('\\', '$#')
            vargs = shlex.split(str(args), posix=True)
            for x in vargs:
                if x.find("=") != -1:
                    k, v = x.split("=",1)
                    options[k] = eval(repr(v).replace('$#', '\\'))
        return options
```

```
    def run(self, conn, tmp, module_name, module_args, inject, complex_
args=None, **kwargs):
        ''' 处理文件传输操作'''

        # load up options
        options = {}
        if complex_args:
            options.update(complex_args)
        # options.update(utils.parse_kv(module_args))
        # 使用改造后的代码解析模块变量
        options.update(self.parse_kv_copy(module_args))
        # ...

    # ...
```

最后在 Ansible 1.1 的虚拟环境中修改相应位置处的代码，然后运行改造后的 ansible 命令，具体操作如下：

```
(ansible1.1) [root@master ~]# vim /root/.pyenv/versions/2.7.18/envs/
ansible1.1/lib/python2.7/site-packages/ansible/runner/action_plugins/
copy.py
(ansible1.1) [root@master ~]# ansible ceph-1 -i hosts  -m copy -a "content=
'xxxxxx\nyyyyyy' dest=/root/test_copy.txt"
ceph-1 | success >> {
    "changed": true,
    "dest": "/root/test_copy.txt",
    "group": "root",
    "md5sum": "00cd67a22f6c5083b24a0388607a57de",
    "mode": "0644",
    "owner": "root",
    "size": 13,
    "src": "/root/.ansible/tmp/ansible-1603000276.75-8087808413471/source",
    "state": "file"
}
```

查看目标节点上/root/test_copy.txt 位置的文件内容，发现已经实现了想要的结果，具体如下：

```
[root@ceph-1 ~]# cat test_copy.txt
xxxxxx
yyyyyy[root@ceph-1 ~]#
```

可以看到，经过简单改造 copy 模块的动作插件代码后，新的 copy 模块已经能满足自定义的需求了。

1.6.3 Ansible 模块的核心步骤小结

通过 1.5 节和 1.6 节的学习，读者应该已经了解了 Ansible 模块的完整运行流程。Ansible 模块的核心代码包括以下两个方面：

- 模块的动作插件代码（位于 lib/ansible/runner/action_plugins/目录下）；
- 模块本身的代码（位于 library 目录下，非完整的模块代码）。

理解了以上两部分代码的核心内容，就等于掌握了该模块的所有功能。前面已经分析了 ping 和 copy 这两个典型的模块，其余模块的分析过程基本一致，这里不再详细介绍。Ansible 提供的 ping 模块和 copy 模块的运行机制，也是 Ansible 工具的核心运行原理。正是由于这样的设计理念，使得开发者可以将主要精力放在模块开发上，不断丰富和完善各种模块，从而扩展 Ansible 的功能。

1.7 本章小结

本章详细介绍了 Ansible 1.1 的核心源码，并对 Ansible 模块的运行原理进行了详尽分析。限于篇幅原因，关于 Ansible 1.1 的 Playbook 的运行流程、对返回结果的统计及打印等部分的代码并没有介绍，感兴趣的读者可以自行学习这部分代码。在完全掌握了 Ansible 1.1 的核心源码后，学习 Ansible 1.9.6 和 Ansible 2.8 的源码时会发现，许多目录及代码命名都似曾相识，而且 Ansible 的模块运行机制在最新的 Ansible 2.9 中也没有发生变化。第 2 章中我们将继续探索 Ansible 1 中的最后一个版本的源码（v1.9.6），并解析其 Playbook 的运行机制。

第 2 章　Ansible 1.9.6 源码剖析

本章主要学习 Ansible 1 的最后一个版本源码，即 Ansible 1.9.6 源码。本章将在 Ansible 1.1 的源码基础上进一步分析 Ansible 1.9.6 的源码，除了再次分析 Ansible 1.9.6 的模块运行机制外，还会重点分析 Ansible 中 Playbook 的执行过程，达到彻底掌握 Ansible 1 源码的目的。

2.1　Ansible 1.1 和 Ansible 1.9.6 源码对比

本节会从一些直观的角度对比 Ansible 1.1 和 Ansible 1.9.6 的源码。从 Ansible 官网的历史版本中可以看到，Ansible 从 1.1 开始经历了 8 个大版本迭代发展到 v1.9.6，耗时近 3 年。Ansible 1.1 和 Ansible 1.9.6 的源码文件对比如图 2-1 所示。

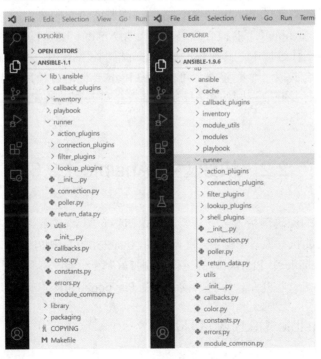

图 2-1　Ansible 1.1 与 Ansible 1.9.6 源码对比

通过简单对比两个版本的 Ansible 源码可以发现，Ansible 1.9.6 中保留了大部分 Ansible 1.1 的源码目录及文件命名，甚至大部分类和方法及 Ansible 核心的模块运行机制也没有发生大的变化，但是新版本的 Ansible 功能更强大，支持的模块更多且 Playbook 的语法越来越多样化。其中最直观的变化如下：

- Ansible 1.1 的模块并没有包括在源码中，而是独立在 library 目录下。Ansible 1.9.6 的模块文件全部移动到了 lib/ansible/modules 目录下，它们和 Ansible 源码一起迭代管理。
- Ansible 1.9.6 中多了缓存模块，支持使用 Redis、Memcached 及内存进行数据缓存，而 Ansible 1.1 并没有这样的功能。
- Ansible 1.9.6 虽然保留了 module_common.py 这个文件，但是其内容和 Ansible 1.1 中的 module_common.py 已经完全不同。其中关于公共模块代码部分已转移至 module_utils 目录下，并且这些公共代码已经不再是字符串的形式，而是真正的 Python 的类和函数。例如 AnsibleModule 类定义在 module_utils/basic.py 文件中，是一个真正的 Python 类。
- Ansible 1.9.6 的 runner 目录下多了一个 shell_plugins 目录，该目录下有一个 powershell.py 文件，由此可知 Ansible 1.9.6 支持远程操作 Windows 主机，而 Ansible 1.1 则只能支持 Linux 主机的操作。

仅从目录和文件的角度来看，比较明显的改变就是以上几点。至于具体的模块执行细节上的优化及支持更复杂的 Playbook 语法等，则需要通过源码来判断。

其实，分析过 Ansible 1.1 的核心源码后，再看 Ansible 1.9.6 的源码并不会感到陌生。在 Ansible 的版本迭代过程中，其模块的核心运行机制没有变化，同样是将模块内容与一个通用的代码部分合并成一个完整的、可执行的 Python 代码，最后通过 SSH 协议远程传输到远端主机上执行并获取返回结果，模块的参数会写入合成的代码中并用于后面解析模块功能。

2.2　快速上手 Ansible 1.9.6

同 Ansible 1.1 的源码剖析方式一样，这里先初步介绍 Ansible 1.9.6 工具的使用，然后再分析其命令背后的执行原理。本节会重点介绍 Ansible 1.9.6 中 Playbook 的语法及使用实例，为后续分析 Ansible 中 Playbook 的运行机制打好基础。此外，本节介绍的 Playbook 语法会在 Ansible 2 中进一步介绍，以帮助读者掌握 Playbook 的编写方法和排查错误的方法。

2.2.1　搭建 Ansible 1.9.6 测试环境

首先使用 pyenv 工具创建一个名为 ansible1.9.6 的虚拟环境，具体操作命令如下：

```
[root@master ~]# pyenv virtualenv 2.7.18 ansible1.9.6
# ...
[root@master ~]# pyenv versions
* system (set by /root/.pyenv/version)
  2.7.18
  2.7.18/envs/ansible1.1
  2.7.18/envs/ansible1.9.6
  ansible1.1
  ansible1.9.6
```

接着激活 ansible 1.9.6 虚拟环境并在该环境下安装 Ansible 1.9.6 工具，具体操作命令如下：

```
[root@master ~]# pyenv activate ansible1.9.6
(ansible1.9.6) [root@master ~]# pip install ansible==1.9.6 -i https://
pypi.tuna.tsinghua.edu.cn/simple
```

可以在命令行调用 Ansible 模块进行测试：

```
>>> import ansible
>>> ansible.__version__
'1.9.6'
```

此外，搭建 Ansible 调试环境并和 Ansible 1.1 源码调试一致：准备 CentOS 的桌面系统并安装好 VSCode 软件，导入 Ansible 1.9.6 的源码并配置好想要的调试参数即可。完成上面这些步骤后，就可以开始 Ansible 1.9.6 的源码阅读和学习了。不过在此之前，先介绍一下 Ansible 1.9.6 中的一些模块及 Playbook 的常见用法，确保后续研究 Ansible 1.9.6 源码时能更好地理解其运行机制。

2.2.2　Ansible 1.9.6 中的部分模块剖析

对于 Ansible 模块的学习，不管是 Ansible 1 还是 Ansible 2，通过学习其模块源码，能够迅速理解并掌握该模块的作用和其支持的参数。通常，模块中有对该模块的完整描述及支持的所有参数说明，这是掌握 Ansible 模块的最佳途径。此外，对于 Ansible 2 而言，其官方网站上有对各个模块及其参数的详细介绍，同时还配有部分案例，方便开发者迅速掌握该模块的用法。不同于 Ansible 1.1，Ansible 1.9.6 中的模块数量已有数百个之多，并且将模块按功能进行了划分，目录层次明确。本节依旧只介绍部分经典模块及其源码，通过源码分析模块功能和实现逻辑，进一步帮助读者掌握该模块的用法。

1. stat模块

stat 模块用于获取文件的信息，其模块文件的路径为：lib/ansible/modules/core/files/stat.py。下面介绍该模块的常用参数：

- path：文件路径。
- follow：是否访问符号链接。
- get_md5：返回结果中是否包含文件的 MD5 值，默认为 yes。

- get_checksum：返回结果中是否包含文件的 checksum 值，默认为 yes。

下面是 stat 模块的几个简单使用示例。

```
(ansible1.9.6) [root@master ~]# ansible ceph-1 -i hosts -m stat -a "path=
/etc/hosts"
ceph-1 | success >> {
    "changed": false,
    "stat": {
        "atime": 1604471285.841277,
        "checksum": "9b067f75a052862d72361fa6c3720e741d98a742",
        "ctime": 1600313827.2628028,
        "dev": 64768,
        "exists": true,
        "gid": 0,
        "gr_name": "root",
        "inode": 16784717,
        "isblk": false,
        "ischr": false,
        "isdir": false,
        "isfifo": false,
        "isgid": false,
        "islnk": false,
        "isreg": true,
        "issock": false,
        "isuid": false,
        "md5": "663c6dda5012ccc93144de679d6c78d9",
        "mode": "0644",
        "mtime": 1600313826.8288481,
        "nlink": 1,
        "path": "/etc/hosts",
        "pw_name": "root",
        "rgrp": true,
        "roth": true,
        "rusr": true,
        "size": 249,
        "uid": 0,
        "wgrp": false,
        "woth": false,
        "wusr": true,
        "xgrp": false,
        "xoth": false,
        "xusr": false
    }
}
(ansible1.9.6) [root@master ~]# ansible ceph-1 -i hosts -m stat -a "path=
/etc/hosts get_checksum=no get_md5=no "
ceph-1 | success >> {
    "changed": false,
    "stat": {
        "atime": 1604471285.841277,
        "ctime": 1600313827.2628028,
        "dev": 64768,
        "exists": true,
        "gid": 0,
        "gr_name": "root",
```

```
        "inode": 16784717,
        "isblk": false,
        "ischr": false,
        "isdir": false,
        "isfifo": false,
        "isgid": false,
        "islnk": false,
        "isreg": true,
        "issock": false,
        "isuid": false,
        "mode": "0644",
        "mtime": 1600313826.8288481,
        "nlink": 1,
        "path": "/etc/hosts",
        "pw_name": "root",
        "rgrp": true,
        "roth": true,
        "rusr": true,
        "size": 249,
        "uid": 0,
        "wgrp": false,
        "woth": false,
        "wusr": true,
        "xgrp": false,
        "xoth": false,
        "xusr": false
    }
}
(ansible1.9.6) [root@master ~]# ansible ceph-1 -i hosts -m stat -a "path=
/usr/bin/python get_checksum=no get_md5=no"
ceph-1 | success >> {
    "changed": false,
    "stat": {
        "atime": 1604477338.7286284,
        "ctime": 1587608110.437001,
        "dev": 64768,
        "exists": true,
        "gid": 0,
        "gr_name": "root",
        "inode": 50544386,
        "isblk": false,
        "ischr": false,
        "isdir": false,
        "isfifo": false,
        "isgid": false,
        "islnk": true,
        "isreg": false,
        "issock": false,
        "isuid": false,
        "lnk_source": "/usr/bin/python2.7",
        "mode": "0777",
        "mtime": 1587608110.436001,
        "nlink": 1,
        "path": "/usr/bin/python",
        "pw_name": "root",
```

```
            "rgrp": true,
            "roth": true,
            "rusr": true,
            "size": 7,
            "uid": 0,
            "wgrp": true,
            "woth": true,
            "wusr": true,
            "xgrp": true,
            "xoth": true,
            "xusr": true
        }
    }
(ansible1.9.6) [root@master ~]# ansible ceph-1 -i hosts -m stat -a "path=
/usr/bin/python get_checksum=no get_md5=no follow=yes"
ceph-1 | success >> {
    "changed": false,
    "stat": {
        "atime": 1604477338.7286284,
        "ctime": 1587608110.437001,
        "dev": 64768,
        "exists": true,
        "gid": 0,
        "gr_name": "root",
        "inode": 50544388,
        "isblk": false,
        "ischr": false,
        "isdir": false,
        "isfifo": false,
        "isgid": false,
        "islnk": false,
        "isreg": true,
        "issock": false,
        "isuid": false,
        "mode": "0755",
        "mtime": 1565139122.0,
        "nlink": 1,
        "path": "/usr/bin/python",
        "pw_name": "root",
        "rgrp": true,
        "roth": true,
        "rusr": true,
        "size": 7216,
        "uid": 0,
        "wgrp": false,
        "woth": false,
        "wusr": true,
        "xgrp": true,
        "xoth": true,
        "xusr": true
    }
}
```

　　上面的操作比较简单，stat 模块会返回文件的一些具体信息，例如其是否存在（exists）、是否为目录（isdir）、文件的 iNode 值等。通常在 Playbook 中会使用该模块获取文件的信息来影响下一个任务的执行。以下是该模块的实现源码：

```python
# 源码位置: lib/ansible/modules/core/files/stat.py
# ...

import os
import sys
from stat import *
import pwd
import grp

def main():
    # ansible 通用模块类
    module = AnsibleModule(
        argument_spec = dict(
            path = dict(required=True),
            follow = dict(default='no', type='bool'),
            get_md5 = dict(default='yes', type='bool'),
            get_checksum = dict(default='yes', type='bool')
        ),
        supports_check_mode = True
    )

    # 获取模块参数
    path = module.params.get('path')
    # 支持~、$HOME 这样的写法
    path = os.path.expanduser(path)
    # 是否访问符号链接
    follow = module.params.get('follow')
    # 是否返回相关值
    get_md5 = module.params.get('get_md5')
    get_checksum = module.params.get('get_checksum')

    try:
        # 根据 follow 参数调用不同的方法
        if follow:
            st = os.stat(path)
        else:
            st = os.lstat(path)
    except OSError, e:
        # 处理文件不存在的情况
        if e.errno == errno.ENOENT:
            d = { 'exists' : False }
            module.exit_json(changed=False, stat=d)

        module.fail_json(msg = e.strerror)

    # 获取 st_mode 信息
    mode = st.st_mode

    # 返回给 ansible 的结果
    d = {
        'exists'   : True,
        'path'     : path,
```

```
        'mode'      : "%04o" % S_IMODE(mode),
        'isdir'     : S_ISDIR(mode),
        'ischr'     : S_ISCHR(mode),
        'isblk'     : S_ISBLK(mode),
        'isreg'     : S_ISREG(mode),
        'isfifo'    : S_ISFIFO(mode),
        'islnk'     : S_ISLNK(mode),
        'issock'    : S_ISSOCK(mode),
        'uid'       : st.st_uid,
        'gid'       : st.st_gid,
        'size'      : st.st_size,
        'inode'     : st.st_ino,
        'dev'       : st.st_dev,
        'nlink'     : st.st_nlink,
        'atime'     : st.st_atime,
        'mtime'     : st.st_mtime,
        'ctime'     : st.st_ctime,
        'wusr'      : bool(mode & stat.S_IWUSR),
        'rusr'      : bool(mode & stat.S_IRUSR),
        'xusr'      : bool(mode & stat.S_IXUSR),
        'wgrp'      : bool(mode & stat.S_IWGRP),
        'rgrp'      : bool(mode & stat.S_IRGRP),
        'xgrp'      : bool(mode & stat.S_IXGRP),
        'woth'      : bool(mode & stat.S_IWOTH),
        'roth'      : bool(mode & stat.S_IROTH),
        'xoth'      : bool(mode & stat.S_IXOTH),
        'isuid'     : bool(mode & stat.S_ISUID),
        'isgid'     : bool(mode & stat.S_ISGID),
        }

# 如果是链接文件，需要找到完整的链接源路径
if S_ISLNK(mode):
    d['lnk_source'] = os.path.realpath(path)

if S_ISREG(mode) and get_md5 and os.access(path,os.R_OK):
    # Will fail on FIPS-140 compliant systems
    try:
        d['md5']        = module.md5(path)
    except ValueError:
        d['md5']        = None

if S_ISREG(mode) and get_checksum and os.access(path,os.R_OK):
    d['checksum']       = module.sha1(path)


try:
    # pwd 模块
    pw = pwd.getpwuid(st.st_uid)

    d['pw_name']   = pw.pw_name

    grp_info = grp.getgrgid(st.st_gid)
    d['gr_name'] = grp_info.gr_name
except:
    pass
```

```
    module.exit_json(changed=False, stat=d)

# import module snippets
from ansible.module_utils.basic import *

main()
```

上面的代码非常简单，这里不再过多介绍。stat 模块中最核心的就是 os.stat() 和 os.lstat() 这两个方法，了解这两个方法的返回结果，对于这个模块的返回结果就会非常清楚了。

2．replace模块

replace 模块用于替换目标文件中所有匹配的文本，其功能类似于 lineinfile 模块。首先介绍 replace 模块支持的参数选项。

- dest：待操作的目标文件。
- regexp：正则表达式，用于寻找目标文件中匹配的文本。
- replace：替换匹配的文本。
- backup：是否备份。
- validate：上传文件到指定位置前是否进行校验。

另外，file 模块支持的参数 replace 模块都支持。

首先准备好相关环境，目标节点为 ceph-1，目标操作文件为/root/passwd，其内容如下：

```
[root@ceph-1 ~]# cp /etc/passwd .
[root@ceph-1 ~]# cat /root/passwd
root:x:0:0:root:/root:/bin/bash
bin:x:1:1:bin:/bin:/sbin/nologin
daemon:x:2:2:daemon:/sbin:/sbin/nologin
adm:x:3:4:adm:/var/adm:/sbin/nologin
lp:x:4:7:lp:/var/spool/lpd:/sbin/nologin
sync:x:5:0:sync:/sbin:/bin/sync
shutdown:x:6:0:shutdown:/sbin:/sbin/shutdown
halt:x:7:0:halt:/sbin:/sbin/halt
mail:x:8:12:mail:/var/spool/mail:/sbin/nologin
operator:x:11:0:operator:/root:/sbin/nologin
games:x:12:100:games:/usr/games:/sbin/nologin
ftp:x:14:50:FTP User:/var/ftp:/sbin/nologin
nobody:x:99:99:Nobody:/:/sbin/nologin
systemd-network:x:192:192:systemd Network Management:/:/sbin/nologin
dbus:x:81:81:System message bus:/:/sbin/nologin
polkitd:x:999:998:User for polkitd:/:/sbin/nologin
sshd:x:74:74:Privilege-separated SSH:/var/empty/sshd:/sbin/nologin
postfix:x:89:89::/var/spool/postfix:/sbin/nologin
shen:x:1000:1000:shen:/home/shen:/bin/bash
store:x:1001:1001::/home/store:/bin/bash
ntp:x:38:38::/etc/ntp:/sbin/nologin
media:x:1002:1002::/home/media:/bin/bash
chrony:x:998:996::/var/lib/chrony:/sbin/nologin
libstoragemgmt:x:997:995:daemon account for libstoragemgmt:/var/run/lsm/:
```

```
sbin/nologin
ceph:x:167:167:Ceph daemons:/var/lib/ceph:/sbin/nologin
dockerroot:x:996:993:Docker User:/var/lib/docker:/sbin/nologin
bird6:x:995:992:BIRD6 daemon user:/var/lib/bird6:/sbin/nologin
tcpdump:x:72:72::/:/sbin/nologin
test:x:1003:1003::/home/test:/bin/bash
halo:x:1004:1004::/home/halo:/bin/bash
```

接着来看几个关于 replace 模块的演示实例。

```
(ansible1.9.6) [root@master ~]# ansible ceph-1 -i hosts -m replace -a "dest=
/root/passwd regexp=^halo.*$ backup=yes"
ceph-1 | success >> {
    "changed": true,
    "msg": "1 replacements made"
}
(ansible1.9.6) [root@master ~]# ansible ceph-1 -i hosts -m replace -a "dest=
/root/passwd regexp=^test.*$ replace=xxx"
ceph-1 | success >> {
    "changed": true,
    "msg": "1 replacements made"
}
(ansible1.9.6) [root@master ~]# ansible ceph-1 -i hosts -m replace -a "dest=
/root/passwd regexp=/sbin/nologin replace=/bin/bash"
ceph-1 | success >> {
    "changed": true,
    "msg": "21 replacements made"
}
```

上面三个操作中，第一个操作是将 ceph-1 节点上/root/passwd 文件中最后一行的 halo 用户清除，这里默认使用了空字符进行替换，第二个操作是将 test 用户所在行的内容替换成 xxx 字符串，第三个操作是将文件中所有出现的/sbin/nologin 替换为/bin/bash。最后得到的文件内容如下：

```
[root@ceph-1 ~]# cat passwd
root:x:0:0:root:/root:/bin/bash
bin:x:1:1:bin:/bin:/bin/bash
daemon:x:2:2:daemon:/sbin:/bin/bash
adm:x:3:4:adm:/var/adm:/bin/bash
lp:x:4:7:lp:/var/spool/lpd:/bin/bash
sync:x:5:0:sync:/sbin:/bin/sync
shutdown:x:6:0:shutdown:/sbin:/sbin/shutdown
halt:x:7:0:halt:/sbin:/sbin/halt
mail:x:8:12:mail:/var/spool/mail:/bin/bash
operator:x:11:0:operator:/root:/bin/bash
games:x:12:100:games:/usr/games:/bin/bash
ftp:x:14:50:FTP User:/var/ftp:/bin/bash
nobody:x:99:99:Nobody:/:/bin/bash
systemd-network:x:192:192:systemd Network Management:/:/bin/bash
dbus:x:81:81:System message bus:/:/bin/bash
polkitd:x:999:998:User for polkitd:/:/bin/bash
sshd:x:74:74:Privilege-separated SSH:/var/empty/sshd:/bin/bash
postfix:x:89:89::/var/spool/postfix:/bin/bash
shen:x:1000:1000:shen:/home/shen:/bin/bash
store:x:1001:1001::/home/store:/bin/bash
```

```
ntp:x:38:38:::/etc/ntp:/bin/bash
media:x:1002:1002::/home/media:/bin/bash
chrony:x:998:996::/var/lib/chrony:/bin/bash
libstoragemgmt:x:997:995:daemon account for libstoragemgmt:/var/run/lsm:/
bin/bash
ceph:x:167:167:Ceph daemons:/var/lib/ceph:/bin/bash
root:x:996:993:Docker User:/var/lib/:/bin/bash
bird6:x:995:992:BIRD6 daemon user:/var/lib/bird6:/bin/bash
tcpdump:x:72:72:::/:/bin/bash
xxx

[root@ceph-1 ~]#
```

接下来学习 replace 模块的源码内容。

```python
# 源码位置: lib/ansible/modules/core/files/replace.py
# ...

def write_changes(module,contents,dest):

    # 创建临时文件
    tmpfd, tmpfile = tempfile.mkstemp()
    f = os.fdopen(tmpfd,'wb')
    # 将 content 内容写入临时文件中
    f.write(contents)
    f.close()

    # 是否进行校验
    validate = module.params.get('validate', None)
    valid = not validate
    if validate:
        # 存在校验，需要在 validate 参数中加上字符串 %s
        if "%s" not in validate:
            module.fail_json(msg="validate must contain %%s: %s" % (validate))
        # 最后生成的 shell 命令将 validate 字符串中的 %s 替换成 tmpfile 后的值
        # 比如 validate="ls %s"，此时允许命令为 ls /tmp/xxxxxx，还可以使用 cat 命
            令做一些其他的校验操作
        (rc, out, err) = module.run_command(validate % tmpfile)
        valid = rc == 0
        if rc != 0:
            module.fail_json(msg='failed to validate: '
                                 'rc:%s error:%s' % (rc,err))
    if valid:
        # 原子移动文件，最终保存在 dest 路径下
        module.atomic_move(tmpfile, dest)

def check_file_attrs(module, changed, message):

    # 获取文件的通用参数
    file_args = module.load_file_common_arguments(module.params)
    if module.set_file_attributes_if_different(file_args, False):

        if changed:
            message += " and "
```

```
        changed = True
        message += "ownership, perms or SE linux context changed"

    return message, changed

def main():
    module = AnsibleModule(
        argument_spec=dict(
            dest=dict(required=True, aliases=['name', 'destfile']),
            regexp=dict(required=True),
            replace=dict(default='', type='str'),
            backup=dict(default=False, type='bool'),
            validate=dict(default=None, type='str'),
        ),
        add_file_common_args=True,
        supports_check_mode=True
    )

    params = module.params
    # 操作文件
    dest = os.path.expanduser(params['dest'])

    if os.path.isdir(dest):
        # 如果 dest 是目录，则直接返回
        module.fail_json(rc=256, msg='Destination %s is a directory !' %
dest)

    if not os.path.exists(dest):
        # 如果 dest 文件不存在，则直接返回
        module.fail_json(rc=257, msg='Destination %s does not exist !' %
dest)
    else:
        # 读取文件内容并保存到 contents 中
        f = open(dest, 'rb')
        contents = f.read()
        f.close()

    # 使用 Python 的 re 模块，先编译正则表达式，允许多行匹配
    mre = re.compile(params['regexp'], re.MULTILINE)
    # 直接使用 re 模块的 subn() 方法进行文本替换，最后一个参数是允许最多替换的次数，不
    #   指定时全部替换。返回值为一个元组，第一个参数为替换后的内容，第二个为替换的次数
    result = re.subn(mre, params['replace'], contents, 0)

    # 如果替换次数大于 0 且替换后的文本和源文本不一致，则表明 replacements 已发送
    if result[1] > 0 and contents != result[0]:
        msg = '%s replacements made' % result[1]
        changed = True
    else:
        msg = ''
        changed = False

    if changed and not module.check_mode:
        # 如果替换文本改变了，则需要进行写入操作
        if params['backup'] and os.path.exists(dest):
```

```
            module.backup_local(dest)
        if params['follow'] and os.path.islink(dest):
            dest = os.path.realpath(dest)
        write_changes(module, result[0], dest)

    msg, changed = check_file_attrs(module, changed, msg)
    module.exit_json(changed=changed, msg=msg)

# this is magic, see lib/ansible/module_common.py
#<<INCLUDE_ANSIBLE_MODULE_COMMON>>

main()
```

之前笔者并没有使用 validate 参数，这里通过代码来了解一下 validate 参数的含义。先看第一个函数 write_changes()，该函数有 3 个参数，即 module、contents 和 dest，分别表示当前模块的 AnsibleModule 对象、要写入的内容和最终写入的文件地址。从上述代码中可以看到，validate 参数校验的逻辑如下：

（1）在校验参数生效之前，将文本内容（其实就是替换后的文本内容）写入一个临时文件中，假设临时文件的全路径为/tmp/tmpLizV6x。

（2）获取校验值，判断是否存在%s，如果不存在则直接返回错误信息。

（3）生成校验命令。如输入校验值为 ls %s，则生成待执行的命令为 ls /tmp/tmpLizV6x，即 Ansible 会将这个%s 替换成生成的临时文件的完整路径。

（4）调用 module 的 run_command()方法执行生成的校验命令。如果执行失败则校验不通过，否则校验通过，继续执行后续的原子移动操作。

理解了上述执行逻辑后，可以简单测试下这个 validate 参数。这里想完成这样一个校验：替换全部的 docker 文本并校验新生成的文件内容中不包含 docker 字符串。具体操作命令如下：

```
(ansible1.9.6) [root@master shencong]# ansible ceph-1 -i hosts -m replace
-a "dest=/root/passwd regexp=docker validate='cat %s | grep -v docker'"
ceph-1 | FAILED >> {
    "failed": true,
    "msg": "failed to validate: rc:1 error:cat: |: No such file or directory\
ncat: grep: No such file or directory\ncat: docker: No such file or
directory\n"
}
```

提示执行失败，是不是前面的分析有问题？仔细研究后发现这个问题出在 module.run_command(validate % tmpfile)这一行代码。还记得第 1 章中介绍的 command 和 shell 模块吗？它们之间的区别也体现在这个 run_command()方法中。

- command 模块不支持管道和重定向这样的写法，它调用 run_command()方法的参数分别为 shlex.split(cmd)和 use_unsafe_shell=False。注意，这里输入的命令最后会被 shlex 模块切割，形成命令数组。
- shell 模块支持管道和重定向这样的写法，犹如在控制台上执行 shell 命令一样顺畅。shell 模块调用 run_command()的参数分别为 cmd 和 use_unsafe_shell=True。这里直接输入字符串命令，不用切割命令字符串。

replace 模块中使用的 shell 参数和 command 模块一样，因此不支持带管道的 shell 命令。为了能让 replace 模块的校验参数支持复杂的写法，需要调整模块代码中 run_command() 的 use_unsafe_shell 参数。修改内容如下：

```
# 源码位置: lib/ansible/modules/core/files/replace.py
# ...

def write_changes(module,contents,dest):

    # ...

    if validate:
        # ...
        (rc, out, err) = module.run_command(validate % tmpfile, use_unsafe_
shell=True)
        # ...
    # ...

# ...
```

对于虚拟环境而言，修改的位置为 ~/.pyenv/versions/ansible1.9.6/lib/python2.7/site-packages/ansible/modules/core/files/replace.py。再次测试之前的命令，结果如下：

```
(ansible1.9.6) [root@master ~]# ansible ceph-1 -i hosts -m replace -a "dest=
/root/passwd regexp=docker validate='cat %s | grep -v docker'"
ceph-1 | success >> {
    "changed": true,
    "msg": "2 replacements made"
}
```

最后查看 ceph-1 节点上的 /root/passwd 文件内容，可以看到其中的 docker 字符串已经没有了。

```
[root@ceph-1 ~]# cat /root/passwd
root:x:0:0:root:/root:
bin:x:1:1:bin:/bin:
daemon:x:2:2:daemon:/sbin:
adm:x:3:4:adm:/var/adm:
lp:x:4:7:lp:/var/spool/lpd:
sync:x:5:0:sync:/sbin:/bin/sync
shutdown:x:6:0:shutdown:/sbin:/sbin/shutdown
halt:x:7:0:halt:/sbin:/sbin/halt
mail:x:8:12:mail:/var/spool/mail:
operator:x:11:0:operator:/root:
games:x:12:100:games:/usr/games:
ftp:x:14:50:FTP User:/var/ftp:
nobody:x:99:99:Nobody:/:
systemd-network:x:192:192:systemd Network Management:/:
dbus:x:81:81:System message bus:/:
polkitd:x:999:998:User for polkitd:/:
sshd:x:74:74:Privilege-separated SSH:/var/empty/sshd:
postfix:x:89:89::/var/spool/postfix:
shen:x:1000:1000:shen:/home/shen:
```

```
store:x:1001:1001::/home/store:
ntp:x:38:38::/etc/ntp:
media:x:1002:1002::/home/media:
chrony:x:998:996::/var/lib/chrony:
libstoragemgmt:x:997:995:daemon account for libstoragemgmt:/var/run/lsm:
ceph:x:167:167:Ceph daemons:/var/lib/ceph:
dockerroot:x:996:993:Docker User:/var/lib/docker:
bird6:x:995:992:BIRD6 daemon user:/var/lib/bird6:
tcpdump:x:72:72::/:
xxx

[root@ceph-1 ~]#
```

关于 replace 模块的探索就介绍到这里，有兴趣的读者可以仔细研究模块的源码，找出一些不符合自己的场景的例子进行改造并验证，从而更好地掌握该模块的用法。

3．cron模块

cron 模块非常好用，它可以帮助开发者远程在各种系统上任意设置定时任务并统一管理，如同手动登录到远端主机上使用 crontab 命令设置定时任务一样方便。该模块支持的重要参数如下：

- name：定时任务名称。这个参数很重要，后续修改或者删除定时任务都是依据该名称完成的。
- user：指定哪个用户设置的定时任务。
- job：指定定时执行的任务。
- state：添加或者移除定时任务。
- cron_file：非必需。一旦设置该参数，则会使用自定义的文件代替系统的 crontab 文件控制定时任务。
- backup：当定时任务对应的 crontab 文件有改动时，设置 backup=yes 会对该文件进行备份。
- minute/hour/day/month/weekday：设置脚本定时执行的分、时、天、月和星期。写法都是一样的，例如对于分钟参数而言，*/2 表示每两分钟，*表示每分钟。
- reboot：设置的定时任务是否需要重启。
- special_time：特殊时间的别名，主要是为了简化定时任务的编写代码。包括 reboot（重启后执行）、yearly（每年执行一次）、annually、monthly、weekly、daily 和 hourly 等选项。

了解了 cron 模块的参数后，下面使用该模块完成几个简单测试，具体操作如下：

```
(ansible1.9.6) [root@master ~]# ansible ceph-1 -i hosts -m cron -a "name=
'sync time' job='/usr/sbin/ntpdate ntp1.aliyun.com' minute='*/10'"
ceph-1 | success >> {
    "changed": true,
    "jobs": [
        "sync time"
    ]
```

```
}
(ansible1.9.6) [root@master ~]# ansible ceph-1 -i hosts -m cron -a "name=
'clear data' job='/bin/sh /root/clear_data.sh' minute='0'"
ceph-1 | success >> {
    "changed": true,
    "jobs": [
        "sync time",
        "clear data"
    ]
}
(ansible1.9.6) [root@master ~]# ansible ceph-1 -i hosts -m cron -a "name=
'test echo hello' job='echo hello >> /root/test.log' special_time='hourly'"
ceph-1 | success >> {
    "changed": true,
    "jobs": [
        "sync time",
        "clear data",
        "test echo hello"
    ]
}
```

上述操作给 ceph-1 节点添加了 3 个定时任务：第一个是每 10 分钟执行一次同步时钟的命令，确保节点的系统时间和外界同步；第二个是每小时整执行清理数据的脚本；第三个是每小时执行一次 echo 语句，将输出的 hello 字符串写入/root/test.log 文件中。最后，ceph-1 节点上的定时任务列表如下：

```
[root@ceph-1 ~]# crontab -l
#Ansible: sync time
*/10 * * * * /usr/sbin/ntpdatentp1.aliyun.com
#Ansible: clear data
0 * * * * /bin/sh /root/clear_data.sh
#Ansible: test echo hello
@hourly echo hello >> /root/test.log
```

注意，定时任务上面的注释非常重要，Ansible 正是通过这行注释实现对该定时任务的增、删、改、查操作的。如果把 Ansible 生成的注释去掉，就无法通过 cron 模块来管理定时任务了。后续分析完 cron 模块的源码后，就能够明白其中的原因了。

在 cron 模块中定义了一个 CronTab 类，专门用于管理节点上的 crontab 文件。这个类非常有用，可以直接拿来管理系统上的定时任务。该类的实现源码如下：

```
# 源码位置：lib/ansible/modules/core/system/cron.py
# ...

CRONCMD = "/usr/bin/crontab"

class CronTab(object):

    def __init__(self, module, user=None, cron_file=None):
        self.module    = module
        self.user      = user
        self.root      = (os.getuid() == 0)
        self.lines     = None
        self.ansible   = "#Ansible: "
```

```
        # 选择是否需要通过 syslog 导出额外的调试信息
        self.syslogging = False

        # 自定义的定时任务文件
        if cron_file:
            self.cron_file = '/etc/cron.d/%s' % cron_file
        else:
            self.cron_file = None
        # 读取系统定时任务配置文件，获取定时任务信息。最后得到的定时任务内容列表会放到
          self.lines 中
        self.read()

    def read(self):
        # 读取系统上的 crontab
        self.lines = []
        if self.cron_file:
            # 如果是自定义的定时文件，读取相应内容
            try:
                f = open(self.cron_file, 'r')
                self.lines = f.read().splitlines()
                f.close()
            except IOError, e:
                # cron 文件不存在
                return
            except:
                raise CronTabError("Unexpected error:", sys.exc_info()[0])
        else:
            # 这里的 self._read_user_execute() 最后组成的命令是获取该用户下的定时
              任务；对于大部分 Linux 系统，返回命令为 "/usr/bin/crontab -l"
            # 如果需要查看其他用户的定时任务，将返回 "/usr/bin/crontab -u 用户 -l"
            (rc, out, err) = self.module.run_command(self._read_user_execute(),
use_unsafe_shell=True)

            if rc != 0 and rc != 1: # 1 can mean that there are no jobs.
                raise CronTabError("Unable to read crontab")

            # out 就是输出定时任务列表
            lines = out.splitlines()
            count = 0
            for l in lines:
                if count > 2 or (not re.match( r'# DO NOT EDIT THIS FILE - edit
the master and reinstall.', l) and
                                 not re.match( r'# \(/tmp/.*installed on.*\)', l) and
                                 not re.match( r'# \(.*version.*\)', l)):
                    # 跳过了一些常见的注释
                    self.lines.append(l)
                count += 1

    # ...

    def is_empty(self):
```

```python
            # 判断获得的定时任务列表是否为空
            if len(self.lines) == 0:
                return True
            else:
                return False

    def write(self, backup_file=None):
        """
        将定时信息写入系统的 crontab 文件中并保存所有信息
        """
        if backup_file:
            fileh = open(backup_file, 'w')
        elif self.cron_file:
            fileh = open(self.cron_file, 'w')
        else:
            filed, path = tempfile.mkstemp(prefix='crontab')
            fileh = os.fdopen(filed, 'w')

        fileh.write(self.render())
        fileh.close()

        # 如果需要备份，则直接返回
        if backup_file:
            return

        if not self.cron_file:
            (rc, out, err) = self.module.run_command(self._write_execute
(path), use_unsafe_shell=True)
            os.unlink(path)

            if rc != 0:
                self.module.fail_json(msg=err)

    def add_job(self, name, job):
        # 添加定时任务注释
        self.lines.append("%s%s" % (self.ansible, name))

        # 添加定时任务
        self.lines.append("%s" % (job))

    def update_job(self, name, job):
        return self._update_job(name, job, self.do_add_job)

    def do_add_job(self, lines, comment, job):
        # 添加一个定时任务（job），需要添加一行注释和一行任务
        lines.append(comment)

        lines.append("%s" % (job))

def remove_job(self, name):
    # 移除定时任务
        return self._update_job(name, "", self.do_remove_job)

    def do_remove_job(self, lines, comment, job):
```

```
            return None

    def remove_job_file(self):
        try:
            # 删除自定义的定时文件
            os.unlink(self.cron_file)
            return True
        except OSError, e:
            # cron file does not exist
            return False
        except:
            raise CronTabError("Unexpected error:", sys.exc_info()[0])

    def find_job(self, name):
        comment = None
        for l in self.lines:
            if comment is not None:
                # 找到对应的任务名称
                if comment == name:
                    # 返回相应的定时任务名称及相应的表达式
                    return [comment, l]
                else:
                    comment = None
            elif re.match( r'%s' % self.ansible, l):
                # 获取注释行，去掉 ansible 定时标识，得到定时任务名称
                comment = re.sub( r'%s' % self.ansible, '', l)

        return []

    def get_cron_job(self,minute,hour,day,month,weekday,job,special):
        if special:
            # 特殊定时时间写法，@reboot、@annually、@yearly 等
            if self.cron_file:
                return "@%s %s %s" % (special, self.user, job)
            else:
                return "@%s %s" % (special, job)
        else:
            # 非特殊时间，返回相应的定时表达式，其中，自定义的 crontab 需要包含用户信
            # 息，而系统的 crontab 则不需要
            if self.cron_file:
                return "%s %s %s %s %s %s %s" % (minute,hour,day,month,weekday,
self.user,job)
            else:
                return "%s %s %s %s %s %s" % (minute,hour,day,month,weekday,
job)

        return None

    def get_jobnames(self):
        jobnames = []

        for l in self.lines:
            # 对于有 Ansible 固定注释的行则认为是由 ansible 添加的定时任务，加入 jobnames
            # 前需要去掉相应的标识，只保留定时任务的名称即可
```

```
            if re.match( r'%s' % self.ansible, l):
                jobnames.append(re.sub( r'%s' % self.ansible, '', l))

        return jobnames

    def _update_job(self, name, job, addlinesfunction):
        # 非常重要，核心函数
        ansiblename = "%s%s" % (self.ansible, name)
        newlines = []
        comment = None

        for l in self.lines:
            if comment is not None:
                addlinesfunction(newlines, comment, job)
                comment = None
            elif l == ansiblename:
                comment = l
            else:
                newlines.append(l)

        self.lines = newlines

        if len(newlines) == 0:
            return True
        else:
            return False

    def render(self):

        # 将 self.lines 中的内容翻译成字符串，每行用 \n 连接
        crons = []
        for cron in self.lines:
            crons.append(cron)

        result = '\n'.join(crons)
        if result and result[-1] not in ['\n', '\r']:
            result += '\n'
        return result

    def _read_user_execute(self):
        """
        返回用于读 crontab 的命令
        """
        user = ''
        if self.user:
            # 为了兼容多种系统做出的判断，对于大部分 Linux 而言，程序最终执行的是 else
            #   分支下的代码
            if platform.system() == 'SunOS':
                return "su %s -c '%s -l'" % (pipes.quote(self.user), pipes.
quote(CRONCMD))
            elif platform.system() == 'AIX':
                return "%s -l %s" % (pipes.quote(CRONCMD), pipes.quote(self.
user))
            elif platform.system() == 'HP-UX':
```

```
                    return "%s %s %s" % (CRONCMD , '-l', pipes.quote(self.user))
            else:
                user = '-u %s' % pipes.quote(self.user)
        return "%s %s %s" % (CRONCMD , user, '-l')

    def _write_execute(self, path):
        """
        返回用于写 crontab 文件的命令
        """
        user = ''
        if self.user:
            if platform.system() in ['SunOS', 'HP-UX', 'AIX']:
                # 特殊系统处理
                # ...
            else:
                user = '-u %s' % pipes.quote(self.user)
        return "%s %s %s" % (CRONCMD , user, pipes.quote(path))
```

上面的 CronTab 类并不复杂，但该类中定义的方法较多。为了能更好地理解 CronTab 类的用途，笔者将在 master 节点上基于该类完成一系列操作。为了能初始化 CronTab 类，需要模拟一个 AnsibleModule 类，并实现该类的 run_command() 和 fail_json() 方法。笔者先给出一个简单的实现，具体代码如下：

```
#   -*- coding: utf-8  -*-

#
# 模拟 AnsibleModule 类，文件名称为 module.py
#
import sys
import subprocess
import shlex

class AnsibleModule:

    def run_command(self, cmd, use_unsafe_shell=False):
        if isinstance(cmd, str) and not use_unsafe_shell:
            # use shell, so we should remove it
            cmd = shlex.split(cmd)
        process = subprocess.Popen(
            cmd,
            stdout=subprocess.PIPE,
            stderr=subprocess.PIPE,
            stdin=subprocess.PIPE,
            close_fds=True,
            shell = use_unsafe_shell,
        )

        stdout_stream = process.stdout.read()
        stderr_stream = process.stderr.read()
        returncode = process.wait()
        if not isinstance(stdout_stream, str):
            stdout_stream = stdout_stream.decode('utf-8')
        if not isinstance(stderr_stream, str):
            stderr_stream = stderr_stream.decode('utf-8')
```

```
        return returncode, stdout_stream, stderr_streamz

    def fail_json(self, msg=''):
        print('抛出异常:{},退出'.format(msg))
        sys.exit(1)
```

将上述代码保存到当前目录下新建的 module.py 文件中，接着在 Python 交互模式下进行如下测试：

```
>>> from module import AnsibleModule
>>> module = AnsibleModule()
>>> module.run_command('ls -l | awk "{print \$1}"', True)
(0, 'total\n-rw-------.\ndrwxr-xr-x.\n-rw-r--r--.\ndrwxr-xr-x.\ndrwxr-
xr-x.\n-rw-r--r--.\n-rw-r--r--.\n-rw-r--r--.\n-rw-r--r--.\n-rw-r--r--.
\n-rw-r--r--.\n-rw-r--r--.\n-rw-r--r--.\ndrwxr-xr-x.\n-rw-r--r--.\n-rw-
r--r--.\n-rw-r--r--.\n-rw-r--r--.\n-rw-r--r--.\n-rw-r--r--.\n-rw-r--r
--.\n', '')
>>> module.fail_json('hello')
抛出异常:hello,退出
```

这样在实例化 CronTab 类时，就可以使用这里实现的 AnsibleModule 对象作为第一个参数了。下面看一下如何操作这个 CronTab 类：

```
# 接着上面的交互模式继续执行，这样可以不用再重复写导入 AnsibleModule 模块
>>> from ansible.modules.core.system.cron import CronTab
{"msg": "this module requires key=value arguments (['<<INCLUDE_ANSIBLE_
MODULE_ARGS>>'])", "failed": true}
```

一开始导入 CronTab 类就报错，这是什么原因呢？这里暂时不进行讲解，后面在解析模块运行流程时就可以知道原因了。为了能跳过这个报错，只需要注释掉 cron 模块代码中的最后几行代码即可。修改虚拟环境下的 cron 模块代码如下（注意，真正使用 cron 模块时，这几行的注释要去掉，否则无法构成完整的 Python 模块代码，导致报错）：

```
# 源码位置：lib/ansible/modules/core/system/cron.py
# 修改虚拟环境的位置：~/.pyenv/versions/ansible1.9.6/lib/python2.7/site-
packages/ansible/modules/core/system/cron.py
# ...

# 注释掉最后两行语句
# import module snippets
# from ansible.module_utils.basic import *

# main()
```

修改完代码后，再次导入 CronTab 类，可以看到本次操作已经正常，具体如下：

```
>>> from module import AnsibleModule
>>> module = AnsibleModule()
>>> from ansible.modules.core.system.cron import CronTab
```

下面通过一个示例介绍如何添加定时任务。首先，使用 crontab-l 命令查看定时任务列表，可以看到，在 master 节点的 root 账号下并没有定时任务，具体如下：

```
[root@master ~]# crontab -l
[root@master ~]#
```

此时，在 Python 交互模式下实例化 CronTab 类（初始化过程中会读取相应用户的定时任务文件内容（self.read()）并按行存入 self.lines 中），具体操作如下：

```
>>> cron = CronTab(module)
>>> cron.lines
[]
```

下面为当前的 master 节点添加一个 root 用户下的定时任务，具体操作如下：

```
>>> new_job = cron.get_cron_job('*/10', '*', '*', '*', '*', '/usr/sbin/
ntpdate ntp1.aliyun.com', None)
>>> new_job
'*/10 * * * * /usr/sbin/ntpdate ntp1.aliyun.com'
>>> cron.find_job('sync time')
[]
>>> cron.add_job('sync time', new_job)
>>> cron.lines
['#Ansible: sync time', '*/10 * * * * /usr/sbin/ntpdate ntp1.aliyun.com']
>>> cron.find_job('sync time')
['sync time', '*/10 * * * * /usr/sbin/ntpdate ntp1.aliyun.com']
```

在上面的操作中，先是调用 get_cron_job() 方法并根据设置的 minute、hour、day、month、weekday 及 job 参数生成完整的定时任务表达式，接着调用 find_job() 方法查看定时任务列表中是否包含名称为 sync time 的定时任务，随后继续调用 add_job() 方法将 sync time 定时任务添加到对象的 lines 值中，最后再次调用 find_job() 方法查看定时任务列表中是否包含名称为 sync time 的定时任务。

注意，CronTab 对象在添加定时任务后，其定时任务信息是写入 lines 属性中，而没有写入系统的定时任务文件里，因此这里添加的定时任务并没有生效。继续进行如下操作：

```
>>> cron.write()
```

使用 crontab -l 命令查看系统的定时任务列表：

```
[root@master ~]# crontab -l
#Ansible: sync time
*/10 * * * * /usr/sbin/ntpdate ntp1.aliyun.com
```

一个完整的添加定时任务的过程就是这么简单。至于删除定时任务，同样非常简单，具体操作示例如下：

```
>>> cron.remove_job('sync time')
True
>>> cron.lines
[]
>>> cron.write()
```

这样就通过 CronTab 对象删除了系统中名为'sync time'的定时任务。最后再次运行 crontab -l 命令，结果如下：

```
[root@master ~]# crontab -l
[root@master ~]#
```

经过以上一系列操作后，读者是否对 cron 模块中定义的 CronTab 类有了清晰的认识呢？最后再来看一下 cron 模块的 main()方法，其核心代码正是通过 CronTab 类实现对系统中的定时任务进行统一管理的。

```python
# 源码位置: lib/ansible/modules/core/system/cron.py
# ...

def main():
    module = AnsibleModule(...)

    # 获取模块参数并初始化一些变量
    # ...
    do_install  = state == 'present'
    # ...

    # 得到 CronTab 对象
    crontab = CronTab(module, user, cron_file)

    # 去掉一些不重要的代码
    # ...

    if backup:
        # 如果需要备份，则生成一个临时文件。backuph 为文件描述符，backup_file 为临时文件路径
        (backuph, backup_file) = tempfile.mkstemp(prefix='crontab')
        crontab.write(backup_file)

    # 异常情况
    # ...

    # 生成相应的定时任务表达式
    job = crontab.get_cron_job(minute, hour, day, month, weekday, job, special_time)
    # 根据名称查询是否已经生成相应的定时任务，根据 name 判断
    old_job = crontab.find_job(name)

    if do_install:
        # 添加定时任务
        if len(old_job) == 0:
            # 没有则添加
            crontab.add_job(name, job)
            changed = True
        if len(old_job) > 0 and old_job[1] != job:
            # 有旧任务则更新
            crontab.update_job(name, job)
            changed = True
    else:
        # 删除定时任务
        if len(old_job) > 0:
            crontab.remove_job(name)
            changed = True
```

```
res_args = dict(
    jobs = crontab.get_jobnames(), changed = changed
)

if changed:
    # 最后将变化写入系统的 crontab 中
    crontab.write()

# 准备返回记录一些参数
# ...

module.exit_json(**res_args)

    # ...

# ...
```

看完上述 main()方法的源码及相关注释后，相信读者对于 cron 模块的作用及控制系统定时任务的原理已经非常清楚了，其核心就是对 crontab 命令的封装与使用。

2.2.3　Ansible 1.9.6 中 Playbook 的常用语法

本节中将介绍 Ansible 中 Playbook 的一些常用语法，其在 Ansible 的 Playbook 项目中应用广泛。

1. register

register 语法在 Ansible 的 Playbook 编写中应用非常普遍，主要用于捕获命令的输出并保存在临时变量中，后续会使用该临时变量进行判断或生成主机变量。假设 master 节点上有一个模板文件 test_register.conf.j2，其内容如下：

```
name: {{ host_name }}
host_ip: {{ host_ip }}
```

现在想将这个模板文件分别下发到 ceph-1、ceph-2 和 ceph-3 节点上，其中，host_name 为下发主机的主机名，host_ip 为主机的 IP。下面是针对这个需求编写的 Playbook 文件。

```
---
- hosts: all
  remote_user: root
  # 如果不设置为 false，第一步会对所有机器执行 setup 模块收集信息，时间比较长
  # 不执行 setup 模块
  gather_facts: false
  # 遇到任何错误即停止
  any_errors_fatal: true

  tasks:
    - name: get host ip and name
      shell: hostname -s
      # 获取以上面命令的输出结果，其结果是一个字典
```

```
    register: shell_name

  - name: set host_name
    set_fact:
      host_name: "{{ shell_name.stdout }}"

  - name: get host ip
    shell: cat /etc/hosts | grep `hostname` | awk '{print $1}'
    register: shell_ip

  - name: set host_ip
    set_fact:
      host_ip: "{{ shell_ip.stdout }}"

  - name: copy template file to host
    template:
      src: test_template.txt.j2
      dest: /tmp/test_template.txt
```

上面的 Playbook 文件中主要是得到两个变量：host_name 和 host_ip。每次使用 shell 模块时，会将该模块返回的结果保存到 register 指定的变量中。这个变量是一个字典形式，常用的有以下 3 个属性值：

- rc：命令执行的退出码。
- stdout：命令的正确输出。
- stderr：命令的错误输出。

以下是 Playbook 文件的运行结果。

```
(ansible1.9.6) [root@master test_playbook]# ansible-playbook -i hosts
test_register.yml

PLAY [all] ********************************************************

TASK: [get host ip and name] ************************************
changed: [master]
changed: [ceph-2]
changed: [ceph-3]
changed: [ceph-1]

TASK: [set host_name] ********************************************
ok: [master]
ok: [ceph-2]
ok: [ceph-3]
ok: [ceph-1]

TASK: [get host ip] *********************************************
changed: [master]
changed: [ceph-2]
changed: [ceph-3]
changed: [ceph-1]

TASK: [set host_ip] *********************************************
ok: [master]
ok: [ceph-1]
```

```
ok: [ceph-3]
ok: [ceph-2]

TASK: [copy template file to host] *********************************
changed: [master]
changed: [ceph-2]
changed: [ceph-3]
changed: [ceph-1]

PLAY RECAP *********************************************************
ceph-1                     : ok=5    changed=3    unreachable=0    failed=0
ceph-2                     : ok=5    changed=3    unreachable=0    failed=0
ceph-3                     : ok=5    changed=3    unreachable=0    failed=0
master                     : ok=5    changed=3    unreachable=0    failed=0

# 在 master 主机上查看结果
(ansible1.9.6) [root@master test_playbook]# cat /tmp/test_register.conf
name: master
host_ip: 192.168.26.110
```

通过这个例子，读者可以了解 register 的用法。

2. delegate_to

delegate_to 语法表示将任务委托到某台主机上执行，以下是关于该语法的一个简单示例。

```
---
- hosts: nodes
  remote_user: root
  gather_facts: false
  any_errors_fatal: true

  tasks:
    - name: run on localhost
      shell: hostname
      delegate_to: master
      register: shell_name

    - name: debug master name
      debug:
        msg: "{{ inventory_hostname }}:get master name = {{ shell_name.
stdout }}"
```

上面的第一个任务是将所有主机上执行 hostname 命令的动作委托到 master 节点上执行，最后保存的变量为 shell_name。此时在第二个任务中使用 debug 模块可以打印 hostname 命令的执行结果，其结果必然均为 master，因为所有的 hostname 命令都是委托到 master 节点上执行的。以下是 Playbook 文件的运行结果。

```
(ansible1.9.6) [root@master test_playbook]# ansible-playbook -i hosts
test_delegate_to.yml

PLAY [nodes] *******************************************************
```

```
TASK: [run on localhost] ****************************************
changed: [ceph-1 -> master]
changed: [ceph-3 -> master]
changed: [ceph-2 -> master]

TASK: [debug master name] ***************************************
ok: [ceph-1] => {
    "msg": "ceph-1:get master name = master"
}
ok: [ceph-2] => {
    "msg": "ceph-2:get master name = master"
}
ok: [ceph-3] => {
    "msg": "ceph-3:get master name = master"
}

PLAY RECAP ******************************************************
ceph-1                     : ok=2    changed=1    unreachable=0    failed=0
ceph-2                     : ok=2    changed=1    unreachable=0    failed=0
ceph-3                     : ok=2    changed=1    unreachable=0    failed=0
```

3．条件判断

条件判断语法中最重要的就是 when 语法了，它类似于编程语言中的 if 语句，在 Playbook 中几乎无处不在。when 语法接收 bool 类型的值，如果当前 when 值为 False，则跳过该任务不予执行；否则会执行该任务。以下是 when 语法的几种典型示例。

```
# 文件名: test_when.yml
---
- hosts: all
  remote_user: root
  gather_facts: false
  any_errors_fatal: true

  tasks:
    - name: get host ip and name
      shell: fail-command
      register: shell_name
      # 失败的时候不报错
      failed_when: false

    - name: set host_name
      set_fact:
        host_name: "{{ shell_name.stdout }}"
      when: shell_name.rc == 0 | bool

    - name: get host ip
      shell: cat /etc/hosts | grep `hostname` | awk '{print $1}'
      register: shell_ip
      failed_when: false

    - name: set host_name
      set_fact:
        host_ip: "{{ shell_ip.stdout }}"
```

```
      when: shell_ip.rc == 0 | bool

    - name: debug host_ip
      debug:
        msg: "we get host_name = {{ host_name | default('(error)') }} host_ip
= {{ host_ip }} on {{ inventory_hostname }}"

      - name: error occured
        debug:
          msg: error occured
      when: shell_name.rc != 0 or shell_ip.rc != 0

      - name: no error occured
        debug:
          msg: no error occured
      when:
        - shell_name.rc == 0
        - shell_ip.rc == 0
```

> 💬 注意：failed_when 语法表示任务执行失败时，并不会停止和抛出错误信息，而是会
> 一如既往地向下执行。

上面的示例演示了 when 的几种判断条件：

- 单个条件判断：可以在判断条件之后添加| bool 进一步过滤，将结果强制转成 bool
  类型。
- 多个条件或关系：可以直接使用 or 进行逻辑连接（参考倒数第二个任务）。
- 多个条件与关系：可以使用 and 写成列表的形式（参考最后一个任务）。

以下是该 Playbook 的执行效果。

```
(ansible1.9.6) [root@master test_playbook]# ansible-playbook -i hosts
test_when.yml

PLAY [all] ***************************************************************

TASK: [get host ip and name] ********************************************
changed: [master]
changed: [ceph-1]
changed: [ceph-3]
changed: [ceph-2]

TASK: [set host_name] ***************************************************
skipping: [master]
skipping: [ceph-1]
skipping: [ceph-2]
skipping: [ceph-3]

TASK: [get host ip] *****************************************************
changed: [master]
changed: [ceph-1]
changed: [ceph-2]
changed: [ceph-3]
```

```
TASK: [set host_name] ********************************************************
ok: [master]
ok: [ceph-1]
ok: [ceph-2]
ok: [ceph-3]

TASK: [debug host_ip] ********************************************************
ok: [master] => {
    "msg": "we get host_name = (error) host_ip = 192.168.26.110 on master"
}
ok: [ceph-1] => {
    "msg": "we get host_name = (error) host_ip = 192.168.26.120 on ceph-1"
}
ok: [ceph-2] => {
    "msg": "we get host_name = (error) host_ip = 192.168.26.121 on ceph-2"
}
ok: [ceph-3] => {
    "msg": "we get host_name = (error) host_ip = 192.168.26.122 on ceph-3"
}

TASK: [error occured] ********************************************************
ok: [master] => {
    "msg": "error occured"
}
ok: [ceph-1] => {
    "msg": "error occured"
}
ok: [ceph-2] => {
    "msg": "error occured"
}
ok: [ceph-3] => {
    "msg": "error occured"
}

TASK: [no error occured] ********************************************************
skipping: [master]
skipping: [ceph-1]
skipping: [ceph-2]
skipping: [ceph-3]

PLAY RECAP ********************************************************
ceph-1                     : ok=5    changed=2    unreachable=0    failed=0
ceph-2                     : ok=5    changed=2    unreachable=0    failed=0
ceph-3                     : ok=5    changed=2    unreachable=0    failed=0
master                     : ok=5    changed=2    unreachable=0    failed=0
```

4. 过滤器的用法

在 Playbook 的编写过程中，过滤器也是常用的语法。例如上面的| bool 语法，就是将上一个结果过滤成 bool 类型变量。Ansible 的模板引擎正是基于 Jinja2 模块实现的，因此它支持所有的 Jinja2 过滤器，如默认值过滤器（| default("")）、大小写字符过滤器（| upper、| lower）、翻转过滤器（| reverse）等。下面是一个简单的过滤器用法示例。

```
---
- hosts: ceph-1
  remote_user: root
  gather_facts: false
  any_errors_fatal: true

  vars:
    nums:
      - 1
      - 2
      - 3
      - 4

  tasks:
    - name: get host ip and name
      shell: hostname -s
      register: shell_name

    - name: set host_name
      set_fact:
        host_name: "{{ shell_name.stdout }}"

    - name: debug upper hostname
      debug:
        msg: "upper hostname = {{ host_name | upper }}"

    - name: debug first letter
      debug:
        msg: "upper hostname = {{ host_name | first | upper }}"

    - name: calculate the sum of nums
      debug:
        msg: "sum(1+2+3+4) = {{ nums | sum | float}}"

    - name: calculate the sum of nums
      debug:
        msg: "join(1+2+3+4) = {{ nums | join('#')}}"
```

上面的示例中使用了一些常见的过滤器并通过 debug 模块直接打印结果，运行效果如下：

```
(ansible1.9.6) [root@master test_playbook]# ansible-playbook -i hosts
test_filter.yml

PLAY [ceph-1] *************************************************

TASK: [get host ip and name] *********************************
changed: [ceph-1]

TASK: [set host_name] ****************************************
ok: [ceph-1]

TASK: [debug upper hostname] *********************************
ok: [ceph-1] => {
    "msg": "upper hostname = CEPH-1"
}
```

```
TASK: [debug first letter] *********************************************
ok: [ceph-1] => {
    "msg": "upper hostname = C"
}

TASK: [calculate the sum of nums] **************************************
ok: [ceph-1] => {
    "msg": "sum(1+2+3+4) = 10.0"
}

TASK: [calculate the sum of nums] **************************************
ok: [ceph-1] => {
    "msg": "join(1+2+3+4) = 1#2#3#4"
}

PLAY RECAP ************************************************************
ceph-1                     : ok=6    changed=1    unreachable=0    failed=0
```

5. 通知触发

在 Ansible 中通常使用 handlers 和 notify 语法实现触发任务并依次执行这些任务。其中，handlers 称为触发器，本质上就是一系列的任务集合，这些任务只有在 notify 通知时才会执行，notify 称为通知，通过该语法可以调用 handlers 中定义的任务。这些任务将在每个剧本的最后被触发，这样可以避免多次改变发生时重复执行指定的操作。

以下是笔者在工作中升级集群的 IPv6 功能时编写的一个 Playbook。在修改了网卡信息及 bird6 服务配置后，需要通知系统重启网卡服务和 bird6 服务，这里正是采用了通知触发的方式。

```
---
- hosts: pxys
  remote_user: store
  gather_facts: false

  tasks:
    - name: get interface
      script: "get_network_interface.sh"
      register: shell

    - name: set network file path
      set_fact:
        interface: "{{ shell.stdout }}"

    - name: debug interface
      debug:
        msg: "now we get {{ interface }}"

    - name: set network file path
      set_fact:
        network_file_path:"/etc/sysconfig/network-scripts/ifcfg-{{interface}}"

    - name: add ipv6 config to ifcfg-bond0
      shell: "grep {{ item }} {{ network_file_path }} > /dev/null || sed -i
```

```
'$a\\{{ item }}' {{ network_file_path }}"
    with_items:
      - "IPV6INIT=yes"
      - "IPV6_AUTOCONF=no"
      - "IPV6ADDR={{ ipv6_in }}/{{ mask_num }}"
      - "IPV6_DEFAULTGW={{ ipv6_gateway }}"
    notify:
      # 通知触发 ifup bond0 任务，注意任务名称和触发器中的任务名称需对应
      - ifup bond0
    become: true

  - name: copy bird file to remote_host
    copy:
      src: "{{ bird_rpm_file }}"
      dest: "/home/store/{{ bird_rpm_file }}"

  - name: install bird6
    shell: "rpm -qa | grep bird6 > /dev/null || rpm -ivh {{ bird_rpm_file }}"
    args:
      chdir: "/home/store"
    become: true

  - name: get current ip
    shell: cat /etc/hosts | grep -i `hostname` | awk '{print $1}'
    register: "shell_out"

  - name: debug current ip info
    debug:
      msg: "get current ip:{{ shell_out.stdout }}"

  - name: set local ip
    set_fact:
      local_ip: "{{ shell_out.stdout }}"

  - name: copy ip6tables
    copy:
      src: "ip6tables"
      dest: "/etc/sysconfig/ip6tables"
    become: true

  - name: set commands
    set_fact:
      commands:
        - "/usr/sbin/ip link add dev {{ out_interface }} type dummy"
        - "/usr/sbin/ip link set {{ out_interface }} up"
        - "/usr/sbin/ip addr add dev {{ out_interface }} {{ ipv6_out }}/128"
        - "/usr/sbin/ip6tables-restore /etc/sysconfig/ip6tables"

  - name: run shell command
    shell: "{{ item }}"
    with_items: "{{ commands }}"
    become: true

  - name: copy command to rc.local
    shell: "grep \"{{ item }}\" /etc/rc.d/rc.local > /dev/null || sed -i
```

```
      '$a\\{{ item }}' /etc/rc.d/rc.local"
        with_items: "{{ commands }}"
        become: true

    - name: copy bird6.conf to remote
      template:
        src: "bird6.conf.j2"
        dest: "/etc/bird6.conf"
        backup: yes
      become: true
      notify:
        # 通知触发启动 bird6 服务的任务
        - start bird6

    - name: change ipv6 config in haproxy.cfg
      script: add_ipv6_support.sh

  # 触发器
  handlers:
    - name: ifup bond0
      shell: "/usr/sbin/ifup {{ interface }}"
      become: true

    - name: start bird6
      systemd: name=bird6 state=started enabled=yes
      become: true
```

执行完毕后，通过 ipconfig 命令可以看到相关的 IPv6 已经成功设置。这里由于实验条件原因，无法展示相关结果，感兴趣的读者可以在有条件的环境下测试该 Playbook。

2.3 再探 Ansible 1.9.6 源码

本节首先会简单分析一下 Ansible 1.9.6 的源码文件，重点介绍和 Ansible 1.1 的源码不同的地方。此外，还会介绍 Ansible 1.9.6 中一些重要的类与方法，为后续分析 Ansible 1.9.6 模块的执行流程及 Playbook 运行原理打好基础。

2.3.1 cache 模块

首先来看 Ansible 1.9.6 中 cache 模块的代码。该模块提供了多种形式的缓存插件，包括外部服务缓存及内存缓存。先看其统一的对外接口：

```
# 源码位置: lib/ansible/cache/__init__.py
# ...

class FactCache(MutableMapping):

    def __init__(self, *args, **kwargs):
```

```
    # 用于选择不同缓存方式的插件
    self._plugin = utils.plugins.cache_loader.get(C.CACHE_PLUGIN)
    if self._plugin is None:
        return

def __getitem__(self, key):
# 魔法函数，当调用形式为"对象[ key]"时，调用这里的方法
    if key not in self:
        raise KeyError
    return self._plugin.get(key)

def __setitem__(self, key, value):
    self._plugin.set(key, value)

def __delitem__(self, key):
    self._plugin.delete(key)

def __contains__(self, key):
    return self._plugin.contains(key)

def __iter__(self):
    return iter(self._plugin.keys())

def __len__(self):
    return len(self._plugin.keys())

def copy(self):
    """ 返回简单包含缓存数据(k, v)的字典 """
    return dict([(k, v) for (k, v) in self.iteritems()])

def keys(self):
    return self._plugin.keys()

def flush(self):
    """ 清空所有值 """
    self._plugin.flush()
```

注意：这里定义的 FactCache 类继承自 MutableMapping，其功能和行为类似于字典（dict）。初始化中设置的 self._plugin 属性值非常重要，它决定了缓存的存取方式，是使用内存还是 Redis 服务等。

在 cache 目录下，base.py 定义了一个缓存模块的基础类（BaseCacheModule），其余文件都是在这个基础类上实现的各种形式的缓存插件，并且都定义了 CacheModule 类，而该类正是继承自 BaseCacheModule。这里主要介绍 memory.py 和 redis.py 中实现的 CacheModule 类。首先来看一下 base.py 文件定义的缓存模块的基础类。

```
# 源码位置：lib/ansible/cache/base.py
import exceptions

class BaseCacheModule(object):

    def get(self, key):
```

```
        raise exceptions.NotImplementedError

    def set(self, key, value):
        raise exceptions.NotImplementedError

    def keys(self):
        raise exceptions.NotImplementedError

    def contains(self, key):
        raise exceptions.NotImplementedError

    def delete(self, key):
        raise exceptions.NotImplementedError

    def flush(self):
        raise exceptions.NotImplementedError

    def copy(self):
        raise exceptions.NotImplementedError
```

上面的代码中只定义了基本缓存模块要实现的方法名，并未实现相关方法。可以看到，一个基本的缓存模块有存（get()）、取（set()）、删除（delete()）、复制（copy()）、判断 key 是否存在（contains()）及清空（flush()）等动作。

下面来看一下 memory.py 中的代码。由于是使用内存实现缓存，可以直接用字典的形式保存数据，因此该 CacheModule 类的实现非常简单，具体代码如下：

```
# 源码位置：lib/ansible/cache/memory.py
from ansible.cache.base import BaseCacheModule

class CacheModule(BaseCacheModule):

    def __init__(self, *args, **kwargs):
        # 直接使用字典形式实现数据缓存
        self._cache = {}

    def get(self, key):
        # 从字典中获取 key 值对应的 value
        return self._cache.get(key)

    def set(self, key, value):
        # 保存数据
        self._cache[key] = value

    def keys(self):
        # 输出缓存字典中的所有 key
        return self._cache.keys()

    def contains(self, key):
        # 直接用 in 语句判断即可
        return key in self._cache

    def delete(self, key):
        del self._cache[key]
```

```
    def flush(self):
        self._cache = {}

    def copy(self):
        return self._cache.copy()
```

再来看一下 redis.py 中实现的 CacheModule 类，它是基于第三方模块 redis-py 来操作
Redis 服务的，其代码如下：

```
# 源码位置：lib/ansible/cache/redis.py
# ...

# 检查是否安装了 redis-py 模块，若未安装则直接退出
try:
    from redis import StrictRedis
except ImportError:
    print "The 'redis' python module is required, 'pip install redis'"
    sys.exit(1)

class CacheModule(BaseCacheModule):
    """
    基于 Redis 服务的缓存类
    """
    def __init__(self, *args, **kwargs):
        if C.CACHE_PLUGIN_CONNECTION:
            # 需要设置 Redis 的连接信息
            connection = C.CACHE_PLUGIN_CONNECTION.split(':')
        else:
            connection = []

        # 设置 Redis 连接的相关信息
        self._timeout = float(C.CACHE_PLUGIN_TIMEOUT)
        self._prefix = C.CACHE_PLUGIN_PREFIX
        # 这里通过前面的 connection 连接 Redis，得到的 self._cache 就可以用来操作
          Redis 了
        self._cache = StrictRedis(*connection)
        self._keys_set = 'ansible_cache_keys'

    def _make_key(self, key):
        # 保存在 Redis 服务中的 key 值
        return self._prefix + key

    def get(self, key):
        value = self._cache.get(self._make_key(key))
        if value is None:
            self.delete(key)
            raise KeyError
        # 默认保存的都是字符串形式，最后要转换为 JSON 形式
        return json.loads(value)

    def set(self, key, value):
        # 先将数据转换成 str
        value2 = jsonify(value)
```

```
        if self._timeout > 0: # a timeout of 0 is handled as meaning 'never
expire'
            self._cache.setex(self._make_key(key), int(self._timeout), value2)
        else:
            self._cache.set(self._make_key(key), value2)

        # 使用 Redis 的有序集合保存 key 值
        self._cache.zadd(self._keys_set, time.time(), key)

    def _expire_keys(self):
        if self._timeout > 0:
            expiry_age = time.time() - self._timeout
            self._cache.zremrangebyscore(self._keys_set, 0, expiry_age)

    def keys(self):
        # 清除过期的 keys
        self._expire_keys()
        # 返回剩余未过期的 keys
        return self._cache.zrange(self._keys_set, 0, -1)

    def contains(self, key):
        # 先清除过期的 keys
        self._expire_keys()
        # 查看 key 是否位于名为 self._keys_set 的有序集合中
        return (self._cache.zrank(self._keys_set, key) >= 0)

    def delete(self, key):
        # 删除相应的值，同时也删除保存了该 key 值的有序集合
        self._cache.delete(self._make_key(key))
        self._cache.zrem(self._keys_set, key)

    def flush(self):
        # 清空全部 key
        for key in self.keys():
            self.delete(key)

    def copy(self):
        # FIXME: 也许有更好的方式实现该方法
        ret = dict()
        for key in self.keys():
            ret[key] = self.get(key)
        return ret
```

上面的代码只是对 redis-py 模块进行了简单封装，没有什么难度。有兴趣的读者可以自行部署 Redis 服务进行测试。

2.3.2 shell_plugins 插件

shell_plugins 插件使 Ansible 能同时操作 Windows 和 Linux 系统。还记得前面分析 Ansible 1.1 模块的运行原理吗？最后在清除上传的模块文件的操作中直接使用了 rm -rf 命

令，然后调用了远程通信插件执行该命令。这样的操作无法用于 Windows 系统。此时，Ansible 的做法是将和系统相关的操作统一封装在下一层，对外提供统一的接口。这种处理方式在 Ansible 中应用广泛，如前面介绍的 cache 模块就是一个典型的例子，还有 runner 模块下的所有插件也是采用了这种处理方式。在 shell_plugins 插件中最重要的两个文件是 powershell.py 和 sh.py，前者支持对 Windows 系统的远程操作，后者支持对 Linux 系统的远程操作。由于大部分生产环境都是使用的 CentOS 系统，所以下面将重点分析 sh.py 文件。sh.py 中定义的 ShellModule 类如下：

```python
# 源码位置: lib/ansible/runner/shell_plugins/sh.py
# ...

_USER_HOME_PATH_RE = re.compile(r'^~[_.A-Za-z0-9][-_.A-Za-z0-9]*$')

class ShellModule(object):

    # Python 脚本中结束一行的标识符
    _SHELL_EMBEDDED_PY_EOL = '\n'

    # ...

    def join_path(self, *args):
        return os.path.join(*args)

    def path_has_trailing_slash(self, path):
        return path.endswith('/')

    def chmod(self, mode, path):
        # def chmod()方法直接封装 Linux 中的 chmod 命令
        path = pipes.quote(path)
        return 'chmod %s %s' % (mode, path)

    def remove(self, path, recurse=False):
        # remove()方法直接封装 Linux 中的 rm 命令
        path = pipes.quote(path)
        if recurse:
            return "rm -rf %s >/dev/null 2>&1" % path  # 递归删除（使用-r 选项）
        else:
            return "rm -f %s >/dev/null 2>&1" % path  # 非递归删除

    def mkdtemp(self, basefile=None, system=False, mode=None):
        # mkdtemp()方法返回创建临时目录的命令
        if not basefile:
            basefile = 'ansible-tmp-%s-%s' % (time.time(), random.randint(0,
2**48))
        basetmp = self.join_path(C.DEFAULT_REMOTE_TMP, basefile)
        if system and basetmp.startswith('$HOME'):
            basetmp = self.join_path('/tmp', basefile)
        cmd = 'mkdir -p %s' % basetmp
        if mode:
            cmd += ' && chmod %s %s' % (mode, basetmp)
        cmd += ' && echo %s' % basetmp
```

```
            return cmd

    def expand_user(self, user_home_path):

        # 将路径中的~转成属主目录，使用 echo 命令
        if user_home_path != '~':
            if not _USER_HOME_PATH_RE.match(user_home_path):
                # pipes.quote will make the shell return the string verbatim
                user_home_path = pipes.quote(user_home_path)
        # 执行该命令，最终输出完整的文件路径
        return 'echo %s' % user_home_path

    def checksum(self, path, python_interp):
        # 返回码
        # 校验：成功!
        # 0: 未知错误
        # 1: 远端待校验文件不存在
        # 2: 远端待校验文件没有读权限
        # 3: 该文件是一个目录
        # 4: 远端没有 Python 解释器
        shell_escaped_path = pipes.quote(path)
        test = "rc=flag; [ -r %(p)s ] || rc=2; [ -f %(p)s ] || rc=1; [ -d
%(p)s ] && rc=3; %(i)s -V 2>/dev/null || rc=4; [ x\"$rc\" != \"xflag\" ]
&& echo \"${rc}  \"%(p)s && exit 0" % dict(p=shell_escaped_path, i=python_
interp)
        csums = [
            "({0} -c 'import hashlib; BLOCKSIZE = 65536; hasher = hashlib.
sha1();{2}afile = open(\"'{1}'\", \"rb\"){2}buf = afile.read(BLOCKSIZE){2}
while len(buf) > 0:{2}\thasher.update(buf){2}\tbuf = afile.read(BLOCKSIZE)
{2}afile.close(){2}print(hasher.hexdigest())' 2>/dev/null)".format(python_
interp, shell_escaped_path, self._SHELL_EMBEDDED_PY_EOL),     # Python >
2.4 (including python3)
            "({0} -c 'import sha; BLOCKSIZE = 65536; hasher = sha.sha();
{2}afile = open(\"'{1}'\", \"rb\"){2}buf = afile.read(BLOCKSIZE){2}while
len(buf) > 0:{2}\thasher.update(buf){2}\tbuf = afile.read(BLOCKSIZE){2}
afile.close(){2}print(hasher.hexdigest())' 2>/dev/null)".format(python_
interp, shell_escaped_path, self._SHELL_EMBEDDED_PY_EOL),     # Python ==
2.4
        ]

        cmd = " || ".join(csums)
        cmd = "%s; %s || (echo \'0  \'%s)" % (test, cmd, shell_escaped_path)
        return cmd

def build_module_command(self, env_string, shebang, cmd, rm_tmp=None):
    # 最终合成完整的可执行命令，这里会附上解释器，从 shebang 中获取
    # 脚本第一行通常是 #! /bin/bash、#!/bin/python 等写法
    cmd_parts = [env_string.strip(), shebang.replace("#!", "").strip(),
cmd]
    new_cmd = " ".join(cmd_parts)
    if rm_tmp:
        new_cmd = '%s; rm -rf %s >/dev/null 2>&1' % (new_cmd, rm_tmp)
    return new_cmd
```

　　从上面的代码中可以看到，ShellModule 类中定义的实例方法主要是生成了相应的 shell 命令。下面是具体的调用示例。

```
(ansible1.9.6) [root@master ~]# python
Python 2.7.18 (default, Oct 13 2020, 23:55:15)
[GCC 4.8.5 20150623 (Red Hat 4.8.5-39)] on linux2
Type "help", "copyright", "credits" or "license" for more information.
>>> from ansible.runner.shell_plugins.sh import ShellModule
>>> shell = ShellModule()
>>> shell.join_path('/root', 'test.txt')
'/root/test.txt'
>>> shell.remove('/root/shencong', recurse=False)
'rm -f /root/shencong >/dev/null 2>&1'
>>> shell.remove('/root/shencong', recurse=True)
'rm -rf /root/shencong >/dev/null 2>&1'
>>> shell.chmod(0o644, "/root/test")
'chmod 420 /root/test'
>>> shell.mkdtemp()
Traceback (most recent call last):
  File "<stdin>", line 1, in <module>
  File "/root/.pyenv/versions/ansible1.9.6/lib/python2.7/site-packages/
ansible/runner/shell_plugins/sh.py", line 58, in mkdtemp
    basefile = 'ansible-tmp-%s-%s' % (time.time(), random.randint(0, 2**48))
NameError: global name 'time' is not defined
```

　　以上调用示例中有一个未导入模块的报错信息。仔细查看 mkdtemp()方法，发现其中用到了 time 和 random 模块且均未导入。这里只需要修改 Ansible 1.9.6 的源码部分，添加这两个模块即可。之后再次调用 mkdtemp()方法，便可以得到一个创建临时目录的命令，具体如下：

```
>>> shell = ShellModule()
>>> shell.mkdtemp()
'mkdir -p $HOME/.ansible/tmp/ansible-tmp-1603146248.73-49426153011417 &&
echo $HOME/.ansible/tmp/ansible-tmp-1603146248.73-49426153011417'
```

　　接下来看一个用于校验远程文件 MD5 值的方法，其中使用了 Shell+Python 的联合实现方式，非常值得借鉴和学习。

```
>>> print(shell.checksum('/root/test', '/bin/python'))
rc=flag; [ -r /root/test ] || rc=2; [ -f /root/test ] || rc=1; [ -d /root/test ]
&& rc=3; /bin/python -V 2>/dev/null || rc=4; [ x"$rc" != "xflag" ] && echo
"${rc} "/root/test && exit 0; (/bin/python -c 'import hashlib; BLOCKSIZE
= 65536; hasher = hashlib.sha1();
afile = open("'/root/test'", "rb")
buf = afile.read(BLOCKSIZE)
while len(buf) > 0:
    hasher.update(buf)
    buf = afile.read(BLOCKSIZE)
afile.close()
print(hasher.hexdigest())' 2>/dev/null) || (/bin/python -c 'import sha;
BLOCKSIZE = 65536; hasher = sha.sha();
afile = open("'/root/test'", "rb")
buf = afile.read(BLOCKSIZE)
while len(buf) > 0:
```

```
        hasher.update(buf)
        buf = afile.read(BLOCKSIZE)
    afile.close()
    print(hasher.hexdigest())' 2>/dev/null) || (echo '0  '/root/test)
```

至此，笔者对 shell_plugins 插件的源码已经讲解完毕。

2.3.3　filter_plugins 插件

filter_plugins 插件是指 lib/ansible/runner/filter_plugins 目录下的所有代码文件（除了__init__.py），其中，每一个代码文件都代表 Ansible 中的一个过滤插件。这些代码文件用于实现 Playbook 中的一些过滤器，如 2.2.3 节中讲过的 bool、max 及 to_json 过滤器等。filter_plugins 插件和 Ansible 中的其他插件的设计模式相同，filter_plugins 目录下的所有插件文件内都定义了一个统一的类：FilterModule 类，该类中的 filter() 方法提供了一个对外使用的过滤器函数列表并在代码文件的开头实现了这些过滤器函数。下面分析一个简单的过滤器插件（mathstuff.py），代码如下：

```
# 源码位置: lib/ansible/runner/filter_plugins/mathstuff.py
# ...

class FilterModule(object):
    ''' Ansible 的数学过滤器 '''

    def filters(self):
        return {
            # 普通过滤器
            'isnan': isnotanumber,
            'min' : min,
            'max' : max,

            # 指数和算术过滤器
            'log': logarithm,
            'pow': power,
            'root': inversepower,

            # set theory
            'unique' : unique,
            'intersect': intersect,
            'difference': difference,
            'symmetric_difference': symmetric_difference,
            'union': union,

        }
```

FilterModule 类中的 filters() 方法返回了一个字典，其中，key 值为对外提供的过滤器，value 为相应过滤器的实现方法。这些方法的实现非常简单，这里就不再继续分析了。下面主要演示部分方法的调用，从实际角度考察这些过滤方法的输入和输出，具体操作如下：

```
>>> from ansible.runner.filter_plugins import mathstuff
>>> mathstuff.unique([1, 1, 4, 5, 4, 7])
[1, 4, 5, 7]
>>> a = [1, 2, 3]
>>> b = [3, 4, 5]
>>> mathstuff.intersect(a, b)
[3]
>>> mathstuff.difference(a, b)
[1, 2]
>>> mathstuff.symmetric_difference(a, b)
[1, 2, 4, 5]
>>> mathstuff.union(a, b)
[1, 2, 3, 4, 5]
>>> mathstuff.max([2, 5, 1, 7, 11, 8])
11
>>> mathstuff.min([2, 5, 1, 7, 11, 8])
1
>>> mathstuff.power(2, 3)
8.0
>>> mathstuff.inversepower(2, 3)              # 计算 2 开 3 次根，即 2^(1/3)
1.2599210498948732
>>> mathstuff.inversepower(16, 4)             # 计算 16 开 4 次根，即 16^(1/4)
2.0
```

通过上面的操作，读者可以了解这些过滤器的作用。下面写一个简单的 Playbook，帮助读者学习这些过滤器的用法，具体代码如下：

```
---
- hosts: ceph-1
  remote_user: root
  gather_facts: false
  any_errors_fatal: true

  vars:
    a:
      - 1
      - 2
      - 3
    b:
      - 3
      - 4
      - 5
  tasks:
    - name: test intersect filter
      debug:
        msg: " intersect(a, b) = {{ a | intersect(b) }}"

    - name: test difference filter
      debug:
        msg: " difference(a, b) = {{ a | difference(b) }}"

    - name: test symmetric_difference filter
      debug:
        msg: " symmetric_difference(a, b) = {{ a | symmetric_difference(b) }}"
```

```
    - name: test union filter
      debug:
        msg: " union(a, b) = {{ a | union(b) }}"

    - name: test max filter
      debug:
        msg: "max(2, 5, 1, 7, 11, 8) = {{ [2, 5, 1, 7, 11, 8] | max }}"

    - name: test min filter
      debug:
        msg: "min(2, 5, 1, 7, 11, 8) = {{ [2, 5, 1, 7, 11, 8] | min }}"

    - name: test pow filter
      debug:
        msg: "power(2, 3) = {{ 2 | pow(3) }}"

    - name: test root filter
      debug:
        msg: "inversepower(2, 3) = {{ 2 | root(3) }}"

    - name: test root filter second
      debug:
        msg: "inversepower(16, 4) = {{ 16 | root(4) }}"
```

运行上面的 Playbook，发现输出结果和之前在 Python 交互模式下的结果一模一样，具体如下：

```
(ansible1.9.6) [root@master test_playbook]# ansible-playbook -i hosts
test_filter2.yml

PLAY [ceph-1] *********************************************************

TASK: [test intersect filter] ****************************************
ok: [ceph-1] => {
    "msg": " intersect(a, b) = [3]"
}

TASK: [test difference filter] ***************************************
ok: [ceph-1] => {
    "msg": " difference(a, b) = [1, 2]"
}

TASK: [test symmetric_difference filter] *****************************
ok: [ceph-1] => {
    "msg": " symmetric_difference(a, b) = [1, 2, 4, 5]"
}

TASK: [test union filter] ********************************************
ok: [ceph-1] => {
    "msg": " union(a, b) = [1, 2, 3, 4, 5]"
}

TASK: [test max filter] **********************************************
ok: [ceph-1] => {
    "msg": "max(2, 5, 1, 7, 11, 8) = 11"
```

```
}

TASK: [test min filter] *********************************************
ok: [ceph-1] => {
    "msg": "min(2, 5, 1, 7, 11, 8) = 1"
}

TASK: [test pow filter] *********************************************
ok: [ceph-1] => {
    "msg": "power(2, 3) = 8.0"
}

TASK: [test root filter] ********************************************
ok: [ceph-1] => {
    "msg": "inversepower(2, 3) = 1.25992104989"
}

TASK: [test root filter second] ************************************
ok: [ceph-1] => {
    "msg": "inversepower(16, 4) = 2.0"
}

PLAY RECAP *********************************************************
ceph-1                     : ok=9    changed=0    unreachable=0    failed=0
```

通过上面的示例，读者应该对 Ansible 的过滤器模块已经有了一定的认识。至于 Ansible 的其他内置过滤器，都可以按照这样的方式去学习。另外还有两个过滤器插件 core.py 和 ipaddr.py，对它们的分析方法也是一样的，限于篇幅这里就不再演示了，留给读者自行学习。

2.3.4　lookup_plugins 插件

lookup_plugins 插件主要用来扩展 Playbook 里面的各种字符串和变量。为了方便理解该插件的功能，先来看几个简单的 Playbook 案例。

```
# 实现 master 机器对 ceph-[1:3] 免密
(ansible1.9.6) [root@master test_playbook]# cat test_sshpass.yml
---
- hosts: all
  remote_user: root
  gather_facts: false

  tasks:
    - name: remove old .ssh directory
      file:
        path: /root/.ssh
        state: absent
        force: yes
      when: inventory_hostname == 'master'

    - name: remove old ssh pass key
```

```
      file:
        path: /root/.ssh/authorized_keys
        state: absent

    - name: generate ssh key
      shell: /usr/bin/ssh-keygen -t rsa -P '' -f ~/.ssh/id_rsa
      when: inventory_hostname == 'master'

    - name: copy secret key to all hosts
      authorized_key:
        user: root
        key: "{{ lookup('file', '/root/.ssh/id_rsa.pub') }}"
        state: present
```

上面的代码中首先去掉了 master 节点上旧的.ssh 目录及其他节点的 authorized_keys 文件，以方便多次运行该 Playbook，接着在 master 节点调用 ssh-keyger 命令生成相应的 ssh 密钥文件，最后调用 authorized_key 模块，将公钥复制到所有主机的~/.ssh/authorized_keys 中。在最后一个模块中使用了 lookup 插件中的 file 插件，该插件的主要作用是读取某个文件的内容。该 Playbook 的执行结果如下：

```
(ansible1.9.6) [root@master test_playbook]# ansible-playbook -i hosts
test_sshpass.yml

PLAY [all] ******************************************************

TASK: [remove old .ssh directory] ******************************
skipping: [ceph-1]
skipping: [ceph-2]
skipping: [ceph-3]
changed: [master]

TASK: [generate ssh key] ***************************************
skipping: [ceph-1]
skipping: [ceph-2]
skipping: [ceph-3]
changed: [master]

TASK: [copy secret key to all hosts] ***************************
changed: [master]
changed: [ceph-1]
changed: [ceph-2]
changed: [ceph-3]

PLAY RECAP *****************************************************
ceph-1                     : ok=1    changed=1    unreachable=0    failed=0
ceph-2                     : ok=1    changed=1    unreachable=0    failed=0
ceph-3                     : ok=1    changed=1    unreachable=0    failed=0
master                     : ok=3    changed=3    unreachable=0    failed=0

(ansible1.9.6) [root@master test_playbook]# ssh ceph-1
Last login: Sun Nov  8 21:41:26 2020 from master
```

执行完上面的 Playbook 后，就可以直接使用 ssh ceph-1 免密登录到 ceph-1 节点了。

这是 lookup 插件中一个非常简单的应用。下面来看一下 lookup 插件的源码，具体如下：

```python
# 源码位置: lib/ansible/runner/lookup_plugins/file.py
from ansible import utils, errors
import os
import codecs

class LookupModule(object):

    def __init__(self, basedir=None, **kwargs):
        self.basedir = basedir

    def run(self, terms, inject=None, **kwargs):
        # 该插件的核心动作
        terms = utils.listify_lookup_plugin_terms(terms, self.basedir,
inject)
        ret = []

        # 将输入转换成列表
        if not isinstance(terms, list):
            terms = [ terms ]

        for term in terms:
            basedir_path  = utils.path_dwim(self.basedir, term)
            relative_path = None
            playbook_path = None

            if '_original_file' in inject:
                relative_path = utils.path_dwim_relative(inject['_original_
file'], 'files', term, self.basedir, check=False)
            if 'playbook_dir' in inject:
                playbook_path = os.path.join(inject['playbook_dir'], term)

            for path in (basedir_path, relative_path, playbook_path):
                if path and os.path.exists(path):
                    # 读取文件内容，找到第一个存在的问题，读取其内容后直接跳出循环
                    ret.append(codecs.open(path, encoding="utf8").read().
rstrip())
                    break
            else:
                raise errors.AnsibleError("could not locate file in lookup:
%s" % term)

        # ret 为找到的第一个存在文件的内容，只有一个元素
        return ret
```

上述插件代码中的核心方法 run()非常简单，其输入可以是一个或者多个文件，而返回的 ret 结果为找到的第一个存在的文件的内容。因此，对于前面测试的 Playbook 来说，最后一个 task 中的 key 值即为/root/.ssh/id_rsa.pub 文件的内容。至于其他 lookup 插件的学习，可以阅读插件中 LookupModule 类的 run()方法了解该插件的含义，然后进行实际测试，或者先找案例测试理解该插件的功能，然后再深入源码，分析看到的现象，这些都是非常

有效的学习方式。

2.3.5 辅助模块代码

位于 lib/ansible/utils 目录下的主要是一些辅助的代码文件，如和导入插件相关的 plugins.py、提供警告和提示删除函数的 display_functions.py，以及用于 Ansible 加密和解密的 valut.py 等。下面挑选几个简单且常用的代码文件进行分析，其余的代码在遇到时再进行详细介绍。

首先来看 utils 目录下的 cmd_functions.py 文件。它的内容非常简单，只定义了一个 run_cmd() 函数，输入是一个 Shell 命令，输出是该命令的执行结果。下面是该函数的调用示例：

```
>>> from ansible.utils.cmd_functions import run_cmd
>>> run_cmd('hostname -s')
(0, 'master\n', '')
>>> run_cmd('pwd')
(0, '/root/test_playbook\n', '')
>>> run_cmd('ps -ef | grep sshd')
(1, '', "error: garbage option\n\nUsage:\n ps [options]\n\n Try 'ps --help
<simple|list|output|threads|misc|all>'\n  or 'ps --help <s|l|o|t|m|a>'\n
for additional help text.\n\nFor more details see ps(1).\n")
```

上面的代码中演示了多个 Shell 命令的执行情况，分析代码后发现，run_cmd() 函数同样不支持带管道的 Shell 命令，原因和之前碰到的情况一样，同样是在 subprocess.Popen() 方法中没有指定 shell 参数值。简单改造一下该方法，具体代码如下：

```
# 源码位置: lib/ansible/utils/cmd_funtions.py
# ...

def run_cmd(cmd, live=False, readsize=10, shell=False):

    #readsize = 10

    # 调整下面的代码
    cmdargs = cmd
    if not shell:
        cmdargs = shlex.split(cmd)
    p = subprocess.Popen(cmdargs, stdout=subprocess.PIPE, stderr=
subprocess.PIPE, shell=shell)

    # ...
```

在虚拟环境中完成上述代码的修改，然后继续在交互模式下进行测试。

```
>>> from ansible.utils.cmd_functions import run_cmd
>>> run_cmd('hostname -s')
(0, 'master\n', '')
>>> run_cmd('ps -ef | grep sshd')
(1, '', "error: garbage option\n\nUsage:\n ps [options]\n\n Try 'ps --help
<simple|list|output|threads|misc|all>'\n  or 'ps --help <s|l|o|t|m|a>'\n
for additional help text.\n\nFor more details see ps(1).\n")
```

```
>>> run_cmd('ps -ef | grep sshd', shell=True)
(0, 'root       1035       1  0 Oct13 ?        00:00:00 /usr/sbin/sshd -D\nroot
    894260    1035  0 12:28 ?        00:00:01 sshd: root@pts/0\nroot
    895952  895914  0 14:54 pts/0    00:00:00 /bin/sh -c ps -ef | grep
sshd\nroot       895954  895952  0 14:54 pts/0    00:00:00 /bin/sh -c ps -ef
| grep sshd\n', '')
```

可以看到，这里的 run_cmd()函数已经成功支持带管道的 Shell 命令了。

第二个要学习的代码文件为 utils 目录下的 display_functions.py。该文件中定义了 3 个函数，分别为 deprecated()、warning()和 system_warning()。这些打印函数只是在打印的文本前加上了一些专有的信息，下面是这些函数的调用示例。

```
# 在控制台中打印的颜色为紫色
>>> display_functions.deprecated('test deprecated()', '1.0.0')
 [DEPRECATION WARNING]: test deprecated(). This feature will be removed in
version 1.0.0. Deprecation warnings can be disabled by setting
deprecation_warnings=False in ansible.cfg.
# 如果设置了 removed=True，内部还会抛出异常
>>> display_functions.deprecated('test deprecated()', '1.0.0', removed=
True)
Traceback (most recent call last):
  File "<stdin>", line 1, in <module>
  File "/root/.pyenv/versions/ansible1.9.6/lib/python2.7/site-packages/
ansible/utils/display_functions.py", line 43, in deprecated
    raise errors.AnsibleError("[DEPRECATED]: %s.  Please update your
playbooks." % msg)
ansible.errors.AnsibleError: [DEPRECATED]: test deprecated().  Please
update your playbooks.
# 在控制台中打印的颜色为紫色
>>> display_functions.warning('test warning()')
 [WARNING]: test warning()
```

继续学习第三个辅助代码文件 hashing.py。其比较重要的两个函数为 md5()和 md5s()，前者传入文件路径，计算文件内容的 MD5 值，后者传入字符串并计算其 MD5。下面是这两个函数的调用示例。

```
>>> from ansible.utils import hashing
>>> hashing.md5('/root/hosts')
'769413a8b44ee5bda8e424c187a10d65'
>>> hashing.md5s('xxxx')
'ea416ed0759d46a8de58f63a59077499
```

接下来分析 plugins.py 中的代码，这个辅助模块与 Ansible 1.1 中的 plugins.py 相比变化并不大，同样定义了 PluginLoader 类并基于该类加载所有的插件。这里，笔者将调用 plugins.py 中加载的部分插件，帮助读者更好地理解所加载的 loader。

```
>>> from ansible.utils import plugins
>>> shell_loader = plugins.shell_loader
>>> shell_loader
<ansible.utils.plugins.PluginLoader object at 0x7fa6a5d6d510>
>>> action_loader = plugins.action_loader
>>> action_loader
<ansible.utils.plugins.PluginLoader object at 0x7fa6a5d6d410>
```

上述操作中，笔者导入了两个 loader，分别用于加载动作插件和 shell 插件，这两个loader 都是 PluginLoader 对象。接着使用这两个 loader 分别加载相应的插件。先来看一下shell_loader 的演示代码，具体如下：

```
>>> shell_loader.find_plugin('sh')
'/root/.pyenv/versions/ansible1.9.6/lib/python2.7/site-packages/ansible
/runner/shell_plugins/sh.py'
>>> shell_loader.has_plugin('sh')
True
>>> shell_loader.has_plugin('shx')
False
>>> shell_loader.has_plugin('powershell')
True
>>> sh = shell_loader.get('sh')
# 导入的 ShellModule 对象
>>> sh
<ansible.runner.shell_plugins.sh.ShellModule object at 0x7fa6b30d6b90>
>>> sh.mkdtemp()
'mkdir -p $HOME/.ansible/tmp/ansible-tmp-1603181973.91-245729687831151 &&
echo $HOME/.ansible/tmp/ansible-tmp-1603181973.91-245729687831151'
```

在上面的示例中，先演示了 find_plugin()和 has_plugin()方法，这两个方法可以用于判断想要的插件是否存在，接着通过 get()方法获取对应 shell 插件的 ShellModule 对象，即shell_plugins/sh.py 中的 ShellModule 类实例。再来看一下 action_loader 的演示代码，具体如下：

```
>>> action_loader.find_plugin('copy')
'/root/.pyenv/versions/ansible1.9.6/lib/python2.7/site-packages/ansible
/runner/action_plugins/copy.py'
>>> action_loader.has_plugin('copy')
True
>>> from ansible.runner import Runner
>>> from ansible import inventory
>>> inventory_manager = inventory.Inventory('/root/hosts')
>>> runner = Runner(inventory=inventory_manager)
>>> action_loader.get('copy', {'runner': runner})
```

这里相比上一个操作的 shell_loader 有所不同，ShellModule 类的实例化并不需要传入参数，而 ActionModule 类则需要传入 Runner 对象进行实例化。因此在上面的操作中需要传入{'runner': runner}才能正确得到 ActionModule 对象。

继续看下一个代码文件 su_prompts.py。在该文件中只有一个函数 check_su_prompt()，这个函数用来判断输入的字符串中是否包含密码字符串。对于不同语言的 Linux 系统而言，其提示的输入密码的字符串不同。为了能兼容这些字符串，该文件中还定义了一个字符串数组，具体如下：

```
>>> action_loader.get('copy', {'runner': runner})
# 源码位置：lib/ansible/utils/su_prompts.py
# ...

SU_PROMPT_LOCALIZATIONS = [
```

```
        'Password',
        '암호',
        'パスワード',
        'Adgangskode',
        'Contraseña',
            # 其他语言
            # ...
            '密码',
            '密碼',
        ]

        # ...
```

下面尝试调用 check_su_prompt()方法，操作示例如下：

```
>>> from ansible.utils.su_prompts import check_su_prompt
>>> check_su_prompt('密码:')
True
>>> check_su_prompt('password:')
True
>>> check_su_prompt('xxx password')
False
```

注意：check_su_prompt()方法主要用于底层的 SSH 通信插件中。如果使用 sudo 形式进行提权，则可能面临要输入 sudo 密码的情况。此时，在连接信道（channel）上会返回要求输入密码的字符串，Ansible 则会调用该方法判断是否需要输入密码并进行相应处理。

继续学习下一个辅助代码文件 unicode.py。在该文件中主要有两个函数，即 to_unicode()和 to_bytes()，分别用于将字符串转换成 unicode 和 bytes 格式。下面在 Python 的交互模式下调用这两个函数，具体代码如下：

```
>>> from ansible.utils.unicode import to_unicode, to_bytes
>>> to_unicode('中文')
u'\u4e2d\u6587'
>>> to_unicode('xxx')
u'xxx'
>>> to_bytes('中文')
'\xe4\xb8\xad\xe6\x96\x87'
>>> to_bytes('xxx')
'xxx'
```

注意：对于最后一个 to_bytes()方法，调用结果仍然为 str 类型，其原因可从 to_bytes()的代码中看到。

```
# 源码位置：lib/ansible/utils/unicode.py
# ...

def to_bytes(obj, encoding='utf-8', errors='replace', nonstring=None):
    # ...
    if isinstance(obj, basestring):
```

```
        if isinstance(obj, str):
            # 发现是普通字符串，直接返回
            return obj
        return obj.encode(encoding, errors)
    # ...
```

再来看一个较少用到的文件 vault.py，该文件主要是辅助 ansible-vault 命令的执行，对编写的 Playbook 进行加密和解密操作。这里只简单演示一下如何使用该文件中的几个核心类。

```
>>> from ansible.utils.vault import VaultLib
>>> with VaultLib('my_password') as v:
...     encrypt_data = v.encrypt('这段文本将被加密，机密文件\n')
...
>>> encrypt_data
'$ANSIBLE_VAULT;1.1;AES256\n62306232666339346335333032656431336463626364
4336463393766383366653439653932323161\n32303838626437326165666538353534
34636232663838380a626362313761386636636536363730\n65393439376664656333563
33737613861363462383932386238633064333764616431363439323332\n6538326639
333566300a316364656561393131633533333536535626133363537653033336643030635\
n61326661393332303933663962363133663636662323331336263336313836663639363862
6335366136\n65336362626165623939623161323234353333663393432323838\n'
>>> print(v.decrypt(encrypt_data))
这段文本将被加密，机密文件

>>>
```

在上面的代码中，笔者基于 VaultLib 类对一段文本进行了加密和解密操作。另外，在 vault.py 中还定义了两个类似功能的类：VaultAES 和 VaultAES256。这几个类的主要区别在于加密算法不同，使用方法大体一致。下面使用 VaultAES 类进行演示，具体示例如下：

```
>>> from ansible.utils.vault import VaultAES
>>> aes_v = VaultAES()
>>> aes_v.encrypt('这是一段即将加密的文本', 'secret key for myself')
'53616c7465645f5f72a1fc8b3664c5d9980b9e72a8276ebd0e41cfdd143b41240a6334
bc921f5c9b676ae1295dfa3b419b53192248c2ad58181367a94ef84b927510ea2182361
75109c6fdc5e849fa5b1f8d6b6e797f7322e87119420a45b34e62be0fb5428252b01c07
863a5d108641f90a5a3e30eed9499b27e666babf2e82'
>>> encrypt_data = aes_v.encrypt('这是一段即将加密的文本', 'secret key for
myself')
# 输入错误的密钥
>>> aes_v.decrypt(encrypt_data, 'error secret')
Traceback (most recent call last):
  File "<stdin>", line 1, in <module>
  File "/root/.pyenv/versions/ansible1.9.6/lib/python2.7/site-packages/
ansible/utils/vault.py", line 471, in decrypt
    raise errors.AnsibleError("Decryption failed")
ansible.errors.AnsibleError: Decryption failed
# 输入正确的密钥
>>> print(aes_v.decrypt(encrypt_data, 'secret key for myself'))
这是一段即将加密的文本
```

在 VaultLib 的基础上，Ansible 又在 vault.py 中实现了 VaultEditor 类，用于直接操作

文件的加密和解密。由于其代码内容较少，比较容易，这里就不再给出其源码内容了。下面同样使用调用的方式来演示该类的功能，具体操作如下：

```
>>> from ansible.utils.vault import VaultEditor
>>> VaultEditor(None, 'hard_password', '/root/hosts')
>>> ve.encrypt_file()
>>> ve.read_data('/root/hosts')
'$ANSIBLE_VAULT;1.1;AES256\n643564356531323831363339373730646565363662
4623065663138663565343230323039353061\n3030376161373966663937383266346
231363535373566380a37636339363139656339613061396365\n36396331373666313
1626139303862333376333306330613964343338626666461626637363335616161\n6
430353835626535630a66663262626235666346361613264626531653265353462366
5323765303530\n303865653336626266623163323376333343861373133373432336
2616566383439306131333653330\n636139383134313235373666636383133626461
6162616365303630626436466623532383364396\n393562663635333336593936464
3432383366663931343133536135653661363038939933383866623134\n3231323539
63353633534653653530343031633639356239656233626433616439323461396138
3632\n3735383735623262623335353730838303938343463363739396132643930656
46166636563366366\n323866326465356526131616430303863366231643763335346
563613165613334666365343538613961\n6233666363637663137633766356565564
623432343636666261326332386566386166326333838\n62393361643262376666386
1363461306133533336133616265356234343235663534353665343338\n3463\n'
>>> ve.decrypt_file()
>>> ve.read_data('/root/hosts')
'[master]\nmaster\n\n[nodes]\nceph-[1:3]\n\n[nodes:vars]\nansible_ssh_
user=root\nansible_ssh_pass=@SHENcong19920522\nansible_ssh_connection=
ssh\n\n[master:vars]\nansible_connection=local\n'
```

看完上面的操作后，读者对 vault.py 中定义的这些类及相应的方法应该有了一定的了解，至于深入研究如何实现加密和解密的细节，会涉及一些密码学的知识，这些就留给读者自行学习了。

最后一个定义了较多辅助函数的文件就是__init__.py 了。该文件中有很多后续会用到的函数，这里先简单介绍几个函数，其余的在后续跟踪模块运行流程和解析 Playbook 原理时遇到了再分析。

```
# 源码位置: lib/ansible/utils/__init__.py
# ...

def is_failed(result):
    ''' 对于一个失败的结果，是否为 JSON 格式的结果？'''

    return ((result.get('rc', 0) != 0) or (result.get('failed', False) in
[ True, 'True', 'true']))

def is_changed(result):
    ''' 一个有改动的结果，是否为 JSON 格式呢？'''

    return (result.get('changed', False) in [ True, 'True', 'true'])

# ...

def parse_kv(args):
```

```
        ''' 将 key=value 这样成对的字符串转换成字典形式 '''
        options = {}
        if args is not None:
            try:
                vargs = split_args(args)
            except ValueError, ve:
                if 'no closing quotation' in str(ve).lower():
                    raise errors.AnsibleError("error parsing argument string, try
quoting the entire line.")
                else:
                    raise
            for x in vargs:
                if "=" in x:
                    k, v = x.split("=",1)
                    options[k.strip()] = unquote(v.strip())
        return options

# ...

def is_executable(path):
    '''给定的 path 文件是否有可执行权限？'''
    return (stat.S_IXUSR & os.stat(path)[stat.ST_MODE]
            or stat.S_IXGRP & os.stat(path)[stat.ST_MODE]
            or stat.S_IXOTH & os.stat(path)[stat.ST_MODE])

def unfrackpath(path):
    '''
    将 path 转换成全路径返回
    '''
    return os.path.normpath(os.path.realpath(os.path.expandvars(os.path.
expanduser(path))))

# ...

def boolean(value):
    # 通过输入字符串，判断是否表达 "是" 的含义
    val = str(value)
    if val.lower() in [ "true", "t", "y", "1", "yes" ]:
        return True
    else:
        return False
# ...
```

上面这些简单的函数在后续的 Ansible 核心模块调用中会经常使用。像这些基础的函数，没什么依赖，内容较少，因此非常适合移植到个人的项目中使用。例如，笔者就将 boolean() 这个判断函数用到了工作项目中，还有路径展开函数（unfrackpath()）、判断是否是可执行文件的函数（is_executable()）等，这些都是 Python 初学者绝佳的入门素材，非常值得借鉴和使用。

2.4　追踪 Ansible 1.9.6 的模块运行机制

本节将介绍 Ansible 1.9.6 的模块运行机制，相比 Ansible 1.1 而言，其核心流程并没有多少变化，但是整体代码却有不少的调整。本节将重点介绍 Ansible 1.9.6 与 Ansible 1.1 的不同之处，梳理 Ansible 1.9.6 中模块执行的完整流程，为 2.5 节剖析 Playbook 的运行机制奠定基础。

2.4.1　inventory 目录

为了更好地理解 Ansible 1.9.6 中模块的运行过程，和 Ansible 1.1 的源码分析过程一样，我们先介绍 inventory 目录下的代码。和前面一样，该目录下的代码用于解析 hosts 文件，获取 Ansible 的主机、组及相应的主机变量等信息。首先来看一下 inventory 目录下的 expand_hosts.py 文件，其内容比较简单，具体如下：

```python
# 源码位置：lib/ansible/inventory/expand_hosts.py
# ...

def detect_range(line = None):
    if 0 <= line.find("[") < line.find(":") < line.find("]"):
        return True
    else:
        return False


def expand_hostname_range(line = None):
    all_hosts = []
    if line:
        # 将第 1 个括号（左括号或者右括号）替换成竖线，并按照竖线切割字符串
        (head, nrange, tail) = line.replace('[','|',1).replace(']','|',1).split('|')
        bounds = nrange.split(":")
        # 只能写成 1:10 或者 1:10:2 这样的形式，第一个数字是 start，第二个数字是 end，
          如果存在第三个数字，则表示 step，其余形式将抛出错误
        if len(bounds) != 2 and len(bounds) != 3:
            raise errors.AnsibleError("host range incorrectly specified")
        # 第一个数字为 begin
        beg = bounds[0]
        # 第二个数字为 end
        end = bounds[1]
        if len(bounds) == 2:
            step = 1
        else:
            step = bounds[2]
        # 没有 beg 则默认为 0
        if not beg:
            beg = "0"
```

```
        # 必须要有 end
    if not end:
        raise errors.AnsibleError("host range end value missing")

        # 这段代码的作用会在后面演示
    if beg[0] == '0' and len(beg) > 1:
        rlen = len(beg)
        if rlen != len(end):
            raise errors.AnsibleError("host range format incorrectly
specified!")
        fill = lambda _: str(_).zfill(rlen)
    else:
        fill = str

        # try...catch 语句用于得到简写表达式中完整的字母或者数字的序列
    try:
        i_beg = string.ascii_letters.index(beg)
        i_end = string.ascii_letters.index(end)
        if i_beg > i_end:
            raise errors.AnsibleError("host range format incorrectly
specified!")
        # 对应的字母序列
        seq = string.ascii_letters[i_beg:i_end+1]
    except ValueError:  # not an alpha range
        seq = range(int(beg), int(end)+1, int(step))

    for rseq in seq:
        # 生成完整的 host 名称
        hname = ''.join((head, fill(rseq), tail))

        if detect_range(hname):
            # 递归实现
            all_hosts.extend( expand_hostname_range( hname ) )
        else:
            all_hosts.append(hname)

    return all_hosts
```

上述代码中定义的 detect_range()和 expand_hostname_range()函数的功能都非常明确，第 2 个函数正是解析诸如 ceph-[1:3]这样的简写形式。其中，第 2 个函数有个 fill 值，表示填充。假设 beg='008', end='010', 则得到的连续序列为：008、009、010。注意需要满足 len(beg) == len(end)，否则就会抛出 AnsibleError 错误。下面来演示这两个函数的使用。

```
(ansible1.9.6) [root@master ~]# python
Python 2.7.18 (default, Oct 13 2020, 23:55:15)
[GCC 4.8.5 20150623 (Red Hat 4.8.5-39)] on linux2
Type "help", "copyright", "credits" or "license" for more information.
>>> from ansible.inventory.expand_hosts import detect_range, expand_
hostname_range
>>> detect_range('ceph-[1:10:2]')
True
>>> expand_hostname_range('ceph-[1:10:2]')
['ceph-1', 'ceph-3', 'ceph-5', 'ceph-7', 'ceph-9']
```

```
>>> expand_hostname_range('ceph-[008:010]')
['ceph-008', 'ceph-009', 'ceph-010']
```

前面在 Ansible 1.1 的源码中也分析过 expand_hostname_range()函数，不过注意到该函数的最后还有一个递归动作，因此在 Ansible 1.9.6 中该函数支持输入的字符串中可以包含多个[]，而这种写法在 Ansible 1.1 中并不支持。下面演示 Ansible 1.9.6 中的多[]写法。

```
>>> expand_hostname_range('ceph-[1:3]-[08:10]')
['ceph-1-08', 'ceph-1-09', 'ceph-1-10', 'ceph-2-08', 'ceph-2-09', 'ceph-
2-10', 'ceph-3-08', 'ceph-3-09', 'ceph-3-10']
```

再来看一下 Ansible 1.1 中多[]写法的测试结果，具体如下：

```
(ansible1.1) [root@master ~]# python
Python 2.7.18 (default, Oct 13 2020, 23:55:15)
[GCC 4.8.5 20150623 (Red Hat 4.8.5-39)] on linux2
Type "help", "copyright", "credits" or "license" for more information.
>>> from ansible.inventory.expand_hosts import expand_hostname_range
>>> expand_hostname_range('ceph-[1:3]-[08:10]')
Traceback (most recent call last):
  File "<stdin>", line 1, in <module>
  File "/root/.pyenv/versions/ansible1.1/lib/python2.7/site-packages/
ansible/inventory/expand_hosts.py", line 73, in expand_hostname_range
    (head, nrange, tail) = line.replace('[','|').replace(']','|').split
('|')
ValueError: too many values to unpack
```

除此之外，在 inventory 目录下的其他代码文件，如 group.py 和 host.py 等相对于 Ansible 1.1 而言变化都不大，只是在相关类中补充了部分方法。有兴趣的读者可以继续研读相关类的实现代码，都比较简单。

2.4.2　Runner 类及其方法

根据 Ansible 1.1 的分析经验，runner 目录下的代码是整个 Ansible 工具的核心。这在 Ansible 1.9.6 中几乎没有变化，就连设置多进程的实现方式都是一样的。不过对于进程数量的设置而言，Ansible 1.9.6 和 Ansible 1.1 明显不同：

```
# 源码位置: lib/ansible/runner/__init__.py
# ...

class Runner(object):
    ''' ansible 的核心 api 接口 '''

    def run(self):
        # ...

        if self.forks == 0 or self.forks > len(hosts):
            self.forks = len(hosts)

        # ...
```

通过以上代码可知，Ansible 1.9.6 对于主机数量少于设置进程数量的情况，会调整启动进程数量为主机数量。此外，在 Ansible 1.9.6 中，多进程的目标函数依旧是_executor_hook()，具体代码如下：

```
# 源码位置: lib/ansible/runner/__init__.py
# ...

class Runner(object):
    # ...

    def _parallel_exec(self, hosts):
        ''' 处理多进程任务 '''

        manager = multiprocessing.Manager()
        job_queue = manager.Queue()
        for host in hosts:
            job_queue.put(host)
        result_queue = manager.Queue()

        # ...

        workers = []
        for i in range(self.forks):
            # ...

            # 进程的目标函数正是_executor_hook()
            prc = multiprocessing.Process(target=_executor_hook,
                args=(job_queue, result_queue, new_stdin))
            prc.start()
            workers.append(prc)

        try:
            # 进程同步
            for worker in workers:
                worker.join()
        except KeyboardInterrupt:
            for worker in workers:
                worker.terminate()
                worker.join()

        results = []
        try:
            # 接收返回队列中的结果
            while not result_queue.empty():
                results.append(result_queue.get(block=False))
        except socket.error:
            raise errors.AnsibleError("<interrupted>")
        return results

        # ...
```

在_executor_hook()函数中依旧是调用 Runner 对象的_executor()方法，具体代码如下：

```
# 源码位置: lib/ansible/runner/__init__.py
# ...

def _executor_hook(job_queue, result_queue, new_stdin):

    # ...

    while not job_queue.empty():
        try:
            # 从任务队列中取目标主机
            host = job_queue.get(block=False)
            # 执行模块
            return_data = multiprocessing_runner._executor(host, new_stdin)
            # 向结果队列中放入结果
            result_queue.put(return_data)
        except Queue.Empty:
            pass
        except:
            traceback.print_exc()

# ...
```

和 Ansible 1.1 一样，multiprocessing_runner 是一个 Runner 对象，具体代码如下：

```
# 源码位置: lib/ansible/runner/__init__.py
# ...

class Runner(object):
    # ...

    def run(self):
        # ...

        global multiprocessing_runner
        multiprocessing_runner = self
        results = None

        # ...

    # 其余方法
    # ...
```

以上代码和 Ansible 1.1 中的源码是一致的，后面对于这些相同的代码笔者就不再过多介绍了，只针对 Ansible 1.9.6 中的一些新的代码和逻辑进行介绍。继续追踪 Runner 类中定义的_executor()方法，具体代码如下：

```
# 源码位置: lib/ansible/runner/__init__.py
# ...

class Runner(object):
    # ...

    def _executor(self, host, new_stdin):
        ''' 多进程中处理的目标方法 '''
```

```
    # ...

    try:
        # 处理新的输入
        # ...

        exec_rc = self._executor_internal(host, new_stdin)

        # 校验结果，抛出异常或者返回
        # ...
    except errors.AnsibleError, ae:
        # 处理异常
        # ...
    except Exception:
        # 处理异常
        # ...
```

又回到了 _executor_internal() 方法中，该方法和 Ansible 1.1 中的逻辑基本一致。因此下面笔者会注释并解析大部分代码，最后找到核心调用的方法，具体代码如下：

```
# 源码位置: lib/ansible/runner/__init__.py
# ...

class Runner(object):
    # ...

    def _executor_internal(self, host, new_stdin):

        # 获取主机变量
        # ...

        hostvars = HostVars(temp_vars, self.inventory, vault_password=
self.vault_pass)
        inject['hostvars'] = hostvars

        # 获取主机的远程通信方式，对于 Paramiko、SSH 和 Accelerate 方式，需要提供远
          端 SSH 服务端口
        host_connection = inject.get('ansible_connection', self.transport)
        if host_connection in [ 'paramiko', 'ssh', 'accelerate' ]:
            port = hostvars.get('ansible_ssh_port', self.remote_port)
            if port is None:
                port = C.DEFAULT_REMOTE_PORT
        else:
            # fireball, local, etc
            port = self.remote_port

        # 检测和加载 items 插件
        # ...

        if items is None:
            complex_args = _safe_template_complex_args(self.complex_args,
inject)
            return self._executor_internal_inner(host, self.module_name,
```

```
        self.module_args, inject, port, complex_args=complex_args)
        elif len(items) > 0:
            # ...

            for x in items:
                # ...

                result = self._executor_internal_inner(
                    host,
                    self.module_name,
                    self.module_args,
                    this_inject,
                    port,
                    complex_args=complex_args
                )

                # 处理结果
                # ...
        else:
            # 其余情况跳过处理
            self.callbacks.on_skipped(host, None)
            return ReturnData(host=host, comm_ok=True, result=dict(changed=
False, skipped=True))
```

这样又到了_executor_internal_inner()方法中。后面一部分代码只是处理 Playbook 中
with_items 这样的语法，于是可知 with_items 语法对应的就是使用 for 循环反复调用
_executor_internal_inner()方法完成相应的任务。

相比 Ansible 1.1 而言，在 Ansible 1.9.6 中，_executor_internal_inner()方法扩展了不少
功能，不过整体的逻辑依旧不变。下面给出新版本中该方法的重要代码段并进行详细注释。

```
# 源码位置: lib/ansible/runner/__init__.py
# ...

class Runner(object):
    # ...

    def _executor_internal_inner(self, host, module_name, module_args,
inject, port, is_chained=False, complex_args=None):

        # ...

        # 1. 加载模块对应的动作插件，和 Ansible 1.1 中类似
        module_name = template.template(self.basedir, module_name, inject)

        if module_name in utils.plugins.action_loader:
            if self.background != 0:
                raise errors.AnsibleError("async mode is not supported with
the %s module" % module_name)
            handler = utils.plugins.action_loader.get(module_name, self)
        elif self.background == 0:
            handler = utils.plugins.action_loader.get('normal', self)
        else:
            handler = utils.plugins.action_loader.get('async', self)
```

```
# 针对 Playbook 中的 task，处理相应的条件判断
# ...

# 从 inject 变量中获取各种参数值。此外，还会处理参数，包含变量、设置默认值等情况
# ...

# 2. 加载远程通信连接插件和 shell 插件
try:
    if self.delegate_to or host != actual_host:
        delegate_host = host
    else:
        delegate_host = None
    # 重要语句
    conn = self.connector.connect(actual_host, actual_port, actual_
user, actual_pass, actual_transport, actual_private_key_file, delegate_
host)

    # ...

    shell_plugin = utils.plugins.shell_loader.get(shell_type)
    if shell_plugin is None:
        shell_plugin = utils.plugins.shell_loader.get('sh')

    # 注意 conn.shell 赋值，非常重要，后续会经常使用
    conn.shell = shell_plugin

except errors.AnsibleConnectionFailed, e:
    # 返回异常情况
    # ...

tmp = ''
if self._early_needs_tmp_path(module_name, handler):
    # 3. 创建临时目录时不会直接使用 shell 命令而会使用前面得到的 conn.shell
    tmp = self._make_tmp_path(conn)

# ...

# 对于模块参数中有变量的，需要渲染得到完整的模块参数值
# ...

# 4. 用 split_args() 方法代替 ansible1.1 中的 parse_kv() 方法
args = split_args(module_args)
final_args = []
for arg in args:
    # 解析模块参数，除了 shell 模块，大部分模块参数都是 "k1=v1 k2=v2" 这样的
    #   形式
    if '=' in arg:
        k,v = arg.split('=', 1)
        if unquote(v) != self.omit_token:
            final_args.append(arg)
    else:
        # 如果不是 k=v 形式的参数，直接添加到 final_args 中
```

```
                    final_args.append(arg)
            module_args = ' '.join(final_args)

            # 5. 核心语句，执行模块的 action_plugins 插件中的 run()方法，获取最后的结果
            result = handler.run(conn, tmp, module_name, module_args, inject,
complex_args)

            # 如果存在 until 语法，那么会根据设置的 retries 值重试
            # ...

            # 处理并返回结果
            # ...
```

相比前面遇到的方法，_executor_internal_inner()方法内容较多且复杂。为了能让读者看懂，笔者去掉了部分赋值和判断异常的语句，以及在 Playbook 中处理 when 和 until 语法的代码段。此外，以下 5 处核心代码段需要特别说明。

- 第 1 处代码段和 Ansible 1.1 中一样，加载相应模块的动作插件，如果没有相应模块名称的插件，则加载 normal 或者 async 插件，最终得到一个 handler（后面可知该handler 为插件中的一个 ActionModule 对象）。
- 第 2 处代码段是加载远程通信连接插件和 shell 插件，将 shell 插件类赋值给conn.shell。其中 shell 插件是 Ansible 1.1 中没有的，它使得 conn 支持 Windows 主机的远程操作，如创建临时目录等。
- 第 3 处代码段是在远端连接主机上创建临时目录。在 Ansible 1.1 版本中是直接使用mkdir 命令创建目录，而继续跟踪_make_tmp_path()方法源码可知，它是通过 shell插件中的 mkdtemp()方法来创建临时目录的。对于远端为 Linux 系统而言，也是使用 mkdir 命令。
- 第 4 处代码段是处理模块参数，去掉一些需要忽略的 k=v 值。
- 第 5 处代码段是整个模块运行的核心，即调用动作插件中的 run()方法，最后获得返回结果。

为了能继续跟踪模块运行的完整流程，我们选择 normal 插件继续跟踪。该插件对应的代码文件为 action_plugins 目录下的 normal.py 文件，具体代码如下：

```
# 源码位置：lib/ansible/runner/action_plugins/normal.py
# ...

class ActionModule(object):

    def __init__(self, runner):
        self.runner = runner

    def run(self, conn, tmp, module_name, module_args, inject, complex_
args=None, **kwargs):
        ''' 传输并执行模块，非 copy 模块，也非 template 模块'''

        # 处理输入的模块参数
        module_args = self.runner._complex_args_hack(complex_args, module_args)
```

```
            if self.runner.noop_on_check(inject):
                # shell 和 command 模块不支持 check 模式
                if module_name in [ 'shell', 'command' ]:
                    return ReturnData(conn=conn, comm_ok=True, result=dict
(skipped=True, msg='check mode not supported for %s' % module_name))
                # 其余模块，对于 check 模式需要在模块参数的最后面加上下面的标识
                module_args += " CHECKMODE=True"

            if self.runner.no_log:
                module_args += " NO_LOG=True"

            # shell 和 command 是同一个模块，只不过 shell 模块的参数后需要加上 #USE_
              SHELL 标识
            if module_name == 'shell':
                module_name = 'command'
                module_args += " #USE_SHELL"

            if self.runner.no_log:
                module_display_args = "(no_log enabled, args censored)"
            else:
                module_display_args = module_args

            # 继续调用 Runner 对象的_execute_module()方法并返回其结果
            return self.runner._execute_module(conn, tmp, module_name, module_
            args, inject=inject, complex_args=complex_args)
```

normal 插件同样只做一件事情：针对部分模块调整模块参数（shell 和 command 模块），然后远程传输并执行模块代码（调用 Runner 对象的_execute_module()方法）。其他模块可能会有不同的动作，这些动作完全依赖于动作插件中的 run()方法的代码逻辑。继续回到 Runner 类中定义的_execute_module()方法，具体代码如下：

```
# 源码位置：lib/ansible/runner/__init__.py
# ...

class Runner(object):
    # ...

    def _execute_module(self, conn, tmp, module_name, args,
        async_jid=None, async_module=None, async_limit=None, inject=None,
persist_files=False, complex_args=None, delete_remote_tmp=True):

        # ...

        # 1. 非常重要，根据模块名称得到完整的模块代码文件（模块代码与公共代码合并）
        (
        module_style,
        shebang,
        module_data
        ) = self._configure_module(conn, module_name, args, inject, complex_
args)

        # 2. 创建临时目录
```

```
        if self._late_needs_tmp_path(conn, tmp, module_style):
            tmp = self._make_tmp_path(conn)

        # 得到远端临时目录路径
        remote_module_path = conn.shell.join_path(tmp, module_name)

        if (module_style != 'new'
            or async_jid is not None
            or not conn.has_pipelining
            or not C.ANSIBLE_SSH_PIPELINING
            or C.DEFAULT_KEEP_REMOTE_FILES
            or self.become_method == 'su'):
            # 3. 上传完整的模块代码到临时路径
            self._transfer_str(conn, tmp, module_name, module_data)

        # ...

        cmd = conn.shell.build_module_command(environment_string, shebang,
cmd, rm_tmp)
        cmd = cmd.strip()

        sudoable = True
        if module_name == "accelerate":
            # always run the accelerate module as the user
            # specified in the play, not the become_user
            sudoable = False

        # 4. 远程执行模块代码文件并获取执行结果
        res = self._low_level_exec_command(conn, cmd, tmp, become=self.
become, sudoable=sudoable, in_data=in_data)

        # 处理返回结果
        # ...
```

这段代码的逻辑和 Ansible 1.1 中的_execute_module()方法一样，只不过实现方式有所变化。还记得 2.1 节中 Ansible 1.1 和 Ansible 1.9.6 的源码结构对比吗？在 Ansible 1.9.6 中，模块的源码已经放到了 modules 目录下，并且 Ansible 的公共模块代码已经不再是字符串形式，而是挪到了 module_utils/basic.py 文件中。下面是_execute_module()方法中 4 处最重要的代码段说明。

- 第 1 处代码非常重要，它根据模块的名称来合成完整的模块代码，相应的代码内容保存在 module_data 中。
- 第 2 处代码段是创建临时目录。_make_tmp_path()方法会创建临时目录并返回创建目录的全路径。该方法同样借助 shell 插件生成相应的创建临时目录的命令，然后由 Runner 对象的_low_level_exec_command()方法远程调用命令并得到返回结果。
- 第 3 处代码段和 Ansible 1.1 一样，将完整的模块代码上传到远端的临时目录下，得到的临时文件名称为相应的模块名称（例如对于 ping 模块而言，会将 module_data 内容写到临时目录下的 ping 文件中）。
- 第 4 处代码非常简单，它和 Ansible 1.1 中的_low_level_exec_command()方法一样，

只负责对连接的远端主机执行 cmd 命令并得到相应的返回结果。

第 2~4 处代码比较简单，主要是调用底层通信插件的相关方法。例如 exec_command() 方法是远程执行命令、put_file() 方法用于远程上传文件等。下面重点介绍第 1 处的_configure_module() 方法，看 Ansible 是如何将模块源码和公共代码部分组合成完整的模块代码的。

```python
# 源码位置: lib/ansible/runner/__init__.py
# ...

class Runner(object):
    # ...

    def _configure_module(self, conn, module_name, module_args, inject,
complex_args=None):
        ''' 找到模块并进行配置 '''

        # Search module path(s) for named module.
        module_suffixes = getattr(conn, 'default_suffixes', None)
        module_path = utils.plugins.module_finder.find_plugin(module_name,
module_suffixes)

        # 对于找不到相应模块的情况，要及时抛出异常
        # ...

        # insert shared code and arguments into the module
        (module_data, module_style, module_shebang) = module_replacer.
modify_module(
            module_path, complex_args, module_args, inject
        )

        return (module_style, module_shebang, module_data)
```

上面的代码中，module_path 指的是 Ansible 模块的源码路径。它不是一个完整的可运行的代码文件。这里可在 Python 交互模式下演示该 module_path 的值，具体如下：

```python
>>> from ansible import utils
>>> utils.plugins.module_finder.find_plugin('ping')
'/root/.pyenv/versions/ansible1.9.6/lib/python2.7/site-packages/ansible
/modules/core/system/ping.py'
>>> utils.plugins.module_finder.find_plugin('copy')
'/root/.pyenv/versions/ansible1.9.6/lib/python2.7/site-packages/ansible
/modules/core/files/copy.py'
```

继续看后面代码的注释：insert shared code and arguments into the module，即将共享的代码段和参数插入模块代码中，这就是 module_replacer.modify_module() 方法的作用。以下是其实现语句：

```python
# 源码位置: lib/ansible/runner/__init__.py
# ...

module_replacer = ModuleReplacer(strip_comments=False)
```

继续追踪 ModuleReplacer 的代码实现，具体如下：

```python
# 源码位置: lib/ansible/module_common.py
# ...

class ModuleReplacer(object):
    # ...

    def modify_module(self, module_path, complex_args, module_args, inject):

        with open(module_path) as f:
            # 读取模块的源码内容
            module_data = f.read()

            # 1. 将原模块源码中需要导入的类插入该模块代码中，最重要的是 AnsibleModule 类
            (module_data, module_style) = self._find_snippet_imports(module_data, module_path)

            # 准备好模块参数，为后续替换做好准备
            complex_args_json = utils.jsonify(complex_args)
            try:
                encoded_args = repr(module_args.encode('utf-8'))
            except UnicodeDecodeError:
                encoded_args = repr(module_args)
            try:
                encoded_complex = repr(complex_args_json.encode('utf-8'))
            except UnicodeDecodeError:
                encoded_complex = repr(complex_args_json.encode('utf-8'))

            # 替换 module_data 中的相关值
            module_data = module_data.replace(REPLACER_VERSION, repr(__version__))
            module_data = module_data.replace(REPLACER_SELINUX, ','.join(C.DEFAULT_SELINUX_SPECIAL_FS))
            module_data = module_data.replace(REPLACER_ARGS, encoded_args)
            module_data = module_data.replace(REPLACER_COMPLEX, encoded_complex)

            if module_style == 'new':
                facility = C.DEFAULT_SYSLOG_FACILITY
                if 'ansible_syslog_facility' in inject:
                    facility = inject['ansible_syslog_facility']
                module_data = module_data.replace('syslog.LOG_USER', "syslog.%s" % facility)

            lines = module_data.split("\n")
            shebang = None
            # 找出模块代码的解释器，一般在第一行
            if lines[0].startswith("#!"):
                shebang = lines[0].strip()
                args = shlex.split(str(shebang[2:]))
                interpreter = args[0]
                interpreter_config = 'ansible_%s_interpreter' % os.path.basename(interpreter)
```

```
            if interpreter_config in inject:
                lines[0] = shebang = "#!%s %s" % (inject[interpreter_
config], " ".join(args[1:]))
                module_data = "\n".join(lines)

        return (module_data, module_style, shebang)
```

以上是合成完整 Ansible 模块的源码，其中最核心的语句是对 self._find_snippet_imports()方法的调用。该方法正是将 Ansible 模块源码中所依赖的相关类语句替换成真正导入的代码，从而得到完整的可执行的 Ansible 模块代码。self._find_snippet_imports()方法的实现源码如下：

```
# 源码位置：lib/ansible/module_common.py
# ...

REPLACER = "#<<INCLUDE_ANSIBLE_MODULE_COMMON>>"
# ...

class ModuleReplacer(object):
    # ...

    def _find_snippet_imports(self, module_data, module_path):

        # 1. 得到module_style值
        module_style = 'old'
        if REPLACER in module_data:
            module_style = 'new'
        elif 'from ansible.module_utils.' in module_data:
            module_style = 'new'
        elif 'WANT_JSON' in module_data:
            module_style = 'non_native_want_json'

        # 2. 将module_data按行切割成列表
        output = StringIO()
        lines = module_data.split('\n')
        snippet_names = []

        # 3. 按行处理模块代码的每一行内容
        for line in lines:

            if REPLACER in line:
                output.write(self.slurp(os.path.join(self.snippet_path,
"basic.py")))
                snippet_names.append('basic')
            if REPLACER_WINDOWS in line:
                ps_data = self.slurp(os.path.join(self.snippet_path,
"powershell.ps1"))
                output.write(ps_data)
                snippet_names.append('powershell')
            # 这个if分支是核心
            elif line.startswith('from ansible.module_utils.'):
                tokens=line.split(".")
```

```
            import_error = False
            # 几种错误情况
            if len(tokens) != 3:
                import_error = True
            if " import *" not in line:
                import_error = True
            # 不符合要求的模块直接抛错
            if import_error:
                raise errors.AnsibleError("error importing module in %s,
expecting format like 'from ansible.module_utils.basic import *'" %
module_path)
            snippet_name = tokens[2].split()[0]
            snippet_names.append(snippet_name)
            output.write(self.slurp(os.path.join(self.snippet_path,
snippet_name + ".py")))

        else:
            if self.strip_comments and line.startswith("#") or line == '':
                pass
            output.write(line)
            output.write("\n")

    # 处理异常情况
    # ...

    return (output.getvalue(), module_style)
```

下面来解读 self._find_snippet_imports()方法中的逻辑。

（1）整体查看模块的代码中是否包含 REPLACER 对应的字符串或者 from ansible.module_utils.字符串。如果存在，则 module_style 的值为 new。

（2）将模块代码按行切割形成列表。

（3）遍历模块源码，对于包含 REPLACER 对应的字符串及以 from ansible.module_utils.开头的行需要特殊处理。在 Ansible 1.9.6 的模块代码中类似，几乎都有 from ansible.module_utils.这样的导入语句。因此，上面代码中的 elif line.startswith('from ansible.module_utils.')分支最为关键，正是在此处，Ansible 将导入语句替换成了完整的导入代码。

为了帮助读者理解上述代码中关于第三部分代码段的分析，下面以 command 模块为例进行实际演示。该模块中有两个导入模块语句，具体如下：

```
# 源码位置：lib/ansible/modules/core/commands/command.py
# ...

from ansible.module_utils.basic import *
from ansible.module_utils.splitter import *

# ...
```

在 Python 交互模式下查看上述代码中的 elif 分支，具体如下：

```
>>> line = "from ansible.module_utils.basic import *"
>>> tokens = line.split(".")
>>> snippet_name = tokens[2].split()[0]
```

```
>>> snippet_name
'basic'
```

此时再看 ModuleReplacer 对象的 slurp()方法及 snippet_path 值就一目了然了，具体代码如下：

```
# 源码位置: lib/ansible/module_common.py
# ...

class ModuleReplacer(object):
    # ...

    def __init__(self, strip_comments=False):
        this_file = inspect.getfile(inspect.currentframe())
        self.snippet_path = os.path.join(os.path.dirname(this_file),
'module_utils')
        self.strip_comments = strip_comments # TODO: implement

    def slurp(self, path):
        if not os.path.exists(path):
            raise errors.AnsibleError("imported module support code does not
exist at %s" % path)
        fd = open(path)
        data = fd.read()
        fd.close()
        return data
```

ModuleReplacer 对象的 snippet_path 属性值正是 module_utils 目录所在的全路径，例如：

```
>>> from ansible.module_common import ModuleReplacer
>>> ModuleReplacer().snippet_path
'/root/.pyenv/versions/ansible1.9.6/lib/python2.7/site-packages/ansible
/module_utils'
```

此外，ModuleReplacer 对象的 slurp()方法就是读取输入文件的内容，例如：

```
>>> mr = ModuleReplacer()
>>> import os
>>> mr.slurp(os.path.join(mr.snippet_path, 'basic' + '.py'))
# 打印 basic.py 文件内容，由于内容太多，故忽略
# ...
```

看到这里，读者应该对 Ansible 1.9.6 中如何组成完整模块的代码有所了解了。如果模块代码依赖 module_utils 中的代码，直接使用 from ansible.module_utils.xxx import *将其完整导入即可。后续模块运行时，Ansible 会将该行用对应的模块代码进行替换。

至此，Ansible 1.9.6 的模块运行机制已经非常清晰了，这也是基于前面 Ansible 1.1 的分析经验。到这里可以看到 Ansible 1.9.6 和 Ansible 1.1 的模块运行流程基本一致，并且大部分方法名称和功能均保持不变，只不过在 Ansible 1.9.6 中做了诸多扩展。另外，模块的合成方式也是一个重大变化。相比 Ansible 1.1 而言，Ansible 1.9.6 中增加了 module_utils 目录，可以将一些辅助代码放在此处，然后在 Ansible 的模块源码中引入即可。这种模式在一定程度上提高了 Ansible 模块代码的复用性，而且该模式也一直保留至今。

2.5　解析 Playbook 的运行逻辑

本节将探索 Ansible 中 Playbook 的完整运行流程。在这里将解释部分 Playbook 语法的执行逻辑，确保读者能完全理解 Playbook 中相关语法的作用，为后续编写自动化部署的 Playbook 项目打好基础。

2.5.1　准备测试的 Playbook 项目

为了能直观理解 Ansible 1.9.6 中和 Playbook 相关的代码，首先需要准备一个测试的 Playbook 项目，后续将在这个项目上进行测试，一步一步理解 Ansible 中处理 Playbook 的核心类和方法。

首先新建一个测试目录：

```
(ansible1.9.6) [root@master ansible-1.9.6]# mkdir test-playbook
(ansible1.9.6) [root@master ansible-1.9.6]# cd test-playbook/
(ansible1.9.6) [root@master test-playbook]# ls
```

准备好 test-role1 和 test-role2 两个角色：

```
(ansible1.9.6) [root@master test-playbook]# mkdir roles/test-role1
(ansible1.9.6) [root@master test-playbook]# mkdir roles/test-role2
```

角色目录通常包含以下目录：

- tasks：必备目录，为角色的任务入口目录，角色任务的入口文件为 tasks 目录下的 main.yml 文件。
- defaults：非必需目录。该目录下的 main.yml 中的内容为该角色的主机变量。
- templates：非必需目录。通常用于放置模板文件。

为了简单起见，这里只新建 tasks 目录，并在每个角色的 tasks 目录下新建一个 main.yml 文件，其内容如下：

```
(ansible1.9.6) [root@master test-playbook]# cat roles/test-role1/tasks/
main.yml
---
- name: get hostname
  shell: hostname
  register: name_out

- name: debug hostname
  debug:
    msg: "hostname = {{ name_out.stdout }}"
(ansible1.9.6) [root@master test-playbook]# cat roles/test-role2/tasks/
main.yml
---
- name: get hostname
```

```
    shell: hostname
    register: name_out

  - name: debug hostname
    debug:
      msg: "hostname = {{ name_out.stdout }}"
```

此外，为了能让两个角色有所区别，笔者给 test-role1 角色添加了一个 vars 目录，并在该目录下准备了一个简单的 main.yml 文件，其内容如下：

```
(ansible1.9.6) [root@master test-playbook]# cat roles/test-role1/vars/
main.yml
test_role_var1: var1
```

接着准备一个全局的变量文件 group_vars/all.yml，其内容如下：

```
(ansible1.9.6) [root@master test-playbook]# cat group_vars/all.yml
global_variable: vars
```

再来准备一个 site.yml 文件（该文件名可以任意，并无规定，一般而言，大型的部署项目喜欢用 site.yml），其内容如下：

```
---
- hosts: group1
  gather_facts: false

  pre_tasks:
    - name: test pre tasks in group1
      debug:
        msg: this is a pre task in group1

  roles:
    - test-role1

  tasks:
    - name: end task
      debug:
        msg: "vars = {{ global_variable }}"

- hosts: group2
  gather_facts: false

  pre_tasks:
    - name: test pre tasks in group2
      debug:
        msg: this is a pre task in group2

  roles:
    - test-role2

  tasks:
    - name: end task
      debug:
        msg: end task in group2
```

最后准备 hosts 文件，其内容如下：

```
(ansible1.9.6) [root@master test-playbook]# cat hosts
[group1]
master

[group2]
ceph-[1:3]

[group2:vars]
ansible_ssh_user=root
ansible_ssh_pass=@SHENcong19920522
ansible_ssh_connection=ssh

[group1:vars]
ansible_connection=local
```

简单运行这个 Playbook，体验带角色、带变量目录的 Playbook 项目的运行效果，具体如下：

```
(ansible1.9.6) [root@master test-playbook]# ansible-playbook -i hosts
site.yml

PLAY [group1] ************************************************************

TASK: [test pre tasks in group1] ***************************************
ok: [master] => {
    "msg": "this is a pre task in group1"
}

TASK: [test-role1 | get hostname] **************************************
changed: [master]

TASK: [test-role1 | debug hostname] ************************************
ok: [master] => {
    "msg": "hostname = master"
}

TASK: [end task] *******************************************************
ok: [master] => {
    "msg": "vars = vars"
}

PLAY [group2] ************************************************************

TASK: [test pre tasks in group2] ***************************************
ok: [ceph-1] => {
    "msg": "this is a pre task in group2"
}
ok: [ceph-2] => {
    "msg": "this is a pre task in group2"
}
ok: [ceph-3] => {
    "msg": "this is a pre task in group2"
}

TASK: [test-role2 | get hostname] **************************************
```

```
changed: [ceph-2]
changed: [ceph-1]
changed: [ceph-3]

TASK: [test-role2 | debug hostname] ***********************************
ok: [ceph-1] => {
    "msg": "hostname = ceph-1"
}
ok: [ceph-2] => {
    "msg": "hostname = ceph-2"
}
ok: [ceph-3] => {
    "msg": "hostname = ceph-3"
}

TASK: [end task] *****************************************************
ok: [ceph-1] => {
    "msg": "end task in group2"
}
ok: [ceph-2] => {
    "msg": "end task in group2"
}
ok: [ceph-3] => {
    "msg": "end task in group2"
}

PLAY RECAP *********************************************************
ceph-1                     : ok=4    changed=1    unreachable=0    failed=0
ceph-2                     : ok=4    changed=1    unreachable=0    failed=0
ceph-3                     : ok=4    changed=1    unreachable=0    failed=0
master                     : ok=4    changed=1    unreachable=0    failed=0
```

以上是一个非常简单的剧本示例，其包含 tasks、pre_tasks 和 roles 等常见语法。后面的测试与实验将完全基于这个简单的剧本，请读者务必按照上面的步骤准备好这个剧本并使其成功运行。

2.5.2　相关类基础

要想了解 Playbook 的运行流程，lib/ansible/playbook 目录下的代码内容至关重要。首先简单介绍该目录下的文件及其定义的核心类。

- __init__.py：该文件中只定义了一个 PlayBook 类，用于代表所有的剧本（play）集。该类中的 run() 方法正是剧本执行的入口，这一点会在后面的源码分析中介绍。
- play.py：该文件中定义了一个 Play 类，表示单个剧本。
- task.py：该文件中定义了一个 Task 类，表示单个剧本中的单个任务。

首先来看 playbook 目录下的 __init__.py 文件，实例化 PlayBook 类至少需要 inventory、stats、playbook_cb 和 runner_cb 这 4 个参数（inventory 一般有默认值，但这里需要特殊指定 hosts 文件）。后面 3 个参数的赋值可以直接参考 bin/ansible-playbook 中实例化 PlayBook

类的语句，示例操作如下：

```
>>> from ansible.inventory import Inventory
>>> from ansible import callbacks
>>> from ansible.playbook import PlayBook
>>> inventory = Inventory('/root/ansible-1.9.6/test-playbook/hosts')
>>> stats = callbacks.AggregateStats()
>>> playbook_cb = callbacks.PlaybookCallbacks(0)
>>> runner_cb = callbacks.PlaybookRunnerCallbacks(stats, verbose=0)
>>> pb = PlayBook(playbook='/root/ansible-1.9.6/test-playbook/site.yml',
inventory=inventory, runner_callbacks=runner_cb, stats=stats, callbacks=
playbook_cb)
>>> pb._load_playbook_from_file('/root/ansible-1.9.6/test-playbook/site.
yml', {})
([{'gather_facts': False, 'tasks': [{'debug': {'msg': 'vars = {{ global_
variable }}'}, 'name': 'end task'}], 'hosts': 'group1', 'pre_tasks':
[{'debug': {'msg': 'this is a pre task in group1'}, 'name': 'test pre tasks
in group1'}], 'roles': ['test-role1']}, {'gather_facts': False, 'tasks':
[{'debug': {'msg': 'end task in group2'}, 'name': 'end task'}], 'hosts':
'group2', 'pre_tasks': [{'debug': {'msg': 'this is a pre task in group2'},
'name': 'test pre tasks in group2'}], 'roles': ['test-role2']}], ['/root/
ansible-1.9.6/test-playbook', '/root/ansible-1.9.6/test-playbook'])
```

从上面的示例中可以看到，PlayBook 类的_load_playbook_from_file()方法会从输入的剧本文件中找出所有的剧本并形成 json 元素的列表。site.yml 文件中有两个 Play，每个 Play 所属的入口目录均为/root/ansible-1.9.6/test-playbook。下面来看 PlayBook 类中的__init__()方法，具体代码如下：

```
# 源码位置: lib/ansible/playbook/__init__.py
# ...

class PlayBook(object):
    # ...

    def __init__(self,
        # 忽略初始化参数
        # ...
    ):
        # ...

        (self.playbook, self.play_basedirs) = self._load_playbook_from_
file(playbook, vars)

        # ...
```

由此可知，在前面代码中初始化 PlayBook 类后，pb 的 Playbook 属性值为该剧本集中所有剧本的数据列表，每个剧本数据包括任务集（tasks）、前置执行任务集（pre_tasks）、后置执行任务集（post_tasks）和角色集（roles）等，而 play_basedirs 表示的是该 Play 所在的根目录：

```
>>> pb.playbook
[{'gather_facts': False, 'tasks': [{'debug': {'msg': 'vars = {{ global_
variable }}'}, 'name': 'end task'}], 'hosts': 'group1', 'pre_tasks':
```

```
[{'debug': {'msg': 'this is a pre task in group1'}, 'name': 'test pre tasks
in group1'}], 'roles': ['test-role1']}, {'gather_facts': False, 'tasks':
[{'debug': {'msg': 'end task in group2'}, 'name': 'end task'}], 'hosts':
'group2', 'pre_tasks': [{'debug': {'msg': 'this is a pre task in group2'},
'name': 'test pre tasks in group2'}], 'roles': ['test-role2']}]
>>> pb.play_basedirs
['/root/ansible-1.9.6/test-playbook', '/root/ansible-1.9.6/test-playbook']
```

再来看 Play 类的初始化方法，它需要 3 个参数才能实例化（之后一个 vault_password
用于加密，可忽略），即 PlayBook 对象、Play 的数据（即上面 pb.playbook 中的元素）及
Play 所在的根目录（即上面 pb.play_basedirs 中的元素）。因此，这里可以使用上面的结果
来实例化一个 Play 类，具体操作如下：

```
>>> from ansible.playbook.play import Play
>>> play = Play(pb, pb.playbook[0], pb.play_basedirs[0])
>>> play
<ansible.playbook.play.Play object at 0x7fc667ed91c0>
>>> play._get_role_path('test-role1')
('/root/ansible-1.9.6/test-playbook/roles/test-role1', {})
```

Play 对象在初始化时会生成一个_tasks 属性，它是 Play 的 tasks 列表，列表中的每个
元素都是一个 Task 对象，具体如下：

```
>>> play._tasks
[<ansible.playbook.task.Task object at 0x7f0f720ea050>, <ansible.playbook.
task.Task object at 0x7f0f720ea750>, <ansible.playbook.task.Task object at
0x7f0f720ea910>, <ansible.playbook.task.Task object at 0x7f0f720eaad0>,
<ansible.playbook.task.Task object at 0x7f0f720eac90>, <ansible.playbook.
task.Task object at 0x7f0f720eae50>, <ansible.playbook.task.Task object at
0x7f0f72121050>]
```

上面的 Task 对象有 7 个，这是怎么回事？在 pb.playbook 中，第一个元素的 tasks 值
中只有 2 个元素，加上 roles 中的任务，第一个 Play 对象共有 4 个 task，那么剩余的 3 个
task 分别代表什么呢？接下来笔者将从 Task 类的源码开始向上分析，最后会在 Play 类的
源码中解答上述疑惑。Task 类的核心代码如下：

```
# 源码位置：lib/ansible/playbook/task.py
# ...

class Task(object):
    # ...

    def __init__(self, play, ds, module_vars=None, play_vars=None, play_
file_vars=None, role_vars=None, role_params=None, default_vars=None,
additional_conditions=None, role_name=None, no_tags=True):
        # ...

        # 添加额外模块查找目录
        # ...

        for x in ds.keys():
            if x in utils.plugins.module_finder:
```

```
                if 'action' in ds:
                    raise errors.AnsibleError("multiple actions specified in
task: '%s' and '%s'" % (x, ds.get('name', ds['action'])))
                if isinstance(ds[x], dict):
                    if 'args' in ds:
                        # 直接抛出异常
                        # ...
                    ds['args'] = ds[x]
                    ds[x] = ''
                elif ds[x] is None:
                    ds[x] = ''
                if not isinstance(ds[x], basestring):
                    raise errors.AnsibleError("action specified for task %s
has invalid type %s" % (ds.get('name', "%s: %s" % (x, ds[x])), type(ds[x])))
                ds['action'] = x + " " + ds[x]
                ds.pop(x)

            elif x.startswith("with_"):
                if isinstance(ds[x], basestring):
                    param = ds[x].strip()

                plugin_name = x.replace("with_","")
                if plugin_name in utils.plugins.lookup_loader:
                    ds['items_lookup_plugin'] = plugin_name
                    ds['items_lookup_terms'] = ds[x]
                    ds.pop(x)
                else:
                    # 异常情况，直接抛出异常
                    # ...

            elif x in [ 'changed_when', 'failed_when', 'when']:
                if isinstance(ds[x], basestring):
                    param = ds[x].strip()
                    # Only a variable, no logic
                    if (param.startswith('{{') and
                        param.find('}}') == len(ds[x]) - 2 and
                        param.find('|') == -1):
                        # 警告提示，在 when 语法中不推荐使用{{ }}
                        # ...
            elif x.startswith("when_"):
                # 不推荐该语法，后续被移除
                # ...
            elif not x in Task.VALID_KEYS:
                # 抛出异常
                # ...

        # 设置 task 的各种属性值，内容太多，可忽略
        # ...
```

　　上面的 Task 类的实例化方法的内容较多，在__init__()方法中主要是设置 Task 对象的属性值。例如，Playbook 中 task 的书写语句是否包含 "become: true"，如果有则对应生成的 Task 对象中将设置其 become 属性值为 True。通过观察上面列出的代码可以发现，比较重要的一个初始化参数为 ds。每次实例化 Task 类时，参数 ds 的值是什么呢？下面在 Task

类的初始化方法的第一行打印 ds 值并将其更新到虚拟环境中，接着再次执行前面的
ansible-playbook 命令，具体代码如下：

```
(ansible1.9.6) [root@master test-playbook]# cat ~/.pyenv/versions/ansible
1.9.6/lib/python2.7/site-packages/ansible/playbook/task.py | sed -n '46,50p'
  def __init__(self, play, ds, module_vars=None, play_vars=None, play_
file_vars=None, role_vars=None, role_params=None, default_vars=None,
additional_conditions=None, role_name=None, no_tags=True):
    ''' constructor loads from a task or handler datastructure '''

    print('Task,ds={}'.format(ds))  # 笔者添加的 print() 语句
(ansible1.9.6) [root@master test-playbook]# ansible-playbook -i hosts
site.yml
Task,ds={'debug': {'msg': 'this is a pre task in group1'}, 'name': 'test
pre tasks in group1'}
Task,ds={'meta': 'flush_handlers'}
Task,ds={'shell': 'hostname', 'name': 'get hostname', 'register': 'name_
out'}
Task,ds={'debug': {'msg': 'hostname = {{ name_out.stdout }}'}, 'name':
'debug hostname'}
Task,ds={'debug': {'msg': 'vars = {{ global_variable }}'}, 'name': 'end
task'}
Task,ds={'meta': 'flush_handlers'}
Task,ds={'meta': 'flush_handlers'}
Task,ds={'debug': {'msg': 'this is a pre task in group2'}, 'name': 'test
pre tasks in group2'}
Task,ds={'meta': 'flush_handlers'}
Task,ds={'shell': 'hostname', 'name': 'get hostname', 'register': 'name_
out'}
Task,ds={'debug': {'msg': 'hostname = {{ name_out.stdout }}'}, 'name':
'debug hostname'}
Task,ds={'debug': {'msg': 'end task in group2'}, 'name': 'end task'}
Task,ds={'meta': 'flush_handlers'}
Task,ds={'meta': 'flush_handlers'}

# 忽略后续的 task 输出
# ...
```

从上面的结果中可以看到第一个 Play 对象中出现了 7 个 task，至于为何会有{'meta':
'flush_handlers'}这样的 task 出现，则需要在后续的 Play 类源码中寻找。上面的结果直观地
展示了 ds 的值，对于一个普通的 task 而言，ds 是该 task 的一个 JSON 形式的数据。例如
下面的一个示例：

```
{'shell': 'hostname', 'name': 'get hostname', 'register': 'name_out'}
```

其中，shell 是任务执行的模块，name 是该任务的名称，register 参数对应的是将结果
写入的变量，其正好对应下面的 task：

```
- name: get hostname
  shell: hostname
  register: name_out
```

此外，Task 类中还有一个_load_tags()方法，主要用于提取任务的 tags 参数。在本次

的测试 Playbook 中并未使用 tags 参数，而且该方法内容较少，功能清晰，这里就不再介绍了。接下来分析 Play 类，同样首先分析__init__()方法，其内容如下：

```
# 源码位置: lib/ansible/playbook/play.py
# ...

class Play(object):
    _pb_common = [
        'accelerate', 'accelerate_ipv6', 'accelerate_port', 'any_errors_
fatal', 'become',
        'become_method', 'become_user', 'environment', 'force_handlers',
'gather_facts',
        'handlers', 'hosts', 'name', 'no_log', 'remote_user', 'roles',
'serial', 'su',
        'su_user', 'sudo', 'sudo_user', 'tags', 'vars', 'vars_files',
'vars_prompt',
        'vault_password',
    ]

    __slots__ = _pb_common + [
        '_ds', '_handlers', '_play_hosts', '_tasks', 'any_errors_fatal',
'basedir',
        'default_vars', 'included_roles', 'max_fail_pct', 'playbook',
'remote_port',
        'role_vars', 'transport', 'vars_file_vars',
    ]

    VALID_KEYS = frozenset(_pb_common + [
        'connection', 'include', 'max_fail_percentage', 'port', 'post_tasks',
        'pre_tasks', 'role_names', 'tasks', 'user',
    ])

    # ...

    def __init__(self, playbook, ds, basedir, vault_password=None):

        # 1. 校验 ds 中的所有 key, 必须是前面合法的值
        for x in ds.keys():
            if not x in Play.VALID_KEYS:
                raise errors.AnsibleError("%s is not a legal parameter of an
Ansible Play" % x)

        # ds 参数初始化 Play 对象的属性值
        # ...
        self.roles           = ds.get('roles', None)

        # ...

        # 2. 加载角色
        self.included_roles = []
        ds = self._load_roles(self.roles, ds)
```

```
        # ...

        self._ds = ds

        # ...

        # 3. 加载 tasks 和 handlers
        self._tasks    = self._load_tasks(self._ds.get('tasks', []),
load_vars)
        self._handlers  = self._load_tasks(self._ds.get('handlers', []),
load_vars)

            # ...
```

上面的代码中简化了大部分赋值语句和加载变量的代码段，同时也对三处比较重要的代码段进行了详细说明。

第 1 处代码校验 ds 中的所有 key。注意这个 keys 就是 playbook 文件中的第一层值名称，例如下面的写法：

```
- hosts: all
  gather_facts: false
  not_exist: true
  remote_user: root

  any_errors_fatal: false

  tasks:
    - name: xxxx
      xxxx: yyyy

  roles:
    - xxxx

  post_tasks:
    - name: zzzz
      mmmm: nnnn
```

第一层的 keys 为 hosts、gather_facts、not_exist、remote_user、any_errors_fatal、tasks、roles 或 post_tasks，这些是 YAML 语法基础。第一处的 for 循环会对这些字段进行校验，如果不在 Play.VALID_KEYS 数组中则直接抛错，这里 not_exist 明显不是有效的 key 值，会导致异常。

第 2 处代码将导入相应角色中的 tasks、handlers、vars 及 defaults 等目录。以下是导入这些目录文件的具体实现代码：

```
# 源码位置: lib/ansible/playbook/play.py
# ...

class Play(object):
    # ...

    def _load_roles(self, roles, ds):
```

```
# roles 值处理，必须为 None 或者列表值，否则抛出异常
# ...

# 1. 准备好相关初始变量
new_tasks       = []
new_handlers    = []
role_vars_files = []
defaults_files  = []

pre_tasks = ds.get('pre_tasks', None)
if type(pre_tasks) != list:
    pre_tasks = []
for x in pre_tasks:
    new_tasks.append(x)

# 2. 第一个出现的额外 task
new_tasks.append(dict(meta='flush_handlers'))

# 3. 获取角色的相关信息
roles = self._build_role_dependencies(roles, [], {})
# 为了理解函数输出，可以在这里加上一个 print() 语句，查看最后的输出结果
# print('roles:{}'.format(roles))

# ...

role_names = []

# 4. 处理角色目录下的所有相关信息，如 tasks、handlers 和 vars 等
for (role, role_path, role_vars, role_params, default_vars) in roles:
    # ...

    task_basepath     = utils.path_dwim(self.basedir, os.path.join
(role_path, 'tasks'))
    handler_basepath  = utils.path_dwim(self.basedir, os.path.join
(role_path, 'handlers'))
    vars_basepath     = utils.path_dwim(self.basedir, os.path.join
(role_path, 'vars'))
    meta_basepath     = utils.path_dwim(self.basedir, os.path.join
(role_path, 'meta'))
    defaults_basepath = utils.path_dwim(self.basedir, os.path.join
(role_path, 'defaults'))

    task      = self._resolve_main(task_basepath)
    handler   = self._resolve_main(handler_basepath)
    vars_file = self._resolve_main(vars_basepath)
    meta_file = self._resolve_main(meta_basepath)
    defaults_file = self._resolve_main(defaults_basepath)

    library   = utils.path_dwim(self.basedir, os.path.join(role_path,
'library'))

    # 处理异常和赋值
    # ...
```

```
        # 更新 new_tasks、new_handlers 等变量，同时添加模块库地址
        # ...

    # 5. 整理最后的变量并更新到 ds 中，这里的代码加上第 2 处的代码将解释上文的 7 个
      Task 对象
    tasks      = ds.get('tasks', None)
    post_tasks = ds.get('post_tasks', None)
    handlers   = ds.get('handlers', None)
    vars_files = ds.get('vars_files', None)

    if type(tasks) != list:
        tasks = []
    if type(handlers) != list:
        handlers = []
    if type(vars_files) != list:
        vars_files = []
    if type(post_tasks) != list:
        post_tasks = []

    new_tasks.extend(tasks)
    # flush handlers after tasks + role tasks
    new_tasks.append(dict(meta='flush_handlers'))
    new_tasks.extend(post_tasks)
    # flush handlers after post tasks
    new_tasks.append(dict(meta='flush_handlers'))

    new_handlers.extend(handlers)

    ds['tasks'] = new_tasks
    ds['handlers'] = new_handlers
    ds['role_names'] = role_names

    self.role_vars = self._load_role_vars_files(role_vars_files)
    self.default_vars = self._load_role_defaults(defaults_files)

    return ds

# ...
```

上述导入角色的代码非常多，笔者对其进行了精简和注释。此外，上述代码中还有几处需要说明：

- _load_roles()方法一开始就是初始化变量，然后获取 Play 对象的 ds 中的 pre_tasks 值并加到 new_tasks 变量的最前端。

- 在 new_tasks 中追加了一个额外的字典元素，其值为{'meta': 'flush_handlers'}，这就是前面第一个 play 中多出的 3 个 task 之中的一个，参考_load_roles()方法中的第 2 处代码。

- _load_roles()方法中的第 3 处代码是获取角色目录下的一些信息。对于函数内容太多的情况，可以考虑使用简单的 print()语句打印相应的变量结果来得到一个直观的印象（上面代码中注释掉了加入的 print()语句）。本次测试的 Playbook 结果如下：

```
(ansible1.9.6) [root@master test-playbook]# ansible-playbook -i hosts
site.yml
roles:[['test-role1', '/root/ansible-1.9.6/test-playbook/roles/test-
role1', {'test_role_var1': 'var1'}, {}, {}]]
roles:[['test-role2', '/root/ansible-1.9.6/test-playbook/roles/test-
role2', {}, {}, {}]]
# 省略 Playbook 运行过程
# ...
```

- _load_roles()方法中针对每个角色会遍历其目录下的 tasks、handlers 和 vars 等目录，以获得该角色的任务、变量等信息。Ansible 会查找这些目录下的 main、main.yml、main.yaml 和 main.json 文件（只能存在一个）并解析该文件内容以获取相应的信息，最后将得到的信息依次添加到最开始准备的初始化变量中。从代码中可以看到 Ansible 1.9.6 所支持的角色目录的子目录名及对应子目录下初始搜索的文件名，这对于理解 Playbook 中角色的写法十分重要，具体参考_load_roles()方法中的第 4 处代码。
- 接下来的第 5 处代码就非常重要了，new_tasks 首先添加了 ds 中的 pre_tasks 的值，然后又添加了 ds 中的 tasks 的值。由于这里得到的 task 的值为部分任务列表，所以使用了 extend()方法。注意后面又添加了两个额外任务，其内容为{'meta': 'flush_handlers'}，其中还穿插了 post_tasks 任务列表（本次测试的 Playbook 中并没有 post_tasks 字段）。

继续回到__init__()方法中的第 3 处代码，前面设置了通过_load_roles()更新 ds 的 tasks、handlers 和 role_names 属性值，接着调用 Play 对象中的_load_tasks()方法得到最后的 Task 对象列表并赋值给_tasks，这里的结果正是前面 play._tasks 的输出结果。将这里的代码结合上面的_load_roles()方法就可以解释前面多出的 3 个 task 对象了。以下是_load_tasks()方法的具体实现源码：

```
# 源码位置: lib/ansible/playbook/play.py
# ...

class Play(object):
    # ...

    def _load_tasks(self, tasks, vars=None, role_params=None, default_
vars=None, become_vars=None,
        additional_conditions=None, original_file=None, role_name=None):
        # ...

        for x in tasks:
            if 'meta' in x:
                if x['meta'] == 'flush_handlers':
                    if role_name and 'role_name' not in x:
                        x['role_name'] = role_name
                    results.append(Task(self, x, module_vars=task_vars, role_
name=role_name, no_tags=False))
                    continue
```

```
            if 'include' in x:
                # 处理其中包含的 include 语句，由于代码太多，比较复杂，可先忽略
                # ...

            elif type(x) == dict:
                task = Task(
                    self, x,
                    module_vars=task_vars,
                    play_vars=self.vars,
                    play_file_vars=self.vars_file_vars,
                    role_vars=self.role_vars,
                    role_params=role_params,
                    default_vars=default_vars,
                    additional_conditions=list(additional_conditions),
                    role_name=role_name
                )
                results.append(task)
            else:
                raise Exception("unexpected task type")

        for x in results:
            if self.tags is not None:
                x.tags.extend(self.tags)

        return results
```

上面的代码逻辑并不复杂，但是在处理 task 中包含 include 字段的情况时，相关代码较多。对于其他普通模块 task 而言，直接实例化 Task 类即可，其中，Task 的 ds 参数正是 tasks 列表本身的元素。这里也可以看到 tasks 获取的核心方法是 _load_roles()。最后再来数一数测试的 Playbook 中第一个 Play 最后的 task 数，具体如下：

```
# 1. 在 _load_roles() 方法中，pre_tasks 中有一个 task，其 name 为 test pre tasks in
  group1，将其加入 new_tasks 列表中
# 2. 在 _load_roles() 方法中，在 new_tasks 列表中追加一个 {'meta': 'flush_handlers'}
  任务
# 3. 在 _load_roles() 方法中，先依次解析角色目录，将角色中的任务添加到 new_tasks 中，
  这里面共有两个 task，name 分别为 get hostname 和 debug hostname
# 4. 在 _load_roles() 方法中，第 5 处代码是添加 play 中的 task 任务到 new_tasks 中，
  此时 task 的 name 为 end task
# 5. 在 _load_roles() 方法中，第 5 处代码会继续向 new_tasks 中追加 {'meta': 'flush_
  handlers'} 任务，然后导入 play 中的 post_tasks 任务，最后再次追加 {'meta': 'flush_
  handlers'} 任务，由于测试的 Playbook 的第一个 play 中没有 post_tasks 任务，于是会
  在 new_tasks 中直接追加两次 {'meta': 'flush_handlers'} 任务

# 综上所述：第一个 Play 最后得到了 7 个 task
```

下面在交互模式下验证上面分析的结果，具体如下：

```
>>> tasks = play._tasks
>>> tasks[0].name
'test pre tasks in group1'
>>> tasks[1].meta
```

```
'flush_handlers'
>>> tasks[2].name
'get hostname'
>>> tasks[3].name
'debug hostname'
>>> tasks[4].name
'end task'
>>> tasks[5].meta
'flush_handlers'
>>> tasks[6].meta
'flush_handlers'
```

通过输出结果可知，角色的任务永远在 Play 中的任务之前执行。也就是说，Play 中直接定义的 task 无论是在 roles 字段之前还是之后，都会先执行 roles 中的 task。下面简单改造 site.yml 文件中相关语法的位置并进行测试，具体如下：

```
(ansible1.9.6) [root@master test-playbook]# cat site.yml
---
- hosts: group1
  gather_facts: false

  pre_tasks:
    - name: test pre tasks in group1
      debug:
        msg: this is a pre task in group1

  # 在 roles 的前面，但是后执行
  tasks:
    - name: before role task1
      debug:
        msg: "hello1"

    - name: before role task2
      debug:
        msg: "hello2"

  roles:
    - test-role1
(ansible1.9.6) [root@master test-playbook]# ansible-playbook -i hosts
site.yml
PLAY [group1] *********************************************************

TASK: [test pre tasks in group1] *************************************
ok: [master] => {
    "msg": "this is a pre task in group1"
}

TASK: [test-role1 | get hostname] ************************************
changed: [master]

TASK: [test-role1 | debug hostname] **********************************
ok: [master] => {
    "msg": "hostname = master"
}
```

```
TASK: [before role task1] ********************************************
ok: [master] => {
    "msg": "hello1"
}

TASK: [before role task2] ********************************************
ok: [master] => {
    "msg": "hello2"
}

PLAY RECAP ***********************************************************
master                  : ok=5   changed=1   unreachable=0   failed=0
```

看到了吗？是不是 roles 中的 task 先执行？这些小知识点都可以从 Ansible 源码中学到，这也说明了源码学习的重要性，能促进学习者更好地掌握该工具的使用。

除了__init__.py 外，再来看看 Play 类中的其他几个辅助方法。下面这两个方法非常简单，就是根据输入的文件列表使用 yaml 模块解析得到相应的字典数据。

```python
# 源码位置: lib/ansible/playbook/play.py
# ...

class Play(object):
    # ...

    def _load_role_vars_files(self, vars_files):
        # process variables stored in vars/main.yml files
        role_vars = {}
        for filename in vars_files:
            if os.path.exists(filename):
                # 使用 yaml 模块解析文件
                new_vars = utils.parse_yaml_from_file(filename, vault_password=
self.vault_password)
                if new_vars:
                    # 必须是字典形式的结果，若不是则直接抛错
                    if type(new_vars) != dict:
                        # 抛出异常
                        # ...
                    role_vars = utils.combine_vars(role_vars, new_vars)

        return role_vars

    def _load_role_defaults(self, defaults_files):
        # process default variables
        default_vars = {}
        for filename in defaults_files:
            if os.path.exists(filename):
                # 使用 yaml 模块解析文件
                new_default_vars = utils.parse_yaml_from_file(filename,
vault_password=self.vault_password)
                if new_default_vars:
                    if type(new_default_vars) != dict:
                        # 抛出异常
                        # ...
```

```
                    # 合并结果
                    default_vars = utils.combine_vars(default_vars, new_
default_vars)

        return default_vars

    # ...
```

学习了这些基础知识之后，就可以跟踪 ansible-playbook 命令的整个运行流程了。

2.5.3　追踪 ansible-playbook 命令的运行流程

本节将追踪 ansible-playbook 命令的运行流程并简化大部分细节，找出核心的运行语句，同时根据代码说明 Playbook 中一些语法背后的运行逻辑。

首先来看一下 bin/ansible-playbook 文件中的代码内容，具体如下：

```
# 源码位置: bin/ansible-playbook
# ...

def main(args):
    # 解析参数
    # ...

    # 校验（校验 playbook 文件）与赋值
    # ...

    inventory = ansible.inventory.Inventory(options.inventory, vault_
password=vault_pass)

    # 校验 Playbook 中的目标主机，并处理无匹配主机的情况
    # ...

    for playbook in args:

        stats = callbacks.AggregateStats()
        playbook_cb = callbacks.PlaybookCallbacks(verbose=utils.VERBOSITY)
        if options.step:
            playbook_cb.step = options.step
        if options.start_at:
            playbook_cb.start_at = options.start_at
        runner_cb = callbacks.PlaybookRunnerCallbacks(stats, verbose=
utils.VERBOSITY)

        pb = ansible.playbook.PlayBook(
            playbook=playbook,
            module_path=options.module_path,
            inventory=inventory,
            forks=options.forks,
            remote_user=options.remote_user,
            remote_pass=sshpass,
            callbacks=playbook_cb,
```

```
                runner_callbacks=runner_cb,
                stats=stats,
                timeout=options.timeout,
                transport=options.connection,
                become=options.become,
                become_method=options.become_method,
                become_user=options.become_user,
                become_pass=becomepass,
                extra_vars=extra_vars,
                private_key_file=options.private_key_file,
                only_tags=only_tags,
                skip_tags=skip_tags,
                check=options.check,
                diff=options.diff,
                vault_password=vault_pass,
                force_handlers=options.force_handlers,
            )

            # ...

            try:
                pb.run()

                # 整理 Playbook 的执行结果并打印
                # ...

            except errors.AnsibleError, e:
                display(u"ERROR: %s" % utils.unicode.to_unicode(e, nonstring=
        'simplerepr'), color='red')
                return 1

            return 0
```

以上代码经过简化之后就非常清晰了：先是解析选项并得到相关选项值，然后进行校验并对变量进行赋值，接下来是循环运行 Playbook 文件并解析。这里最核心的代码就是 PlayBook 对象的 run() 方法，之后是整理并展示运行结果。继续深入研究 PlayBook 对象的 run() 方法，代码如下：

```
# 源码位置：lib/ansible/playbook/__init__.py
# ...

class PlayBook(object):
    # ...

    def run(self):
        ''' 针对 Playbook 中所有匹配的主机执行 '''
        plays = []
        matched_tags_all = set()
        unmatched_tags_all = set()

        self.callbacks.on_start()
        for (play_ds, play_basedir) in zip(self.playbook, self.play_basedirs):
            # 生成 Play 对象
            play = Play(self, play_ds, play_basedir, vault_password=self.
```

```
vault_password)
        assert play is not None

        # 处理 tags
        # ...

        if (len(matched_tags) > 0 or len(play.tasks()) == 3):
            plays.append(play)

    # 处理 tags 相关的信息
    # ...

    for play in plays:
        ansible.callbacks.set_play(self.callbacks, play)
        ansible.callbacks.set_play(self.runner_callbacks, play)
        # 核心语句
        if not self._run_play(play):
            break

    # ...

    # summarize the results
    results = {}
    for host in self.stats.processed.keys():
        results[host] = self.stats.summarize(host)
    return results
```

2.5.2 节中使用测试的 Playbook 实例化并得到了 self.playbook 和 self.play_basedirs 两个列表，这两个列表中的元素表示剧本的数据和剧本所在的根目录，接着在 for 循环中依次执行这些剧本。通过上面的代码可知，调用了_run_play()方法执行各剧本。如果返回值为 False，则直接停止运行后面的 Play 对象。执行完所有的剧本后，会总结和打印相应的执行结果。接下来研究_run_play()方法的实现源码，具体如下：

```
# 源码位置: lib/ansible/playbook/__init__.py
# ...

class PlayBook(object):
    # ...

    def _run_play(self, play):

        # 检查 hosts 等信息
        # ...

        # 1. 根据 play，获取目标主机的系统信息，如果设置了 "gather_facts=true"
        self._do_setup_step(play)

        # 2. 非常有意思的一个参数——serial
        all_hosts = self._trim_unavailable_hosts(play._play_hosts)
        play.update_vars_files(all_hosts, vault_password=self.vault_
password)
        hosts_count = len(all_hosts)
```

```
    if play.serial.endswith("%"):
        serial_pct = int(play.serial.replace("%",""))
        serial = int((serial_pct/100.0) * len(all_hosts))

        serial = max(serial, 1)
    else:
        serial = int(play.serial)

    serialized_batch = []
    if serial <= 0:
        serialized_batch = [all_hosts]
    else:
        # do N forks all the way through before moving to next
        while len(all_hosts) > 0:
            play_hosts = []
            for x in range(serial):
                if len(all_hosts) > 0:
                    play_hosts.append(all_hosts.pop(0))
            serialized_batch.append(play_hosts)

    # 3. 循环处理节点和相应的 task
    task_errors = False
    for on_hosts in serialized_batch:

        play._play_hosts = self._trim_unavailable_hosts(on_hosts)
        self.inventory.also_restrict_to(on_hosts)

        for task in self.tasks_to_run_in_play(play):

            if task.meta is not None:
                # meta tasks can force handlers to run mid-play
                if task.meta == 'flush_handlers':
                    self.run_handlers(play)

                continue

            if not self._run_task(play, task, False):
                return False

            # ...

        # 处理 handlers，需要根据配置决定是否执行 handlers
        # ...

    return True
```

　　_run_play()方法中有三处非常重要的地方需要进行说明。第一处代码段是获取远端主机的系统信息，它使用的是 Ansible 中的 setup 模块来探测远端主机的各种信息，例如网卡、磁盘、内存、系统和 CPU 等。其中，_do_setup_step()方法的具体实现源码如下：

```
# 源码位置：lib/ansible/playbook/__init__.py
# ...

class PlayBook(object):
```

```
    # ...

    def _do_setup_step(self, play):
        ''' 从远端节点获取一些基本信息 '''

        host_list = self._trim_unavailable_hosts(play._play_hosts)

        # 不调用 setup 模块获取结果的方法: gather_facts 未设置且 C.DEFAULT_GATHERING
            设置为 smart, 或者是 gather_facts 未设置且 C.DEFAULT_GATHERING 设置为
            explicit, 又或者是 gather_facts 设置为 False
        if play.gather_facts is None and C.DEFAULT_GATHERING == 'smart':
            host_list = [h for h in host_list if h not in self.SETUP_CACHE
or 'module_setup' not in self.SETUP_CACHE[h]]
            if len(host_list) == 0:
                return {}
        elif play.gather_facts is False or (play.gather_facts is None and
C.DEFAULT_GATHERING == 'explicit'):
            return {}

        self.callbacks.on_setup()
        self.inventory.restrict_to(host_list)

        ansible.callbacks.set_task(self.callbacks, None)
        ansible.callbacks.set_task(self.runner_callbacks, None)

        # push any variables down to the system
        setup_results = ansible.runner.Runner(
            basedir=self.basedir,
            pattern=play.hosts,
            module_name='setup',
            module_args={},
            inventory=self.inventory,
            forks=self.forks,
            module_path=self.module_path,
            timeout=self.timeout,
            remote_user=play.remote_user,
            remote_pass=self.remote_pass,
            remote_port=play.remote_port,
            private_key_file=self.private_key_file,
            setup_cache=self.SETUP_CACHE,
            vars_cache=self.VARS_CACHE,
            callbacks=self.runner_callbacks,
            become=play.become,
            become_method=play.become_method,
            become_user=play.become_user,
            become_pass=self.become_pass,
            vault_pass=self.vault_password,
            transport=play.transport,
            is_playbook=True,
            module_vars=play.vars,
            play_vars=play.vars,
            play_file_vars=play.vars_file_vars,
            role_vars=play.role_vars,
            default_vars=play.default_vars,
```

```
                check=self.check,
                diff=self.diff,
                accelerate=play.accelerate,
                accelerate_port=play.accelerate_port,
            ).run()
        self.stats.compute(setup_results, setup=True)

        self.inventory.lift_restriction()

        # now for each result, load into the setup cache so we can
        # let runner template out future commands
        setup_ok = setup_results.get('contacted', {})
        for (host, result) in setup_ok.iteritems():
            utils.update_hash(self.SETUP_CACHE, host, {'module_setup': True})
            utils.update_hash(self.SETUP_CACHE, host, result.get('ansible_
facts', {}))
        return setup_results
```

_do_setup_step()方法非常好理解，通过上面的代码示例可以看到，_do_setup_step()方法可以直接调用 Ansible 模块。禁止调用 setup 进行远程探测的方式如下：

- Play 对象中 gather_facts 的值未设置且 C.DEFAULT_GATHERING 设置为 smart。
- Play 对象中 gather_facts 的值明确设置为 False。
- Play 对象中 gather_facts 的值未设置且 C.DEFAULT_GATHERING 设置为 explicit。

调用模块的方式也非常简单，直接实例化 ansible.runner.Runner 类，其中需要的参数部分来自 PlayBook 对象本身，部分来自 Play 对象，而模块名称固定为 setup。最后链式调用 Runner 对象的 run()方法，更新部分参数后返回相应的结果，函数执行完毕。

第 2 处代码非常有意思，如果不是分析代码，笔者一直不知道有 serial 这个参数可用。先来看一种情况，笔者准备了一个示例，内容如下：

```
(ansible1.9.6) [root@master test-playbook]# cat site2.yml
---
- hosts: group2
  gather_facts: false
  serial: 2

  pre_tasks:
    - name: test pre tasks in group2
      debug:
        msg: this is a pre task in group2

  roles:
    - test-role2

  tasks:
    - name: end task
      debug:
        msg: end task in group2
(ansible1.9.6) [root@master test-playbook]# ansible-playbook -i hosts
site2.yml

PLAY [group2] ***********************************************************
```

```
TASK: [test pre tasks in group2] ********************************************
ok: [ceph-1] => {
    "msg": "this is a pre task in group2"
}
ok: [ceph-2] => {
    "msg": "this is a pre task in group2"
}

TASK: [test-role2 | get hostname] ********************************************
changed: [ceph-2]
changed: [ceph-1]

TASK: [test-role2 | debug hostname] ******************************************
ok: [ceph-1] => {
    "msg": "hostname = ceph-1"
}
ok: [ceph-2] => {
    "msg": "hostname = ceph-2"
}

TASK: [end task] *************************************************************
ok: [ceph-1] => {
    "msg": "end task in group2"
}
ok: [ceph-2] => {
    "msg": "end task in group2"
}

TASK: [test pre tasks in group2] ********************************************
ok: [ceph-3] => {
    "msg": "this is a pre task in group2"
}

TASK: [test-role2 | get hostname] ********************************************
changed: [ceph-3]

TASK: [test-role2 | debug hostname] ******************************************
ok: [ceph-3] => {
    "msg": "hostname = ceph-3"
}

TASK: [end task] *************************************************************
ok: [ceph-3] => {
    "msg": "end task in group2"
}

PLAY RECAP ******************************************************************
ceph-1                     : ok=4    changed=1    unreachable=0    failed=0
ceph-2                     : ok=4    changed=1    unreachable=0    failed=0
ceph-3                     : ok=4    changed=1    unreachable=0    failed=0
```

　　发现这种情况了吗？如果设置 serial=2，则对于有 3 台目标主机的情况而言，会按照每次使用 2 台主机的规定执行相应的 Play。如果设置 serial=1，则 Play 会被主机一台台地执行，达到串行执行的效果，具体代码如下：

```
(ansible1.9.6) [root@master test-playbook]# cat site2.yml
---
- hosts: group2
  gather_facts: false
  serial: 1

  # 其余任务不变
  # ...

(ansible1.9.6) [root@master test-playbook]# ansible-playbook -i hosts
site2.yml

PLAY [group2] ********************************************************

TASK: [test pre tasks in group2] ************************************
ok: [ceph-1] => {
    "msg": "this is a pre task in group2"
}

TASK: [test-role2 | get hostname] ***********************************
changed: [ceph-1]

TASK: [test-role2 | debug hostname] *********************************
ok: [ceph-1] => {
    "msg": "hostname = ceph-1"
}

TASK: [end task] ****************************************************
ok: [ceph-1] => {
    "msg": "end task in group2"
}

TASK: [test pre tasks in group2] ************************************
ok: [ceph-2] => {
    "msg": "this is a pre task in group2"
}

TASK: [test-role2 | get hostname] ***********************************
changed: [ceph-2]

TASK: [test-role2 | debug hostname] *********************************
ok: [ceph-2] => {
    "msg": "hostname = ceph-2"
}

TASK: [end task] ****************************************************
ok: [ceph-2] => {
    "msg": "end task in group2"
}

TASK: [test pre tasks in group2] ************************************
ok: [ceph-3] => {
    "msg": "this is a pre task in group2"
```

```
}

TASK: [test-role2 | get hostname] **************************************
changed: [ceph-3]

TASK: [test-role2 | debug hostname] ************************************
ok: [ceph-3] => {
    "msg": "hostname = ceph-3"
}

TASK: [end task] ******************************************************
ok: [ceph-3] => {
    "msg": "end task in group2"
}

PLAY RECAP ************************************************************
ceph-1                     : ok=4    changed=1    unreachable=0    failed=0
ceph-2                     : ok=4    changed=1    unreachable=0    failed=0
ceph-3                     : ok=4    changed=1    unreachable=0    failed=0
```

那么代码是如何实现的呢？本次测试 Play 对象的目标主机共有 3 台，分别为 ceph-1、
ceph-2 和 ceph-3。如果 serial=2，那么就会得到一个具有两个元素的数组[[ceph-1,ceph-2],
[ceph-3]]；如果 serial=0，则得到的批量执行数组为[[ceph-1,ceph-2,ceph-3]]。这样只需要 for
循环遍历这个数组，每次使用数组中的 hosts 执行该 Play 对象即可。这个逻辑对应下面的
代码段：

```
serialized_batch = []
if serial <= 0:
    serialized_batch = [all_hosts]
else:
    # do N forks all the way through before moving to next
    while len(all_hosts) > 0:
        play_hosts = []
        for x in range(serial):
            if len(all_hosts) > 0:
                play_hosts.append(all_hosts.pop(0))
        serialized_batch.append(play_hosts)
```

为了更容易理解，笔者将这段代码封装成一个函数并进行调用，具体代码如下：

```
(ansible1.9.6) [root@master test-playbook]# cat test_serial_segment.py
import copy

def test_serial(all_hosts, serial=0):
    serialized_batch = []
    if serial <= 0:
        serialized_batch = [all_hosts]
    else:
        # do N forks all the way through before moving to next
        while len(all_hosts) > 0:
            play_hosts = []
            for x in range(serial):
                if len(all_hosts) > 0:
                    play_hosts.append(all_hosts.pop(0))
```

```
            serialized_batch.append(play_hosts)

        return serialized_batch

all_hosts = ['ceph-1', 'ceph-2', 'ceph-3']
print(test_serial(copy.copy(all_hosts), 1))
print(test_serial(copy.copy(all_hosts), 2))
print(test_serial(copy.copy(all_hosts)))
(ansible1.9.6) [root@master test-playbook]# python test_serial_segment.py
[['ceph-1'], ['ceph-2'], ['ceph-3']]
[['ceph-1', 'ceph-2'], ['ceph-3']]
[['ceph-1', 'ceph-2', 'ceph-3']]
```

此外，serial 还支持%的写法，例如将 Play 的目标节点作为总数取其百分数，得到相应的 serial 的整数值，代码如下：

```
if play.serial.endswith("%"):
    serial_pct = int(play.serial.replace("%",""))
    serial = int((serial_pct/100.0) * len(all_hosts))

    serial = max(serial, 1)
else:
    serial = int(play.serial)
```

遍历得到的 serialized_batch 数组，并分别执行 Play，对应注释中的第 3 处代码段。这段代码中有两个语句非常重要：

- self.tasks_to_run_in_play(play)：获取 Play 中对应的 task 列表并遍历执行。
- self._run_task(play, task, False)：该语句对应执行 Play 中的 task，如果返回 False，则运行结束，直接返回 False。

下面来看一下这两条语句中调用的方法。第一个为 tasks_to_run_in_play()方法，代码如下：

```
# 源码位置：lib/ansible/playbook/__init__.py
# ...

class PlayBook(object):
    # ...

    def tasks_to_run_in_play(self, play):

        tasks = []

        for task in play.tasks():
            # 处理 tags，最后设置 should_run 变量
            # ...

            if should_run:
                tasks.append(task)

        return tasks
```

tasks_to_run_in_play()方法非常简单，基本上依赖 Play 对象中的 tasks()方法，其中，

tasks()方法的代码如下：

```
# 源码位置：lib/ansible/playbook/play.py
# ...

class Play(object):
    # ...

    def tasks(self):
        ''' 返回该剧本中的 task 对象 '''
        return self._tasks

    # ...
```

又回到了前面分析的 Play 对象的_tasks 属性值问题。它是该 Play 对象的所有任务列表，包括角色任务（roles）、前置任务（pre_tasks）和后置任务（post_tasks），每个元素都是一个 Task 对象。再来看一下_run_task()方法，代码如下：

```
# 源码位置：lib/ansible/playbook/__init__.py
# ...

class PlayBook(object):
    # ...

    def _run_task(self, play, task, is_handler):
        # ...

        # 1. 先将条件变量中的参数进行翻译
        cond = template(play.basedir, task.ignore_errors, task.module_vars,
expand_lists=False)
        # 2. 检查执行条件
        task.ignore_errors = utils.check_conditional(cond, play.basedir,
task.module_vars, fail_on_undefined=C.DEFAULT_UNDEFINED_VAR_BEHAVIOR)

        # 3. 核心调用，最后调用 self._run_task_internal()方法运行 task
        include_failed = is_handler and play.force_handlers
        results = self._run_task_internal(task, include_failed=include_
failed)

        # ...

        contacted = results.get('contacted', {})
        self.stats.compute(results, ignore_errors=task.ignore_errors)

        def _register_play_vars(host, result):
            if 'stdout' in result and 'stdout_lines' not in result:
                result['stdout_lines'] = result['stdout'].splitlines()
            # 注意更新的值，task.register 就是 register 语法指定的变量，保存的是
              result
            utils.update_hash(self.VARS_CACHE, host, {task.register: result})

        # ...

        for host, result in contacted.iteritems():
```

```
        # ...

            # 4．将变量写入全局变量中
            if task.register:
                _register_play_vars(host, result)

        # 5．处理包含register语法的task执行失败且设置忽略错误的情况
        if task.ignore_errors and task.register:
            failed = results.get('failed', {})
            for host, result in failed.iteritems():
                _register_play_vars(host, result)

        # 6．任务中存在回调通知的情况：如果有改变的任务则设置好相应的回调列表
        if task.notify and len(task.notify) > 0:
            for host, results in results.get('contacted',{}).iteritems():
                if results.get('changed', False):
                    for handler_name in task.notify:
                        self._flag_handler(play, template(play.basedir, handler_
name, task.module_vars), host)

        # ...

        return hosts_remaining
    # ...
```

　　_run_task()方法的代码也非常多，同样，为了能说清楚该方法的核心流程，笔者注释掉了部分代码。从上面保留的代码来看，_run_task()方法首先是渲染 task 中的条件语句，将所有模板变量进行替换，然后再对该条件语句进行判断。紧接着调用 self._run_task_internal()方法执行 task 任务并得到该 task 的执行结果。随后迭代 task 执行结果，并在 for 循环中解析 register 语法。对于存在 register 语法的 task，将调用内嵌方法_register_play_vars()设置register 指定的变量。最后是处理一些失败及设置回调的动作，这些并未在本次测试 Playbook 中涉及，故对其细节不再深究。这里找到的最重要的方法就是 PlayBook 对象的_run_task_internal()方法，下面是其具体实现代码：

```
# 源码位置: lib/ansible/playbook/__init__.py
# ...

class PlayBook(object):
    # ...

    def _run_task_internal(self, task, include_failed=False):
        ''' 执行 Playbook 中一个特定的 task '''

        hosts = self._trim_unavailable_hosts(self.inventory.list_hosts
(task.play._play_hosts), keep_failed=include_failed)
        self.inventory.restrict_to(hosts)

        runner = ansible.runner.Runner(
            pattern=task.play.hosts,
            inventory=self.inventory,
```

```
            module_name=task.module_name,
            module_args=task.module_args,
            forks=self.forks,
            remote_pass=self.remote_pass,
            module_path=self.module_path,
            timeout=self.timeout,
            remote_user=task.remote_user,
            remote_port=task.play.remote_port,
            module_vars=task.module_vars,
            play_vars=task.play_vars,
            play_file_vars=task.play_file_vars,
            role_vars=task.role_vars,
            role_params=task.role_params,
            default_vars=task.default_vars,
            extra_vars=self.extra_vars,
            private_key_file=self.private_key_file,
            setup_cache=self.SETUP_CACHE,
            vars_cache=self.VARS_CACHE,
            basedir=task.play.basedir,
            conditional=task.when,
            callbacks=self.runner_callbacks,
            transport=task.transport,
            is_playbook=True,
            check=self.check,
            diff=self.diff,
            environment=task.environment,
            complex_args=task.args,
            accelerate=task.play.accelerate,
            accelerate_port=task.play.accelerate_port,
            accelerate_ipv6=task.play.accelerate_ipv6,
            error_on_undefined_vars=C.DEFAULT_UNDEFINED_VAR_BEHAVIOR,
            vault_pass = self.vault_password,
            run_hosts=hosts,
            no_log=task.no_log,
            run_once=task.run_once,
            become=task.become,
            become_method=task.become_method,
            become_user=task.become_user,
            become_pass=task.become_pass,
        )


        runner.module_vars.update({'play_hosts': hosts})
        runner.module_vars.update({'ansible_version': self._ansible_version})

        if task.async_seconds == 0:
            # 非异步任务，直接调用 Runner 对象的 run() 方法执行
            results = runner.run()
        else:
            results, poller = runner.run_async(task.async_seconds)
            self.stats.compute(results)
            if task.async_poll_interval > 0:
                # if not polling, playbook requested fire and forget, so don't
poll
                results = self._async_poll(poller, task.async_seconds,
task.async_poll_interval)
```

```
            else:
                for (host, res) in results.get('contacted', {}).iteritems():
                    self.runner_callbacks.on_async_ok(host, res, poller.
runner.vars_cache[host]['ansible_job_id'])

        contacted = results.get('contacted',{})
        dark      = results.get('dark', {})

        self.inventory.lift_restriction()

        if len(contacted.keys()) == 0 and len(dark.keys()) == 0:
            return None

        return results
```

从上面的代码中可以看出，PlayBook 对象中的所有 task 的执行都离不开 Runner 对象，因为它们是通过调用 Runner 对象中的 run()或者 run_async()来运行的。由于这里的 task 需要考虑可能设置的一些异步任务，所以使用了 if 来区分模块的两种执行方式，最后统一返回执行的结果 results。

到这里，Playbook 运行的大致流程已经分析清楚了，至于一些细节，如具体到每个语法的实现代码、模板变量的替换、include 的 YAML 文件中继续包含 include 语句的处理、返回结果的分析等，限于篇幅就不再过多分析了。最后，总结 Playbook 的大致运行流程如下：

（1）解析 Playbook 文件，得到相应的 Playbook 的 ds 结果，其包含待运行的 Play 数据列表及相应 Play 所属的 Playbook 文件的根目录所组成的列表。

（2）调用 for 循环执行每个 Play 对象。通过该 Play 对象中的核心方法_load_roles()可以得到剧本中所有的 task。

（3）循环执行剧本中的每个 task 并得到相关结果。最终是调用 PlayBook 类中的_run_task()或者_run_task_internal()方法执行剧本中的 task。

在 PlayBook 和 Play 类中还有部分函数未介绍，有兴趣的读者可以自行学习这些函数在 Playbook 运行过程中的作用，进一步了解 Playbook 运行的细节。

2.6　本　章　小　结

本章分析了 Ansible 1.9.6 的源码，介绍了几个在 Ansible 1.1 源码分析中未介绍的模块及部分插件，如缓存插件和 shell 插件等。整体来看，两个版本之间的源码差异并不是特别大，甚至有很多代码及调用方式都被保留了下来。通过 2.4 节的分析可知，Ansible 1.9.6 中模块的核心运行流程并没有变化，只不过实现方式相比早期版本更加复杂。此外，2.5 节中详细介绍了 Playbook 文件被解析和执行的完整过程，最终发现 Playbook 中 task 的执行仍旧是依靠 runner 目录下的那些核心代码。到这里，整个 Ansible 1 的源码已经分析完毕，对于 Ansible 1 版本的源码，读者应该不难理解了。

第 3 章　Ansible 2 常用模块与 Playbook 语法

本章内容将聚焦于目前较新的 Ansible 2.8 版本，通过深入分析 Ansible 2.8 的源码，帮助读者熟悉 Ansible 2.8 工具的使用及日常 Playbook 的编写。与分析 Ansible 1 一样，本章同样从模块的使用和 Playbook 的编写两个方面介绍 Ansible 2.8，其内容会比 Ansible 1 更加复杂，需要读者认真学习。

3.1　快速上手 Ansible 2

Ansible 2 同时支持 Python 2 和 Python 3，考虑到 Python 2 已经处于非维护状态，为了跟上主流的技术，这里在 Python 3 的环境下安装 Ansible 2.8。具体操作如下：

```
[root@master ~]# yum -y install gcc gcc-c++ zlib zlib-devel bzip2-devel
openssl-devel ncurses-devel sqlite-devel readline-devel tk-devel gdbm-
devel db4-devel libpcap-devel libffi-devel xz-devel
[root@master ~]# v=3.8.6;wget http://mirrors.sohu.com/python/$v/Python-
$v.tar.xz -P ~/.pyenv/cache/;pyenv install $v
```

🔔注意：以上操作和前面安装 Python 2.7 的虚拟环境类似，只不过安装版本变成了 3.8.6。

接着创建一个名为 ansible2.8.16 的虚拟环境（目前最新的 Ansible 2.8 的版本为 v2.8.16）并激活该环境。具体操作如下：

```
[root@master ~]# pyenv virtualenv 3.8.6 ansible2.8.16
[root@master ~]# pyenv activate ansible2.8.16
pyenv-virtualenv: prompt changing will be removed from future release.
configure `export PYENV_VIRTUALENV_DISABLE_PROMPT=1' to simulate the behavior.
(ansible2.8.16) [root@master ~]#
```

在虚拟环境下使用 pip install ansible==2.8.16 命令安装 Ansible 2.8.16。执行命令后，使用 ansible --version 命令查看虚拟环境中安装的 Ansible 版本。具体操作如下：

```
(ansible2.8.16) [root@master ~]# pip install ansible==2.8.16 -i https://
pypi.tuna.tsinghua.edu.cn/simple
(ansible2.8.16) [root@master ~]# ansible --version
ansible 2.8.16
  config file = None
```

```
configured module search path = ['/root/.ansible/plugins/modules', '/usr/
share/ansible/plugins/modules']
  ansible python module location = /root/.pyenv/versions/3.8.6/envs/
ansible2.8.16/lib/python3.8/site-packages/ansible
  executable location = /root/.pyenv/versions/ansible2.8.16/bin/ansible
  python version = 3.8.6 (default, Oct 18 2020, 15:33:08) [GCC 4.8.5 20150623
(Red Hat 4.8.5-39)]
```

接下来借助 Ansible 1.1 中的 hosts 文件简单测试一下 Ansible 2.8.16。具体操作如下：

```
(ansible2.8.16) [root@master ~]# cat /root/hosts
[master]
master

[nodes]
ceph-[1:3]

[nodes:vars]
ansible_ssh_user=root
ansible_ssh_pass=@SHENcong19920522
ansible_ssh_connection=ssh

[master:vars]
ansible_connection=local

(ansible2.8.16) [root@master ~]# ansible all -i hosts -m ping
[WARNING]: Found both group and host with same name: master

master | SUCCESS => {
    "ansible_facts": {
        "discovered_interpreter_python": "/usr/bin/python"
    },
    "changed": false,
    "ping": "pong"
}
ceph-3 | SUCCESS => {
    "ansible_facts": {
        "discovered_interpreter_python": "/usr/bin/python"
    },
    "changed": false,
    "ping": "pong"
}
ceph-2 | SUCCESS => {
    "ansible_facts": {
        "discovered_interpreter_python": "/usr/bin/python"
    },
    "changed": false,
    "ping": "pong"
}
ceph-1 | SUCCESS => {
    "ansible_facts": {
        "discovered_interpreter_python": "/usr/bin/python"
    },
    "changed": false,
    "ping": "pong"
}
```

至此，简单的实验环境就搭建好了。上面的操作与 Ansible 1.1 和 Ansible 1.9.6 没有任何区别。Ansible 2 虽然在代码上进行了大量改造和重构，但是其整体模块的运行机制及大部分模块的命名和功能依旧沿用以前的规定。因此，熟悉了 Ansible 1 的使用后，对 Ansible 2 也能快速上手。其中，主要介绍的就是 Ansible 2 中的一些常用模块和相关参数的使用，以及新的 Playbook 语法。读者在掌握了 Ansible 2 中的部分常用模块的用法及 Playbook 的基本语法后，就可以使用 Ansible 工具进行日常的运维操作和自动化部署等工作了。

接下来在前面创建的虚拟环境下学习 Ansible 2 相关模块的用法及 Playbook 的语法。目前，Ansible 的最新版本是 2.10，但官方网站上依旧保存了 2.8 版本的文档说明，这是 Ansible 2 最权威的学习材料之一。所有模块的用法及 Playbook 语法等几乎都可以在官网上找到对应的说明，这也是所有的 Ansible 入门者和进阶者经常学习的地方。Ansible 官网的文档页面如图 3-1 所示。

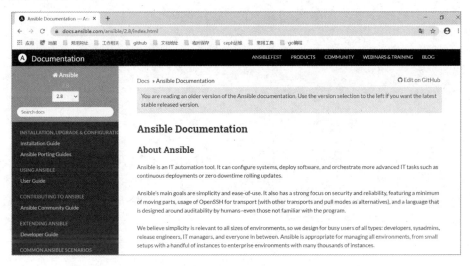

图 3-1　Ansible 2.8 的官方文档

3.2　Ansible 2 的命令行使用

本节将介绍 Ansible 2 的命令行用法，主要包括两种：Ansible 工具支持的选项；数个常用的 Ansible 模块（与 Ansible 1 中介绍的模块不同）。

3.2.1　Ansible 2 支持的常见选项

本节将从源码角度分析 Ansible 2 所支持的选项，并对部分常用选项进行实例演示。与 Ansible 1 相同，Ansible 2 命令行的入口代码也是从 bin/ansible 文件中开始执行。

```
# 源码位置: bin/ansible
# ...

if __name__ == '__main__':
    # ...

    cli = None
    me = os.path.basename(sys.argv[0])

    try:
        # 打印开始运行信息
        # ...

        sub = None
        target = me.split('-')
        if target[-1][0].isdigit():
            target = target[:-1]

        if len(target) > 1:
            sub = target[1]
            myclass = "%sCLI" % sub.capitalize()
        elif target[0] == 'ansible':
            sub = 'adhoc'
            myclass = 'AdHocCLI'
        else:
            raise AnsibleError("Unknown Ansible alias: %s" % me)

        try:
            mycli = getattr(__import__("ansible.cli.%s" % sub, fromlist=
[myclass]), myclass)
        except ImportError as e:
            # 处理导入异常
            # ...

        try:
            args = [to_text(a, errors='surrogate_or_strict') for a in sys.
argv]
        except UnicodeError:
            # 处理编码异常
            # ...
        else:
            cli = mycli(args)
            exit_code = cli.run()

    # ...

    except Exception as e:
        # 处理其他异常
        # ...

    sys.exit(exit_code)
```

上面的代码展示了一种非常经典的编程模式，这在许多热门的 Python 项目中很常见，如 Django 和 Scrapy 框架。在 Ansible 项目源码的 bin 目录下，ansible、ansible-playbook、 ansible-config 和 ansible-console 等命令文件共用 ansible 文件中的代码。代码中的 me 表示的正是执行的命令名，对于执行的 ansible 格式的命令，则是 me = 'ansible'，而对于执行的 ansible- playbook 命令而言，me = 'ansible-playbook'。接着对 me 参数按-符号进行切割，最后得到 sub 和 myclass 变量：

- 对于执行的 ansible 命令，sub='adhoc'且 myclass='AdHocCLI'。
- 对于执行的 ansible-xxx 格式的命令，sub='xxx'且 myclass='XXXCLI'。

代码中的__import__()方式是 Python 中常用的导入类的方式。一定要带上 fromlist 参数，这样最后将导入 sub 对应的模块。接着使用 getattr()方法从 sub 模块中获取 myclass 对应的类，看下面的示例：

```
>>> mycli = __import__('ansible.cli.adhoc', fromlist=['AdHocCLI'])
>>> mycli
<module 'ansible.cli.adhoc' from
'/root/.pyenv/versions/ansible2.8.16/
lib/python3.8/site-packages/ansible/cli/adhoc.py'>
>>> getattr(mycli, 'AdHocCLI')
<class 'ansible.cli.adhoc.AdHocCLI'>
>>> mycli = __import__('ansible.cli.playbook', fromlist=['PlaybookCLI'])
>>> mycli
<module 'ansible.cli.playbook' from
'/root/.pyenv/versions/ansible2.8.16/
lib/python3.8/site-packages/ansible/cli/playbook.py'>
>>> getattr(mycli, 'PlaybookCLI')
<class 'ansible.cli.playbook.PlaybookCLI'>
```

最后，实例化 mycli 对应的类并调用该对象的 run()方法。ansible 和 ansible-playbook 等命令的执行流程及命令代码都相同。

为了能找到 ansible 命令所支持的命令行参数，下面继续追踪 ansible.cli.adhoc 模块中的代码，具体如下：

```
# 源码位置：lib/ansible/cli/adhoc.py
# ...

class AdHocCLI(CLI):

    def init_parser(self):
        # ansible 命令支持的选项全部可以在这个函数中找到
        super(AdHocCLI, self).init_parser(usage='%prog <host-pattern> [options]',
                                          desc="...",
                                          epilog="....")

        opt_help.add_runas_options(self.parser)
        opt_help.add_inventory_options(self.parser)
        opt_help.add_async_options(self.parser)
        opt_help.add_output_options(self.parser)
        opt_help.add_connect_options(self.parser)
        opt_help.add_check_options(self.parser)
        opt_help.add_runtask_options(self.parser)
        opt_help.add_vault_options(self.parser)
        opt_help.add_fork_options(self.parser)
```

```
        opt_help.add_module_options(self.parser)
        opt_help.add_basedir_options(self.parser)

        # options unique to ansible ad-hoc
        self.parser.add_option('-a', '--args', dest='module_args',
                        help="module arguments", default=C.DEFAULT_
MODULE_ARGS)
        self.parser.add_option('-m', '--module-name', dest='module_name',
                        help="...",
                        default=C.DEFAULT_MODULE_NAME)

    # ...
```

通过分析 AdHocCLI 类中的 init_parser()方法基本上可以确定 ansible 命令所支持的选项。最常用的是-a 和-m 选项，前者表示模块参数，后者表示指定运行的模块。没有指定-m 参数时，使用默认的模块 command。下面首先分析父类 CLI 中的 init_parser()方法。其代码如下：

```
# 源码位置: lib/ansible/cli/__init__.py
# ...
from ansible.cli.arguments import optparse_helpers as opt_help
# ...

class CLI(with_metaclass(ABCMeta, object)):
    # ...

    def __init__(self, args, callback=None):
        """
        所有命令行程序基本的初始化方法
        """

        self.args = args
        self.parser = None
        self.action = None
        self.callback = callback

    # ...

    @abstractmethod
    def init_parser(self, usage="", desc=None, epilog=None):
        self.parser = opt_help.create_base_parser(usage=usage, desc=desc,
epilog=epilog)

    # ...
```

由上述代码可知，ansible 命令所支持的全部选项可在 lib/ansible/cli/arguments 目录下的 optparse_helpers.py 文件中找到。继续看该文件中定义的相关函数，具体代码如下：

```
# 源码位置: lib/ansible/cli/arguments/optparse_helpers.py
# ...

def create_base_parser(usage="", desc=None, epilog=None):
    """
    所有 ansible 命令支持的选项
```

```
    """
    parser = SortedOptParser(usage, version=to_native(version("%prog")),
                        description=desc, epilog=epilog)
    parser.remove_option('--version')
    version_help = "..."
    parser.add_option('--version', action="version", help=version_help)
    parser.add_option('-v', '--verbose', dest='verbosity', default=
C.DEFAULT_VERBOSITY, action="count",help="...")
    return parser


def add_async_options(parser):
    """ 支持异步操作的选项 """
    parser.add_option('-P', '--poll', default=C.DEFAULT_POLL_INTERVAL,
type='int', dest='poll_interval',help="...")
    parser.add_option('-B', '--background', dest='seconds', type='int',
default=0, help='...')


def add_fork_options(parser):
    """Add options for commands that can fork worker processes"""
    parser.add_option('-f', '--forks', dest='forks', default=C.DEFAULT_
FORKS, type='int', help="...")


def add_inventory_options(parser):
    """ 添加与 inventory 相关的选项 """
    parser.add_option('-i', '--inventory', '--inventory-file', dest=
'inventory', action="append", help="...")
    parser.add_option('--list-hosts', dest='listhosts', action='store_true',
                help='...')
    parser.add_option('-l', '--limit', default=C.DEFAULT_SUBSET, dest='subset',
                help='...')


def add_runas_options(parser):
    runas_group = optparse.OptionGroup(parser, "...")
    # consolidated privilege escalation (become)  su/sudo 提权选项
    runas_group.add_option("-b", "--become", default=C.DEFAULT_BECOME,
                    action="store_true", dest='become', help="...")
    # 选择使用 sudo 或者是 su 提权
    runas_group.add_option('--become-method', dest='become_method',
                    default=C.DEFAULT_BECOME_METHOD, help="...")
    # 如果是提权用户
    runas_group.add_option('--become-user', default=None, dest='become_user',
                    type='string', help='...')

    add_runas_prompt_options(parser, runas_group=runas_group)
```

```
def add_runtask_options(parser):
    """ 添加与执行任务相关的选项 """
    parser.add_option('-e', '--extra-vars', dest="extra_vars", action=
"append", help="...", default=[])

    # ...
```

上面列出了 Ansible 涉及的部分选项函数，Ansible 2 中对这些选项进行了分类，封装成了不同的函数，这样多个相关的选项和同类型的选项可以更好地被复用，只需要添加对应的函数即可。以下是上述选项的使用与说明。

（1）-a 和-m 选项可以说是必用选项，前面在 Ansible 1 中也使用过多次。当然-m 可以忽略，默认使用的是 command 模块。

（2）-f 选项用于指定启动的进程数，默认的进程数为 5。

（3）-v 和--version 选项中，-v 并不是--version 的简写，它们的含义不同。--version 用于查看 Ansible 版本等信息，调用 optparse_helpers.py 文件中的 version()函数。可查看输出结果。

version()函数的代码如下：

```
# 源码位置: lib/ansible/cli/arguments/optparse_helpers.py
# ...

def version(prog=None):
    """ 返回 Ansible 的版本信息 """
    if prog:
        result = " ".join((prog, __version__))
    else:
        result = __version__

    gitinfo = _gitinfo()
    if gitinfo:
        result = result + " {0}".format(gitinfo)
    result += "\n  config file = %s" % C.CONFIG_FILE
    if C.DEFAULT_MODULE_PATH is None:
        cpath = "Default w/o overrides"
    else:
        cpath = C.DEFAULT_MODULE_PATH
    result = result + "\n  configured module search path = %s" % cpath
    result = result + "\n  ansible python module location = %s" % ':'.
join(ansible.__path__)
    result = result + "\n  executable location = %s" % sys.argv[0]
    result = result + "\n  python version = %s" % ''.join(sys.version.
splitlines())
    return result
```

查看一下与实际打印的结果是否一致，具体如下：

```
(ansible2.8.16) [root@master ~]# ansible --version
ansible 2.8.16
```

```
    config file = None
    configured module search path = ['/root/.ansible/plugins/modules',
'/usr/share/ansible/plugins/modules']
    ansible python module location = /root/.pyenv/versions/3.8.6/envs/
ansible2.8.16/lib/python3.8/site-packages/ansible
    executable location = /root/.pyenv/versions/ansible2.8.16/bin/ansible
    python version = 3.8.6 (default, Oct 18 2020, 15:33:08) [GCC 4.8.5 20150623
(Red Hat 4.8.5-39)]
```

注意，在使用 add_option() 方法添加 -v 选项时使用的 action 值为 count，下面给出一段测试代码看一下 count 的功能。

```
"""
测试选项中 action="count" 的情况
"""
import sys

from ansible.module_utils._text import to_text
from ansible.cli.arguments import optparse_helpers

# 这一行语句出自 Ansible 2.8.16 源码目录下 bin/ansible 文件中的第 103 行
args = [to_text(a, errors='surrogate_or_strict') for a in sys.argv]
# 这一行添加 -v 和 --version 选项
parser = optparse_helpers.create_base_parser(usage='test action=count')
# 解析上面的选项，得到相应的结果
option, args = parser.parse_args(args[1:])
print(option)
```

将上面的代码文件命名为 test_action.py，然后在虚拟环境中的 Python 交互模式下进行测试，具体操作如下：

```
(ansible2.8.16) [root@master ansible-2.8.16]# python test_count.py
{'verbosity': 0}
(ansible2.8.16) [root@master ansible-2.8.16]# python test_count.py -v
{'verbosity': 1}
(ansible2.8.16) [root@master ansible-2.8.16]# python test_count.py -vv
{'verbosity': 2}
(ansible2.8.16) [root@master ansible-2.8.16]# python test_count.py -vvv
{'verbosity': 3}
(ansible2.8.16) [root@master ansible-2.8.16]# python test_count.py -vvvv
{'verbosity': 4}
```

从上面的结果中可以看到，-v 选项中 v 重复的次数将会保存到 option.verbosity 变量中，该变量的值会影响 ansible 命令行的输出信息。option.verbosity 的值越大，则命令行的输出信息越详细。verbosity 的最大值为 5，超过这个值，ansible 命令也不会有更详细的输出。下面是 verbosity 的值为 0 和 3 时的简单对比。

```
(ansible2.8.16) [root@master ansible-2.8.16]# cat hosts
[center]
master

[nodes]
ceph-[1:3]
```

```
[nodes:vars]
ansible_ssh_user=root
ansible_ssh_pass=@SHENcong19920522
ansible_ssh_connection=ssh

[master:vars]
ansible_connection=local
(ansible2.8.16) [root@master ansible-2.8.16]# ansible center -i hosts -m
ping
master | SUCCESS => {
    "ansible_facts": {
        "discovered_interpreter_python": "/usr/bin/python"
    },
    "changed": false,
    "ping": "pong"
}
(ansible2.8.16) [root@master ansible-2.8.16]# ansible center -i hosts -m
ping -vvv
ansible 2.8.16
  config file = None
  configured module search path = ['/root/.ansible/plugins/modules', '/usr/
share/ansible/plugins/modules']
  ansible python module location = /root/.pyenv/versions/3.8.6/envs/
ansible2.8.16/lib/python3.8/site-packages/ansible
  executable location = /root/.pyenv/versions/ansible2.8.16/bin/ansible
  python version = 3.8.6 (default, Oct 18 2020, 15:33:08) [GCC 4.8.5 20150623
(Red Hat 4.8.5-39)]
No config file found; using defaults
host_list declined parsing /root/ansible-2.8.16/hosts as it did not pass
it's verify_file() method
script declined parsing /root/ansible-2.8.16/hosts as it did not pass it's
verify_file() method
auto declined parsing /root/ansible-2.8.16/hosts as it did not pass it's
verify_file() method
Parsed /root/ansible-2.8.16/hosts inventory source with ini plugin
META: ran handlers
<master> ESTABLISH LOCAL CONNECTION FOR USER: root
<master> EXEC /bin/sh -c 'echo ~root && sleep 0'
<master> EXEC /bin/sh -c '( umask 77 && mkdir -p `  echo /root/.ansible/tmp
`"&& mkdir /root/.ansible/tmp/ansible-tmp-1605971879.701043-183950-3546
045909594 && echo ansible-tmp-1605971879.701043-183950-3546045909594=
"` echo /root/.ansible/tmp/ansible-tmp-1605971879.701043-183950-35460459
09594 `" ) && sleep 0'
<master> Attempting python interpreter discovery
<master> EXEC /bin/sh -c 'echo PLATFORM; uname; echo FOUND; command -v
'"'"'/usr/bin/python'"'"'; command -v '"'"'python3.7'"'"'; command -v
'"'"'python3.6'"'"'; command -v '"'"'python3.5'"'"'; command -v '"'"'
python2.7'"'"'; command -v '"'"'python2.6'"'"'; command -v '"'"'/usr/
libexec/platform-python'"'"'; command -v '"'"'/usr/bin/python3'"'"';
command -v '"'"'python'"'"'; echo ENDFOUND && sleep 0'
<master> EXEC /bin/sh -c '/usr/bin/python && sleep 0'
Using module file /root/.pyenv/versions/3.8.6/envs/ansible2.8.16/lib/
python3.8/site-packages/ansible/modules/system/ping.py
<master> PUT /root/.ansible/tmp/ansible-local-183912bdjk01cv/tmp5jmuml8q
```

```
TO /root/.ansible/tmp/ansible-tmp-1605971879.701043-183950-3546045909594/
AnsiballZ_ping.py
<master> EXEC /bin/sh -c 'chmod u+x /root/.ansible/tmp/ansible-tmp-
1605971879.701043-183950-3546045909594/ /root/.ansible/tmp/ansible-tmp-
1605971879.701043-183950-3546045909594/AnsiballZ_ping.py && sleep 0'
<master> EXEC /bin/sh -c '/usr/bin/python /root/.ansible/tmp/ansible-tmp-
1605971879.701043-183950-3546045909594/AnsiballZ_ping.py && sleep 0'
<master> EXEC /bin/sh -c 'rm -f -r /root/.ansible/tmp/ansible-tmp-
1605971879.701043-183950-3546045909594/ > /dev/null 2>&1 && sleep 0'
master | SUCCESS => {
    "ansible_facts": {
        "discovered_interpreter_python": "/usr/bin/python"
    },
    "changed": false,
    "invocation": {
        "module_args": {
            "data": "pong"
        }
    },
    "ping": "pong"
}
META: ran handlers
META: ran handlers
```

以上结果在终端显示时有明显的颜色区别，效果比较好。通过打印详细信息可以看到
Ansible 内部执行模块时的基本步骤是：首先是查找配置文件 ansible.cfg，若没有则使用默
认配置，接着远程执行一些命令，探测远端主机 Python 解释器的位置，然后执行 PUT 上
传 ping 模块文件并执行，最后返回相应的结果并封装成固定格式打印出来。至于打印方
式如何实现，将在第 4 章介绍 Ansible 2.8 的核心源码时进行解读。

（4）-i、--list-hosts 和-l 选项中，-i 选项用于指定 inventory 文件，--list-hosts 选项用于
列出匹配的主机，-l 选项是在前面匹配的主机的基础上再次匹配。以下是这 3 个选项的
用法示例：

```
(ansible2.8.16) [root@master ansible-2.8.16]# ansible all -i hosts -list
-hosts
  hosts (4):
    master
    ceph-1
    ceph-2
    ceph-3
(ansible2.8.16) [root@master ansible-2.8.16]# ansible ceph* -i hosts -list
-hosts
  hosts (3):
    ceph-1
    ceph-2
    ceph-3
(ansible2.8.16) [root@master ansible-2.8.16]# ansible ceph* -i hosts -m ping
-l ceph-1
ceph-1 | SUCCESS => {
    "ansible_facts": {
        "discovered_interpreter_python": "/usr/bin/python"
    },
```

```
    "changed": false,
    "ping": "pong"
}
(ansible2.8.16) [root@master ansible-2.8.16]# ansible all -i hosts -m ping
-l master
master | SUCCESS => {
    "ansible_facts": {
        "discovered_interpreter_python": "/usr/bin/python"
    },
    "changed": false,
    "ping": "pong"
}
```

（5）-b/-become、--become-method 和--become-user 选项用于 ansible 命令执行时切换远程登录身份。-b 或者-become 表示使用 sudo 提权，如果对应的--become-user 没有设置 sudo 免密；则需要在配置或者运行时手动输入相关的 sudo 密码；--become-method 是指提权的方式，在 Linux 系统中常见的是 sudo 和 su，后者一般用于切换用户。为了能进行相关测试，在第 1 章中为了演示 Ansible 1.1 的底层通信代码原理，建立了两个用户，即 store 和 halo，其中，store 为 sudo 免密用户，而 halo 则为 sudo 不免密用户。此外，halo 的密码为：halo.1234!。现在进行如下测试：

```
(ansible2.8.16) [root@master ansible-2.8.16]# cat hosts_store
[nodes]
ceph-1

[nodes:vars]
ansible_ssh_user=store
ansible_ssh_pass=store.1234!
(ansible2.8.16) [root@master ansible-2.8.16]# ansible ceph-1 -i hosts_
store -m shell -a "whoami"
ceph-1 | CHANGED | rc=0 >>
store

(ansible2.8.16) [root@master ansible-2.8.16]# ansible ceph-1 -i hosts_
store -m shell -a "whoami" -b
ceph-1 | CHANGED | rc=0 >>
root

ansible: error: --become-user option requires 1 argument
(ansible2.8.16) [root@master ansible-2.8.16]# ansible ceph-1 -i hosts_
store -m shell -a "whoami" -b --become-user halo
ceph-1 | CHANGED | rc=0 >>
halo
```

由于 store 是 sudo 免密用户，所以直接使用-b 可以提权为 root，类似于在 ceph-1 节点直接执行 sudo su -命令。此外，使用--become-user 可以切换为任意指定的其他用户。下面来看 halo 用户的测试。

```
(ansible2.8.16) [root@master ansible-2.8.16]# cat hosts_halo
[nodes]
ceph-1
```

```
[nodes:vars]
ansible_ssh_user=halo
ansible_ssh_pass=halo.1234!
(ansible2.8.16) [root@master ansible-2.8.16]# ansible ceph-1 -i hosts_halo
-m shell -a "whoami"
ceph-1 | CHANGED | rc=0 >>
halo

(ansible2.8.16) [root@master ansible-2.8.16]# ansible ceph-1 -i hosts_halo
-m shell -a "whoami" -b
ceph-1 | FAILED | rc=-1 >>
Missing sudo password
```

由于 halo 用户是 sudo 非免密的用户，使用 sudo 提权时会要求输入相应的提权密码，而这里并没有设置相应的提权密码，导致报错。因此需要设置 ansible_sudo_pass 变量，确保在提权时能应对需要输入密码的情况。具体操作如下：

```
(ansible2.8.16) [root@master ansible-2.8.16]# cat hosts_halo
[nodes]
ceph-1

[nodes:vars]
ansible_ssh_user=halo
ansible_ssh_pass=halo.1234!
ansible_sudo_pass=halo.1234!
(ansible2.8.16) [root@master ansible-2.8.16]# ansible ceph-1 -i hosts_halo
-m shell -a "whoami" -b
ceph-1 | CHANGED | rc=0 >>
root

(ansible2.8.16) [root@master ansible-2.8.16]# ansible ceph-1 -i hosts_halo
-m shell -a "whoami" -b --become-user=store --become-method=sudo
ceph-1 | CHANGED | rc=0 >>
store
```

（6）-e 和--extra-vars 选项用于给执行的命令添加额外的变量。下面是一个简单的测试示例：

```
(ansible2.8.16) [root@master ansible-2.8.16]# ansible center -i hosts -m
ping -a "data={{ msg }}" -e "msg=hello,world"
master | SUCCESS => {
    "ansible_facts": {
        "discovered_interpreter_python": "/usr/bin/python"
    },
    "changed": false,
    "ping": "hello,world"
}
```

（7）-P 和-B 选项用于启动异步任务，即将 ansible 命令行指定的任务放到远端主机的后台执行，其中，-P 和-B 选项分别指定探测时间和任务执行的超时时间。如果-P 选项设置为 0，表示直接返回，不再进行探测。为了测试该选项，先准备一个测试脚本，其内容如下：

```
(ansible2.8.16) [root@master test_playbook]# cat test.sh
#! /bin/bash

echo "[`date +'%Y-%m-%d %H:%M:%S'`]run backgroud, sleep 3s"
sleep 3
echo "[`date +'%Y-%m-%d %H:%M:%S'`]wake up, on `hostname`, sleep again"
sleep 5
echo "[`date +'%Y-%m-%d %H:%M:%S'`]stop here"
```

接着将以上脚本复制到 ceph-1 节点上，直接使用 Ansible 的 copy 模块即可，具体内容如下：

```
(ansible2.8.16) [root@master test_playbook]# ansible ceph-1 -i hosts -m copy
-a "src=test.sh dest=/tmp/test.sh"
ceph-1 | CHANGED => {
    "ansible_facts": {
        "discovered_interpreter_python": "/usr/bin/python"
    },
    "changed": true,
    "checksum": "e4364a24518c851c4e9f7a25a5d1878eb278281f",
    "dest": "/tmp/test.sh",
    "gid": 0,
    "group": "root",
    "md5sum": "0064ef2e6dc5ac9273f29c5435681063",
    "mode": "0644",
    "owner": "root",
    "size": 210,
    "src": "/root/.ansible/tmp/ansible-tmp-1613796727.4250772-33171-
78738762372252/source",
    "state": "file",
    "uid": 0
}
```

由脚本内容可知，该脚本执行结束耗时至少 8s。因此当设置-B 选项的值为 5 时，执行脚本的任务必定超时，ansible 命令返回超时异常。测试结果如下：

```
(ansible2.8.16) [root@master test_playbook]# ansible ceph-1 -i hosts -m
shell -a "sh /tmp/test.sh" -P 4 -B 5
ceph-1 | FAILED | rc=-1 >>
async task did not complete within the requested time - 5s
```

如果设置-P 为 4，则 Ansible 内部会给该任务生成一个 jid 值用于标识该任务，Ansible 将每隔 4s 去远端查询一次该任务的执行情况。如果任务执行完毕，则立即返回并输出结果，否则返回后等待下次探测或者直到超时。以下是测试-P 选项的命令：

```
(ansible2.8.16) [root@master test_playbook]# ansible ceph-1 -i hosts -m
shell -a "sh /tmp/test.sh" -P 4 -B 20 -vvv
# ...
```

🔔注意：普通情况下无法看到 Ansible 内部的探测动作，需要加上-vvv 选项才能看到探测的详细过程。由于-vvv 选项的输出内容较多，笔者没有在上面展示出来，读者可以自行测试并观察探测结果。

3.2.2　Ansible 2 中的常用模块及其源码解析

和学习 Ansible 1 一样，在介绍 Ansible 2 中的模块时首先将介绍模块的相关参数及用法示例，然后从源码角度分析该模块的实现原理。比较幸运的是，Ansible 2.8 的所有模块可以在官网上找到非常详细的参数说明及用法示例，可以帮助初学者很快地掌握 Ansible 中各模块的用法。本节将以 find、lineinfile 和 systemd 模块为例，介绍这 3 个模块的用法及其源码。

1. find模块

首先在官网上找到 find 模块的介绍、参数说明及相应的示例。以下是官网对该模块的介绍：

```
Return a list of files based on specific criteria. Multiple criteria are
AND'd together
```

简单翻译过来就是，按照搜索的准则找出一系列符合要求的文件并返回查找结果。这是不是和 Linux 中的 find 命令有些相似？但其实 find 模块的实现和 find 命令并无关系，它是直接使用 Python 搜索文件并进行匹配。下面介绍该模块的部分重要参数。

- age：选择对应时间等于或者大于设置时间的文件。
- age_stamp：比较的时间属性，选项有 atime、ctime 和 mtime，默认为 mtime。
- contains：一个或者多个正则表达式，搜索内容匹配该表达式的文件。
- depth：搜索深度。
- excludes：排除某些结果，和 patterns 参数一样，是一个或多个 shell 或者正则形式的 pattern。
- file_type：搜索的文件类型。选项值有 any、directory、file 和 link，默认为 file。
- follow：该参数值有两个，即 yes 和 no，默认值为 no。若设置为 yes，则对于 Python 2.6 以上的系统，将追踪其符号链接地址。
- get_checksum：该参数值有两个，即 yes 和 no，默认值为 no。设置为 yes 时将获取文件的 SHA1 校验和。
- hidden：该参数值有两个，即 yes 和 no，默认值为 no。该参数决定是否搜索隐藏文件。
- paths：必要的参数，表示搜索的目录列表，所有路径必须完全限定。
- patterns：一个或者多个表达式（shell 或者是正则表达式），patterns 的类型取决于 use_regex 参数值。
- recurse：该参数值有两个，即 yes 和 no，默认值为 no。如果目标文件是目录，yes 表示递归搜索。
- size：选择容量大于或者等于该指定值的文件。单位有 b、k、m、g 和 t，分别表示

B、KB、MB、GB 和 TB。

- use_regex：该参数值有 yes 和 no 两个，默认值为 no。no 表示 shell 模式的匹配，而 yes 表示 patterns 参数值为 Python 的正则表达式。

下面来看看官方给出的部分用法示例。

```
# 递归搜索/tmp目录下两天前的文件
- name: Recursively find /tmp files older than 2 days
  find:
    paths: /tmp
    age: 2d
    recurse: yes

# 递归搜索/tmp目录下4周前且容量大于等于1MB的文件
- name: Recursively find /tmp files older than 4 weeks and equal or greater
  than 1 megabyte
  find:
    paths: /tmp
    age: 4w
    size: 1m
    recurse: yes

# 递归搜索/var/tmp目录中上一次访问时间大于3600s的文件
- name: Recursively find /var/tmp files with last access time greater than
  3600 seconds
  find:
    paths: /var/tmp
    age: 3600
    age_stamp: atime
    recurse: yes

# 搜索/var/log目录下大于10MB且为.old和.log.gz形式的文件
- name: Find /var/log files equal or greater than 10 megabytes ending with .old
  or .log.gz
  find:
    paths: /var/log
    patterns: '*.old,*.log.gz'
    size: 10m

# Note that YAML double quotes require escaping backslashes but yaml single
  quotes do not.
- name: Find /var/log files equal or greater than 10 megabytes ending with .old
  or .log.gz via regex
  find:
    paths: /var/log
    patterns: "^.*?\\.(?:old|log\\.gz)$"
    size: 10m
    use_regex: yes

# 搜索/var/log下的所有目录，非递归，排除nginx和mysql目录
- name: Find /var/log all directories, exclude nginx and mysql
  find:
    paths: /var/log
    recurse: no
```

```
      file_type: directory
      excludes: 'nginx,mysql'

# When using patterns that contain a comma, make sure they are formatted
as lists to avoid splitting the pattern
- name: Use a single pattern that contains a comma formatted as a list
  find:
    paths: /var/log
    file_type: file
    use_regex: yes
    patterns: ['^_[0-9]{2,4}_.*.log$']

- name: Use multiple patterns that contain a comma formatted as a YAML list
  find:
    paths: /var/log
    file_type: file
    use_regex: yes
    patterns:
      - '^_[0-9]{2,4}_.*.log$'
      - '^[a-z]{1,5}_.*log$'
```

以上代码中展示了几个非常好的用法示例，如第一个任务使用 find 模块递归查找/tmp
目录下两天前的文件，patterns 可以写成多个表达式，上面的示例中也给出了几种写法。
此外，size、use_regex 和 file_type 等参数的示例也值得参考、学习。

由于 find 模块的返回内容较多，在此给出一个示例后将直接开始模块的源码学习。下
面的示例是搜索 ceph-1 上/root 目录下一天前的 shell 脚本。

```
(ansible2.8.16) [root@master ansible-2.8.16]# ansible ceph-1 -i hosts -m
find -a "paths=/root age=1d recurse=no age_stamp=atime patterns='*.sh'"
ceph-1 | SUCCESS => {
    "ansible_facts": {
        "discovered_interpreter_python": "/usr/bin/python"
    },
    "changed": false,
    "examined": 21,
    "files": [
        {
            "atime": 1604936966.9505732,
            "ctime": 1604936964.8875732,
            "dev": 64768,
            "gid": 0,
            "gr_name": "root",
            "inode": 33622494,
            "isblk": false,
            "ischr": false,
            "isdir": false,
            "isfifo": false,
            "isgid": false,
            "islnk": false,
            "isreg": true,
            "issock": false,
            "isuid": false,
            "mode": "0644",
            "mtime": 1604936964.8875732,
```

```
            "nlink": 1,
            "path": "/root/clear_lvm_system.sh",
            "pw_name": "root",
            "rgrp": true,
            "roth": true,
            "rusr": true,
            "size": 641,
            "uid": 0,
            "wgrp": false,
            "woth": false,
            "wusr": true,
            "xgrp": false,
            "xoth": false,
            "xusr": false
        },
        {
            "atime": 1605345551.6399262,
            "ctime": 1605345538.0908296,
            "dev": 64768,
            "gid": 0,
            "gr_name": "root",
            "inode": 33680358,
            "isblk": false,
            "ischr": false,
            "isdir": false,
            "isfifo": false,
            "isgid": false,
            "islnk": false,
            "isreg": true,
            "issock": false,
            "isuid": false,
            "mode": "0644",
            "mtime": 1605345538.0828285,
            "nlink": 1,
            "path": "/root/make_osds.sh",
            "pw_name": "root",
            "rgrp": true,
            "roth": true,
            "rusr": true,
            "size": 1414,
            "uid": 0,
            "wgrp": false,
            "woth": false,
            "wusr": true,
            "xgrp": false,
            "xoth": false,
            "xusr": false
        }
    ],
    "matched": 2,
    "msg": ""
}
```

下面直接进行 find 模块的源码学习，相应的源码文件位于 lib/ansible/modules/file 目录下。具体源码内容如下：

```
# lib/ansible/modules/files/find.py
# ...

def main():
    # 省略了实例化的参数
    module = AnsibleModule(...)

    params = module.params

    filelist = []

    # 将 m|h|d|w 的单位调整成秒
    if params['age'] is None:
        age = None
    else:
        # convert age to seconds:
        m = re.match(r"^(-?\d+)(s|m|h|d|w)?$", params['age'].lower())
        seconds_per_unit = {"s": 1, "m": 60, "h": 3600, "d": 86400, "w":
604800}
        if m:
            age = int(m.group(1)) * seconds_per_unit.get(m.group(2), 1)
        else:
            module.fail_json(age=params['age'], msg="failed to process age")

    # 将 b|k|m|g|t 的单位调整成字节
    if params['size'] is None:
        size = None
    else:
        # convert size to bytes:
        m = re.match(r"^(-?\d+)(b|k|m|g|t)?$", params['size'].lower())
        bytes_per_unit = {"b": 1, "k": 1024, "m": 1024**2, "g": 1024**3, "t":
1024**4}
        if m:
            size = int(m.group(1)) * bytes_per_unit.get(m.group(2), 1)
        else:
            module.fail_json(size=params['size'], msg="failed to process
size")

    # 遍历所有输入目录
    now = time.time()
    msg = "
    looked = 0
    for npath in params['paths']:
        npath = os.path.expanduser(os.path.expandvars(npath))
        if os.path.isdir(npath):
            ''' ignore followlinks for python version < 2.6 '''
            for root, dirs, files in (sys.version_info < (2, 6, 0) and
os.walk(npath)) or os.walk(npath, followlinks=params['follow']):
                # 处理递归深度配置
                if params['depth']:
                    depth = root.replace(npath.rstrip(os.path.sep), '').
count(os.path.sep)
                    if files or dirs:
                        depth += 1
```

```
                if depth > params['depth']:
                    del(dirs[:])
                    continue
            looked = looked + len(files) + len(dirs)
            for fsobj in (files + dirs):
                fsname = os.path.normpath(os.path.join(root, fsobj))

                # 处理 hidden 参数，如果设置 hidden=no，对于以'.'开头的隐藏文
                  件直接跳过
                if os.path.basename(fsname).startswith('.') and not params
['hidden']:
                    continue

                try:
                    # 获取文件信息
                    st = os.lstat(fsname)
                except Exception:
                    msg += "%s was skipped as it does not seem to be a valid
file or it cannot be accessed\n" % fsname
                    continue

                r = {'path': fsname}
                # 第一层 if 及其分支处理文件类型
                if params['file_type'] == 'any':
                    # 如果 file_type 参数是 any，则调用模式过滤方法（pfilter()）
                      和时间过滤方法（agefilter()）搜索符合条件的文件
                    if pfilter(fsobj, params['patterns'], params['excludes'],
params['use_regex']) and agefilter(st, now, age, params['age_stamp']):

                        r.update(statinfo(st))
                        if stat.S_ISREG(st.st_mode) and params['get_checksum']:
                            r['checksum'] = module.sha1(fsname)
                        filelist.append(r)

                elif stat.S_ISDIR(st.st_mode) and params['file_type'] ==
'directory':
                    # 目录也是模式过滤+时间过滤
                    if pfilter(fsobj, params['patterns'], params['excludes'],
params['use_regex']) and agefilter(st, now, age, params['age_stamp']):

                        r.update(statinfo(st))
                        filelist.append(r)

                elif stat.S_ISREG(st.st_mode) and params['file_type'] ==
'file':
                    # 普通文件是模式过滤+时间过滤+大小过滤+内容过滤
                    if pfilter(fsobj, params['patterns'], params['excludes'],
params['use_regex']) and \
                        agefilter(st, now, age, params['age_stamp']) and \
                        sizefilter(st, size) and contentfilter(fsname,
params['contains']):

                        r.update(statinfo(st))
                        if params['get_checksum']:
```

```
                    r['checksum'] = module.sha1(fsname)
                  filelist.append(r)

            elif stat.S_ISLNK(st.st_mode) and params['file_type'] ==
'link':
                # 链接文件是模式过滤+时间过滤
                if pfilter(fsobj, params['patterns'], params['excludes'],
params['use_regex']) and agefilter(st, now, age, params['age_stamp']):

                    r.update(statinfo(st))
                  filelist.append(r)

            # 非递归则直接返回
            if not params['recurse']:
               break
        else:
            # 非目录。不处理，记录异常信息
            msg += "%s was skipped as it does not seem to be a valid directory
or it cannot be accessed\n" % npath

    matched = len(filelist)
    module.exit_json(files=filelist, changed=False, msg=msg, matched=
matched, examined=looked)

if __name__ == '__main__':
    main()
```

上面的代码并不复杂，逻辑也非常清晰。首先，实例化 AnsibleModule 类，接着从实例化的对象中获取相应的参数值。其次，对时间和大小参数的单位进行修正，统一换算成秒和字节单位，然后遍历 paths 参数提供的搜索目录。对每个搜索目录而言，会遍历该目录下的所有文件，最后针对参数 file_type 组合不同的过滤方式。例如针对 file_type=directory 的情况，执行如下搜索语句：

```
elif stat.S_ISDIR(st.st_mode) and params['file_type'] == 'directory':
    # 目录也是模式过滤+时间过滤
    if pfilter(fsobj, params['patterns'], params['excludes'], params
['use_regex']) and agefilter(st, now, age, params['age_stamp']):
        r.update(statinfo(st))
        filelist.append(r)
```

首先判断搜索的文件类型是否符合 file_type 参数值，然后对于目录的搜索使用模式过滤+时间过滤的组合方式筛选出符合条件的目录，然后添加到 filelist 列表中。其余情况类似，对于普通文件的过滤还额外包含大小过滤和内容过滤。因此核心的过滤函数有以下几个。

- pfilter()：模式过滤，根据 patterns 匹配文件或者目录名，此外还接收 excludes 参数用于排除文件。
- agefilter()：时间过滤。
- sizefilter()：大小过滤。
- contentfilter()：内容过滤。

下面是 pfilter() 函数的具体实现代码，该函数主要使用了 pattern、excludes 和 use_regex 这 3 个参数。

```python
# lib/ansible/modules/files/find.py
# ...

def pfilter(f, patterns=None, excludes=None, use_regex=False):
    '''filter using glob patterns'''
    if patterns is None and excludes is None:
        return True

    if use_regex:
        # 使用正则表达式
        if patterns and excludes is None:
            for p in patterns:
                r = re.compile(p)
                if r.match(f):
                    return True

        elif patterns and excludes:
            for p in patterns:
                r = re.compile(p)
                if r.match(f):
                    for e in excludes:
                        r = re.compile(e)
                        if r.match(f):
                            # 排除 excludes 匹配结果
                            return False
                    return True

    else:
        if patterns and excludes is None:
            for p in patterns:
                if fnmatch.fnmatch(f, p):
                    return True

        elif patterns and excludes:
            for p in patterns:
                if fnmatch.fnmatch(f, p):
                    for e in excludes:
                        if fnmatch.fnmatch(f, e):
                            return False
                    return True

    return False
```

从上面的代码中可以看出，pfilter() 函数的实现逻辑比较清晰：如果 use_regex=yes，则使用 Python 的正则表达式进行匹配，如果 use_regex=no，则调用 fnmatch 模块的 fnmatch() 方法进行模式匹配。针对 excludes 参数，前面两种情况的处理都是一致的，一旦发现匹配 patterns 又匹配 excludes，则直接返回 False。

接下来是 agefilter() 方法，其过滤逻辑非常简单，直接根据获取的文件状态信息中的时间参数与 age 进行比较，其代码如下：

```
# lib/ansible/modules/files/find.py
# ...

def agefilter(st, now, age, timestamp):
    ''' 过滤早于 age 的文件 '''
    if age is None:
        return True
    elif age >= 0 and now - st.__getattribute__("st_%s" % timestamp) >=
abs(age):
        return True
    elif age < 0 and now - st.__getattribute__("st_%s" % timestamp) <=
abs(age):
        return True
    return False
```

从代码中可以看出 age 参数值可以是正数或负数。例如，age=1m 表示过滤在 1ms 之前更新过的文件，而 age=-1m 表示过滤在 1ms 内更新过的文件。

接下来是 sizefilter()方法，其代码如下：

```
# lib/ansible/modules/files/find.py
# ...

def sizefilter(st, size):
    '''filter files greater than size'''
    if size is None:
        return True
    elif size >= 0 and st.st_size >= abs(size):
        return True
    elif size < 0 and st.st_size <= abs(size):
        return True
    return False
```

size 参数和 age 参数一样，可以为正数或负数，含义也类似。通过代码可以理解，size值为正时则匹配比 size 大的文件，否则匹配比 size 小的文件。

最后一个是 contentfilter()函数，其逻辑非常简单，即按行读取文件内容，一行行地匹配，一旦匹配到就直接返回 True，其余情况则返回 False。其代码如下：

```
# lib/ansible/modules/files/find.py
# ...

def contentfilter(fsname, pattern):
    if pattern is None:
        return True

    prog = re.compile(pattern)

    try:
        # 读取文件内容，一行行地匹配，如果匹配成功，则直接返回 True
        with open(fsname) as f:
            for line in f:
                if prog.match(line):
                    return True
```

```
except Exception:
    pass

# 如果没有匹配成功或者读取文件异常，则返回 False
return False
```

2. lineinfile模块

lineinfile 模块用于管理文本文件中的行，简单来说就是对文本内容按行进行管理。该模块的参数众多，这里只介绍部分常用的参数，更多参数及其细节请参考官方网站。

- backup：该参数值有两个，即 yes 和 no，默认值为 no。当设置为 yes 时，将对被操作的文件进行备份。
- create：该参数值有两个，即 yes 和 no，默认值为 no。当设置为 yes 时，对于 state=present 的情况，如果管理的文件不存在，则直接创建；否则模块执行失败。
- firstmatch：该参数值有两个，即 yes 和 no，默认值为 no。配合 insertafter 或 insertbefore 参数一起使用，如果设置为 yes，则在匹配到的第一行之前（insertbefore）或者之后（insertafter）插入。
- insertafter：在匹配行之后插入。如果设置了 firstmatch=yes，则只在匹配到的第一行之后插入。
- insertbefore：在匹配行之前插入，同样可配合 firstmatch 参数只操作匹配到的第一行。
- line：准备插入或者替换文件中的一行内容，配合 state=present 使用。
- mode：修改文件或者目录的权限。
- owner：修改文件或者目录的属主。
- path：准备修改的文件。
- regexp：正则表达式，用于查找文件中的匹配行。
- state：当 state=present 时，需要 line 参数的值替换匹配的行或者在匹配行的上下（由 inserbefore 和 inserafter 参数控制）插入 line 内容；当 state=absent 时，模块参数中不需要 line 参数，并且会删除文件中匹配的行。
- validate：在复制行到文件中的对应位置之前，先运行校验命令。

为了测试 lineinfile 模块，笔者在 ceph-1 节点上创建了两个文件，分别为 test_lineinfile. txt 和 test_lineinfile_absent.txt，其内容一致，具体如下：

```
[root@ceph-1 test_file]# cat test_lineinfile.txt
hello world, there are some apples.
this is a number line: 12345 xxxxxxx
test.1234!
we will have a greate day.
who do you think you are? joke
匹配特殊符号:$xy.1234z$
[root@ceph-1 test_file]# cp test_lineinfile.txt test_lineinfile_absent.
txt
```

示例 1：测试 state=present 的情况。请看下面的操作示例：

```
(ansible2.8.16) [root@master ansible-2.8.16]# ansible ceph-1 -i hosts -m
lineinfile -a "backup=yes regex='.*[0-9]+.*' line='####' path=/root/test_
file/test_lineinfile.txt"
ceph-1 | CHANGED => {
    "ansible_facts": {
        "discovered_interpreter_python": "/usr/bin/python"
    },
    "backup": "/root/test_file/test_lineinfile.txt.102740.2258-07-24@19:
46:54~",
    "changed": true,
    "msg": "line replaced"
}
```

登录到 ceph-1 上查看相应位置上的文件内容，具体如下：

```
[root@ceph-1 test_file]# ls
test_lineinfile_absent.txt
test_lineinfile.txt
test_lineinfile.txt.102740.2258-07-24@19:46:54~
[root@ceph-1 test_file]# cat test_lineinfile.txt
hello world, there are some apples.
this is a number line: 12345 xxxxxxx
test.1234!
we will have a greate day.
who do you think you are? joke
####
```

这里可以看到，只有最后一行匹配成功并进行了替换，前面没有匹配到吗？接下来再试试加上 firstmatch 的效果。首先根据备份文件将原文件还原，然后再做下面的实验：

```
[root@ceph-1 test_file]# mv -f test_lineinfile.txt.102740.2258-07-24\@19\
:46\:54~ test_lineinfile.txt
```

示例 2：测试 firstmatch=yes 的情况。请看下面的操作示例：

```
(ansible2.8.16) [root@master ansible-2.8.16]# ansible ceph-1 -i hosts -m
lineinfile -a "backup=yes regex='.*[0-9]+.*' line='####' firstmatch=yes
path=/root/test_file/test_lineinfile.txt"
ceph-1 | CHANGED => {
    "ansible_facts": {
        "discovered_interpreter_python": "/usr/bin/python"
    },
    "backup": "/root/test_file/test_lineinfile.txt.102956.2258-07-24@19:
55:51~",
    "changed": true,
    "msg": "line replaced"
}
```

再来看 ceph-1 上对应文件的内容，具体如下：

```
[root@ceph-1 test_file]# cat test_lineinfile.txt
hello world, there are some apples.
####
test.1234!
we will have a greate day.
```

```
who do you think you are? joke
匹配特殊符号:$xy.1234z$
```

可以看到，第一个匹配到有数字的行是第二行。在设置了 firstmatch=yes 后，匹配到这一行并替换文本后，程序结束。看到这个现象，读者可以思考一下：是不是 lineinfile 模块默认只会找到最后一个匹配的行并进行替换，而不是对所有匹配的行进行替换？如果是，那么该如何改造才能满足替换多个匹配行的情况呢？出现这个现象的原因及后续的改造将在后面对模块的源码分析中再讲，这里继续第 3 个示例。

示例 3：测试 state=absent 的情况。请看下面的操作示例：

```
(ansible2.8.16) [root@master ansible-2.8.16]# ansible ceph-1 -i hosts -m
lineinfile -a "backup=yes regexp='.*[0-9]+.*' path=/root/test_file/test_
lineinfile_absent.txt state=absent"
ceph-1 | CHANGED => {
    "ansible_facts": {
        "discovered_interpreter_python": "/usr/bin/python"
    },
    "backup": "/root/test_file/test_lineinfile_absent.txt.101828.2258-
07-24@18:39:11~",
    "changed": true,
    "found": 3,
    "msg": "3 line(s) removed"
}
```

使用 ssh 登录到 ceph-1 上查看结果，具体如下：

```
[root@ceph-1 test_file]# cat test_lineinfile_absent.txt
hello world, there are some apples.
we will have a greate day.
who do you think you are? joke
```

通过查看 lineinfile 模块的源码，可以帮助理解上面的代码执行过程。lineinfile 模块文件的源码路径为 lib/ansible/modules/files/lineinfile.py，其 main()方法内容如下：

```
# 源码位置: lib/ansible/modules/files/lineinfile.py
# ...

def main():
    # 简化了这里的实例化参数
    module = AnsibleModule(...)

    # 提取模块参数
    params = module.params
    create = params['create']
    backup = params['backup']
    backrefs = params['backrefs']
    path = params['path']
    firstmatch = params['firstmatch']
    regexp = params['regexp']
    line = params['line']

    if regexp == '':
        # 打印警告提示信息
```

```
    # ...

    # 转换成字节形式
    b_path = to_bytes(path, errors='surrogate_or_strict')
    if os.path.isdir(b_path):
        # 如果 path 参数输入的是目录，直接返回，模块运行失败
        module.fail_json(rc=256, msg='Path %s is a directory !' % path)

    if params['state'] == 'present':
        # 增加和修改 line，开头是对两种异常情况的处理
        if backrefs and regexp is None:
            module.fail_json(msg='regexp is required with backrefs=true')

        if line is None:
            module.fail_json(msg='line is required with state=present')

        ins_bef, ins_aft = params['insertbefore'], params['insertafter']
        if ins_bef is None and ins_aft is None:
            ins_aft = 'EOF'

        # 调用 present() 函数
        present(module, path, regexp, line,
                ins_aft, ins_bef, create, backup, backrefs, firstmatch)
    else:
        # 删除 line
        if regexp is None and line is None:
            module.fail_json(msg='one of line or regexp is required with
state=absent')

        # 调用 absent() 函数
        absent(module, path, regexp, line, backup)
if __name__ == '__main__':
    main()
```

上面的代码非常简洁，逻辑也比较简单。首先从 AnsibleModule 对象中取得模块的参数值，然后检查一遍异常情况，如 path 的值是否为目录等。接着针对 state 参数的值分别调用 present()函数或者 absent()函数进行处理。下面分别解析 present()和 absent()函数的代码。首先来解析 present()函数的源码，具体如下：

```
# 源码位置：lib/ansible/modules/files/lineinfile.py
# ...

def present(module, dest, regexp, line, insertafter, insertbefore, create,
        backup, backrefs, firstmatch):
    # ...

    # 1. 读取文件内容，按行形成列表。如果文件不存在且设置了 create 标识，则创建该文件，
    #    同时将列表置空
    b_dest = to_bytes(dest, errors='surrogate_or_strict')
    if not os.path.exists(b_dest):
        if not create:
            module.fail_json(rc=257, msg='Destination %s does not exist !'
```

```
% dest)
        b_destpath = os.path.dirname(b_dest)
        if b_destpath and not os.path.exists(b_destpath) and not module.
check_mode:
            try:
                os.makedirs(b_destpath)
            except Exception as e:
                # 创建目录异常，直接返回模块执行失败的提示信息
                # ...

        b_lines = []
    else:
        with open(b_dest, 'rb') as f:
            b_lines = f.readlines()

    # ...

    if regexp is not None:
        bre_m = re.compile(to_bytes(regexp, errors='surrogate_or_strict'))

    # 2．如果没有设置 insertafter 或者 insertbefore 参数，则 bre_ins 为 None
    if insertafter not in (None, 'BOF', 'EOF'):
        bre_ins = re.compile(to_bytes(insertafter, errors='surrogate_or_
strict'))
    elif insertbefore not in (None, 'BOF'):
        bre_ins = re.compile(to_bytes(insertbefore, errors='surrogate_or_
strict'))
    else:
        bre_ins = None

    # index 数组的第一个元素表示根据 regexp 找到匹配行的行号
    # index 数组的第二个元素表示根据 insertafter/insertbefore 找到匹配行的行号
    index = [-1, -1]
    match = None
    exact_line_match = False
    b_line = to_bytes(line, errors='surrogate_or_strict')

    # 3. regexp 在前面，如果 regexp、insertafter 或者 insertbefore 都设置了，则优
    #    先考虑 regexp 匹配的行
    if regexp is not None:
        for lineno, b_cur_line in enumerate(b_lines):
            match_found = bre_m.search(b_cur_line)
            if match_found:
                # index 数组第 1 个元素的含义，lineno 为匹配行的行号
                index[0] = lineno
                match = match_found
                # 如果设置了第一次匹配，则第一次匹配后直接跳出即可
                if firstmatch:
                    break

    # 4．当 regexp 没有匹配到结果时才会考虑 insertafter|insertbefore 匹配的行
    if not match:
        for lineno, b_cur_line in enumerate(b_lines):
            if b_line == b_cur_line.rstrip(b'\r\n'):
```

```python
            index[0] = lineno
            exact_line_match = True

        elif bre_ins is not None and bre_ins.search(b_cur_line):
            if insertafter:
                # + 1 for the next line
                index[1] = lineno + 1
                if firstmatch:
                    break

            if insertbefore:
                # index[1] for the previous line
                index[1] = lineno
                if firstmatch:
                    break

# 5. 根据各种情况插入行数据
msg = ''
changed = False
b_linesep = to_bytes(os.linesep, errors='surrogate_or_strict')
if index[0] != -1:
    # 对存在的 regexp 匹配行或者 line 精确匹配行进行处理
    # ...
elif backrefs:
    pass
elif insertbefore == 'BOF' or insertafter == 'BOF':
    # 在第一行插入
    b_lines.insert(0, b_line + b_linesep)
    msg = 'line added'
    changed = True
elif insertafter == 'EOF' or index[1] == -1:
    # 在最后一行插入
    if b_lines and not b_lines[-1][-1:] in (b'\n', b'\r'):
        b_lines.append(b_linesep)

    b_lines.append(b_line + b_linesep)
    msg = 'line added'
    changed = True
elif insertafter and index[1] != -1:
    # 存在匹配 insertafter 的行
    if len(b_lines) == index[1]:
        if b_lines[index[1] - 1].rstrip(b'\r\n') != b_line:
            b_lines.append(b_line + b_linesep)
            msg = 'line added'
            changed = True
    elif b_line != b_lines[index[1]].rstrip(b'\n\r'):
        b_lines.insert(index[1], b_line + b_linesep)
        msg = 'line added'
        changed = True
else:
    # 匹配 insertbefore 的行
    b_lines.insert(index[1], b_line + b_linesep)
    msg = 'line added'
    changed = True
```

```
# 6. 将最后结果写入 dest 路径下的文本中
backupdest = ""
if changed and not module.check_mode:
    if backup and os.path.exists(b_dest):
        backupdest = module.backup_local(dest)
    write_changes(module, b_lines, dest)

# ...
```

present()函数的实现代码较多，略显复杂，但这是整个 lineinfile 模块的核心部分。笔者整理了该函数中的 6 处重要的代码段，每处代码段的含义如下：

- 第 1 处代码是读取文件内容并按行形成列表。如果文件不存在且设置了 create 标识，则创建该文件，同时将文件内容的列表置空。
- 第 2 处代码表示如果没有设置 insertafter 或者 insertbefore 参数，则 bre_ins 为 None。此外，注意后面定义的 index 数组，它只有两个元素，第一个元素值为根据 regexp 找到的匹配行的行号，第二个元素值为根据 insertafter/insertbefore 找到的匹配行的行号。
- 第 3 处代码是根据 regexp 表达式找到匹配的行，同时更新 index 的第一个元素值。这里看到 firstmatch 参数的作用，其在第一次匹配到行后，直接跳出循环。前面的案例 1 和案例 2 的原因就在这里，在多行匹配到 regexp 表达式后，index[0]的值是最后一次匹配的行号，而后续将根据这个行号执行替换和插入的动作。
- 第 4 处代码是如果根据 regexp 表达式没有匹配到行，则根据 line 值或者 insertbefore|insertafter 表达式来匹配文本的行。如果 line 全量匹配到行，则更新 index 的第一个元素，如果是后者能匹配到相应的行，则更新 index 的第二个元素。同样，如果设置了 firstmatch 参数，则可在第一次匹配后跳出循环。
- 第 5 处代码是综合所有的参数以及匹配到的行号等信息进行判断，得到完整的新增或者替换后的行列表。
- 第 6 处代码则是将得到的完整的行列表内容写入最后的文本中。

在大致理解了 present()函数的实现逻辑后，读者可以思考下如何改造该函数，以实现多行替换功能。只实现多行替换功能，其实还是比较简单的，通过上面的 present()方法可以看到，代码中只有一处会进行 line 替换。因此修改对应的逻辑即可，具体如下：

```
# 源码位置: lib/ansible/modules/files/lineinfile.py
# ...

def present(module, dest, regexp, line, insertafter, insertbefore, create,
            backup, backrefs, firstmatch):
    # ...

    if index[0] != -1:
        # ...

        if regexp is None and match is None and not exact_line_match:
```

```
        # ...
        elif b_lines[index[0]] != b_new_line:
            # 改造这里的代码即可
            b_lines[index[0]] = b_new_line
            msg = 'line replaced'
            changed = True
    elif backrefs:
        pass
    elif insertbefore == 'BOF' or insertafter == 'BOF':
        # ...
    elif insertafter == 'EOF' or index[1] == -1:
        # ...
    elif insertafter and index[1] != -1:
        # ...
    else:
        # ...

    # ...
```

首先要给模块添加一个参数 matchnum，表示要替换哪些匹配行，参数含义如下（当没有设置 matchnum 参数时，一切照旧）：

- all：替换全部匹配行。
- 数字 N：表示替换前 N 行匹配行。
- 数字-N：表示替换后 N 行匹配行。

接下来改造模块的 main()方法，添加相应的接收参数。此外，还需要将该参数传至 present()函数中，对应的改动如下：

```
# 源码位置：lib/ansible/modules/files/lineinfile.py
# ...

def main():
    module = AnsibleModule(
        argument_spec=dict(
            path=dict(type='path', required=True, aliases=['dest', 'destfile',
'name']),
            state=dict(type='str', default='present', choices=['absent',
'present']),
            regexp=dict(type='str', aliases=['regex']),
            line=dict(type='str', aliases=['value']),
            insertafter=dict(type='str'),
            insertbefore=dict(type='str'),
            backrefs=dict(type='bool', default=False),
            create=dict(type='bool', default=False),
            backup=dict(type='bool', default=False),
            firstmatch=dict(type='bool', default=False),
            validate=dict(type='str'),
            # 添加新的模块参数 matchnum
            matchnum=dict(type='str')
        ),
        mutually_exclusive=[['insertbefore', 'insertafter']],
        add_file_common_args=True,
        supports_check_mode=True,
```

```
    )
    # ...

    # 接收模块参数值
    matchnum = params['matchnum']

    if params['state'] == 'present':
        # ...

        # 在 present 函数中添加 matchnum 参数
        present(module, path, regexp, line,
                ins_aft, ins_bef, create, backup, backrefs, firstmatch,
matchnum)
    else:
        # ...
```

接着来改造 present()函数，分为以下 3 步：

（1）获取全部匹配行的位置，得到匹配列表 match_index。

（2）解析 matchnum，例如对于 all 和-N 这样的解析并生成新的 match_index。

（3）修改替换行的代码，使用 for 循环替换 match_index 中的行。

以下是经过改造后的 present()函数，这里只给出了改动部分的代码。

```
# 源码位置: lib/ansible/modules/files/lineinfile.py
# ...

def present(module, dest, regexp, line, insertafter, insertbefore, create,
        backup, backrefs, firstmatch, matchnum=None):
    # ...

    # 1. 获取全部匹配行的位置，得到匹配列表 match_index
    match_index = []
    if regexp is not None:
        for lineno, b_cur_line in enumerate(b_lines):
            match_found = bre_m.search(b_cur_line)
            if match_found:
                # 添加匹配行
                match_index.append(lineno)
                index[0] = lineno
                match = match_found
                if firstmatch:
                    break
    # 2. 解析 matchnum 参数值
    match_len = len(match_index)
    if matchnum == 'all':
        # 设置 all 参数
        matchnum = match_len
    elif matchnum and matchnum.isdigit():
        # 对于数字情况
        matchnum = min(int(matchnum), match_len)
    elif matchnum and matchnum.startswith("-") and matchnum[1:].isdigit():
        # 对于"-数字"情况
```

```
        matchnum = min(int(matchnum[1:]), match_len)
        # 重新更新 match_index
        match_index = match_index[match_len - matchnum: match_len]
    else:
        # 其余情况
        matchnum = 0

# ...

if index[0] != -1:
    # ...
    if regexp is None and match is None and not exact_line_match:
        # ...
    elif b_lines[index[0]] != b_new_line:
        # 3. 改造这里的替换行逻辑, 只有设置了 matchnum 且不为 0 的情况才会执行新的
        #    逻辑, 旧逻辑不变
        if matchnum:
            for i in range(matchnum):
                b_lines[match_index[i]] = b_new_line
        else:
            b_lines[index[0]] = b_new_line
        replace_num = "{} ".format(matchnum) if matchnum else ""
        msg = '{}line replaced'.format(replace_num)
        changed = True
elif backrefs:
    pass
elif insertbefore == 'BOF' or insertafter == 'BOF':
    # ...
elif insertafter == 'EOF' or index[1] == -1:
    # ...
elif insertafter and index[1] != -1:
    # ...
else:
    # ...

# ...
```

上面的 3 处代码已做好标注, 分别对应前面描述的 3 处改造点, 对应虚拟环境下的完整路径为:

```
(ansible2.8.16) [root@master ansible-2.8.16]# ls ~/.pyenv/versions/ansible
2.8.16/lib/python3.8/site-packages/ansible/modules/files/lineinfile.py
/root/.pyenv/versions/ansible2.8.16/lib/python3.8/site-packages/ansible
/modules/files/lineinfile.py
```

最后在测试环境下查看改造效果, 执行以下 Ansible 操作 3 次, 每次执行完操作后就登录到 ceph-1 节点查看 test_lineinfile.txt 内容的变化情况。

注意: 每次执行完命令后, 需要根据备份文件将 ceph-1 上的 test_lineinfile.txt 文件还原, 因为 lineinfile 模块直接更改了原来的文件, 否则会影响下一次操作。

首先来看第 1 次操作, 请求替换前两个匹配行, 具体如下:

```
(ansible2.8.16) [root@master ansible-2.8.16]# ansible ceph-1 -i hosts -m
```

```
lineinfile -a "backup=yes regexp='.*[0-9]+.*' line='####' path=/root/
test_file/test_lineinfile.txt matchnum=2"
ceph-1 | CHANGED => {
    "ansible_facts": {
        "discovered_interpreter_python": "/usr/bin/python"
    },
    "backup": "/root/test_file/test_lineinfile.txt.105617.2258-07-24@21:
53:37~",
    "changed": true,
    "msg": "2 line replaced"
}
```

在 ceph-1 节点上可以看到 test_lineinfile.txt 文件中包含数字的前两行被替换掉了，具体代码如下：

```
[root@ceph-1 test_file]# cat test_lineinfile.txt
hello world, there are some apples.
####
####
we will have a greate day.
who do you think you are? joke
匹配特殊符号:$xy.1234z$
```

再来看第 2 次操作，这次设置 matchnum=-2，用于替换最后两个匹配行，具体代码如下：

```
(ansible2.8.16) [root@master ansible-2.8.16]# ansible ceph-1 -i hosts -m
lineinfile -a "backup=yes regexp='.*[0-9]+.*' line='####' path=/root/test_
file/test_lineinfile.txt matchnum=-2"
ceph-1 | CHANGED => {
    "ansible_facts": {
        "discovered_interpreter_python": "/usr/bin/python"
    },
    "backup": "/root/test_file/test_lineinfile.txt.105727.2258-07-24@21:
55:11~",
    "changed": true,
    "msg": "2 line replaced"
}
```

再看 ceph-1 上的目标文件，可以发现，最后两个匹配行被####字符串替换了，具体代码如下：

```
[root@ceph-1 test_file]# cat test_lineinfile.txt
hello world, there are some apples.
this is a number line: 12345 xxxxxxx
####
we will have a greate day.
who do you think you are? joke
####
```

最后测试 matchnum=all，将替换所有的匹配行（一定要先还原文件后再执行该操作），具体代码如下：

```
(ansible2.8.16) [root@master ansible-2.8.16]# ansible ceph-1 -i hosts -m
lineinfile -a "backup=yes regexp='.*[0-9]+.*' line='####' path=/root/test_
```

```
file/test_lineinfile.txt matchnum=all"
ceph-1 | CHANGED => {
    "ansible_facts": {
        "discovered_interpreter_python": "/usr/bin/python"
    },
    "backup": "/root/test_file/test_lineinfile.txt.105826.2258-07-24@21:
56:11~",
    "changed": true,
    "msg": "3 line replaced"
}
```

最后再查看 test_lineinfile.txt 文件的内容，可以看到实现了全量替换，具体代码如下：

```
[root@ceph-1 test_file]# cat test_lineinfile.txt
hello world, there are some apples.
####
####
we will have a greate day.
who do you think you are? joke
####
```

上述改造并没有影响 lineinfile 模块的原有功能且实现了新的需求，这要求对 Ansible
模块的执行逻辑要有清楚的认识。

接下来解析 absent() 函数的源码，具体代码如下：

```python
# 源码位置: lib/ansible/modules/files/lineinfile.py
# ...

def absent(module, dest, regexp, line, backup):

    b_dest = to_bytes(dest, errors='surrogate_or_strict')
    if not os.path.exists(b_dest):
        # 文件不存在，直接返回运行错误的提示信息
        module.exit_json(changed=False, msg="file not present")

    msg = ''
    diff = {'before': '',
            'after': '',
            'before_header': '%s (content)' % dest,
            'after_header': '%s (content)' % dest}

    # 读取文件内容，按行生成列表
    with open(b_dest, 'rb') as f:
        b_lines = f.readlines()

    if module._diff:
        diff['before'] = to_native(b''.join(b_lines))

    if regexp is not None:
        # 编译正则表达式
        bre_c = re.compile(to_bytes(regexp, errors='surrogate_or_strict'))
    # 记录匹配的行
    found = []

    b_line = to_bytes(line, errors='surrogate_or_strict')
```

```
def matcher(b_cur_line):
    if regexp is not None:
        # 调用 search()方法搜索匹配行
        match_found = bre_c.search(b_cur_line)
    else:
        # 如果没有设置正则表达式，则使用行全量比较
        match_found = b_line == b_cur_line.rstrip(b'\r\n')
    if match_found:
        # 找到了就将这行加入 found 列表中
        found.append(b_cur_line)

    # 返回的是 match_found 的反向值
    return not match_found

# 对于不匹配的行，matcher()方法返回 True，于是 b_lines 表示的就是文本中所有不匹
    配 regexp 的行
b_lines = [l for l in b_lines if matcher(l)]
changed = len(found) > 0

if module._diff:
    diff['after'] = to_native(b''.join(b_lines))

backupdest = ""
if changed and not module.check_mode:
    if backup:
        # 备份文件
        backupdest = module.backup_local(dest)
    # 更新 dest 内容，全部为不匹配 regexp 的行，删除所有匹配 regexp 的行
    write_changes(module, b_lines, dest)

if changed:
    msg = "%s line(s) removed" % len(found)

attr_diff = {}
msg, changed = check_file_attrs(module, changed, msg, attr_diff)

attr_diff['before_header'] = '%s (file attributes)' % dest
attr_diff['after_header'] = '%s (file attributes)' % dest

difflist = [diff, attr_diff]

module.exit_json(changed=changed, found=len(found), msg=msg, backup=
backupdest, diff=difflist)
```

上述的 absent()函数相比 present()函数要简单很多，其运行逻辑如下：

（1）判断目标文件是否存在，若不存在，则直接返回运行错误的提示信息。

（2）读取文件内容，按行形成列表。

（3）根据 regexp 对每行进行正则匹配（如果 regexp 未设置，则每行内容和 line 全量
比较）。如果匹配到该行则加入 found 列表，剩余不匹配的行全部归入 b_lines 中。

（4）将不匹配的行重新写入目标文件中，等于删除了匹配 regexp 的行。

3．systemd模块

systemd 模块用于管理远端系统上的服务。该模块使用简单，操作方便，在远端节点的服务启/停上使用频繁。下面介绍该模块的常用参数。

- daemon_reload：该参数值有两个，即 yes 和 no，默认值为 no。yes 表示在做任何操作前执行 daemon-reload 命令，确保服务配置加载成功。
- enabled：该参数值有两个，即 yes 和 no，用于设定是否将操作的服务加入自启动列表中。
- force：该参数值有两个，即 yes 和 no，用于设定是否替换已存在的系统链接。
- name：管理的服务名称。
- scope：该参数值有 3 个，分别是 system、user 和 global，用于指定运行 systemctl 命令的作用域，如默认系统作用域（system）、当前用户的作用域（user）或所有用户的作用域（global）。
- state：该参数值 reloaded、restarted、started 和 stopped，分别表示服务重新加载、重启、启动和停止。

首先来看官方给出的示例，具体代码如下：

```
- name: Make sure a service is running
  systemd:
    state: started
    name: httpd

- name: stop service cron on debian, if running
  systemd:
    name: cron
    state: stopped

- name: restart service cron on centos, in all cases, also issue daemon-reload
to pick up config changes
  systemd:
    state: restarted
    daemon_reload: yes
    name: crond

- name: reload service httpd, in all cases
  systemd:
    name: httpd
    state: reloaded
```

上面是对服务（如 httpd、crond 等）执行启动、停止、重启和重新加载的操作，使用起来非常方便。相应的 ansible 命令如下：

```
(ansible2.8.16) [root@master ansible-2.8.16]# ansible ceph-1 -i hosts -m
systemd -a "name=httpd state=started enabled=yes" -b
(ansible2.8.1ss6) [root@master ansible-2.8.16]# ansible ceph-1 -i hosts -m
systemd -a "name=cron state=stopped" -b
```

⌂注意：对服务的启动和停止等操作需要 root 权限，因此普通用户（具有 sudo 权限）访问时需要带上 -b 选项进行提权。对于非 sudo 免密的用户来说，需要指定 ansible_sudo_pass 变量的值以处理需要输入 sudo 密码的情况。

systemd 模块对应的文件为 lib/ansible/modules/system/systemd.py，该文件的内容如下：

```python
# 源码位置：lib/ansible/modules/system/systemd.py
# ...

def main():
    # 实例化 AnsibleModule 类
    module = AnsibleModule(...)

    # 获取系统中 systemctl 命令的全路径，服务的控制正是通过该命令完成的
    systemctl = module.get_bin_path('systemctl', True)

    # ...

    # 将一些参数转换成 systemctl 命令的选项
    if module.params['scope'] not in (None, 'system'):
        systemctl += " --%s" % module.params['scope']

    if module.params['no_block']:
        systemctl += " --no-block"

    if module.params['force']:
        systemctl += " --force"

    # 获取管理服务名称
    unit = module.params['name']
    rc = 0
    out = err = ''
    # 返回结果
    result = dict(
        name=unit,
        changed=False,
        status=dict(),
    )

    # 如果设置了 daemon_reload 的值为 yes 且不是 check 模式，则需要先执行重新加载服务
      配置的命令
    if module.params['daemon_reload'] and not module.check_mode:
        (rc, out, err) = module.run_command("%s daemon-reload" % (systemctl))
        if rc != 0:
            module.fail_json(msg='failure %d during daemon-reload: %s' %
(rc, err))

    # ...

    if unit:
        found = False
        is_initd = sysv_exists(unit)
        is_systemd = False
```

```
    # 先执行 systemctl 的 show，然后执行 list-unit-files 命令
    (rc, out, err) = module.run_command("%s show '%s'" % (systemctl,
unit))

    if request_was_ignored(out) or request_was_ignored(err):
        (rc, out, err) = module.run_command("%s list-unit-files '%s'" %
(systemctl, unit))
        if rc == 0:
            is_systemd = True

    elif rc == 0:
        # ...
    else:
        # Check for systemctl command
        module.run_command(systemctl, check_rc=True)

# 判断服务在系统中是否存在
found = is_systemd or is_initd

# 处理 masked 参数
# ...

# Enable/disable service startup at boot if requested
if module.params['enabled'] is not None:

    if module.params['enabled']:
        action = 'enable'
    else:
        action = 'disable'

    fail_if_missing(module, found, unit, msg='host')

    enabled = False
    # 先执行 systemctl is-enabled service_name 命令判断服务是否已经是自
      启动模式
    (rc, out, err) = module.run_command("%s is-enabled '%s'" %
(systemctl, unit))

    # 继续检查服务是否处于自启动状态
    if rc == 0:
        enabled = True
    elif rc == 1:
        if module.params['scope'] in (None, 'system') and \
                not module.params['user'] and \
                is_initd and \
                not out.strip().endswith('disabled') and \
                sysv_is_enabled(unit):
            enabled = True

    # 通过代码得到当前服务是否为自启动模式
    result['enabled'] = enabled

    if enabled != module.params['enabled']:
```

```
                result['changed'] = True
                if not module.check_mode:
                    # 如果是非检查模式且检查服务处于非自启动状态，则需要执行命令设置服
                        务为自启动状态
                    (rc, out, err) = module.run_command("%s %s '%s'" %
(systemctl, action, unit))
                    # 处理失败情况
                    # ...

                # 其实就是调整成参数中设置的 enabled 值, yes 对应 True, no 对应 False
                result['enabled'] = not enabled

    if module.params['state'] is not None:
        fail_if_missing(module, found, unit, msg="host")

        # 获取 state 参数值
        result['state'] = module.params['state']

        if 'ActiveState' in result['status']:
            # 重新调整 action, 以配合 systemctl 命令控制服务
            action = None
            if module.params['state'] == 'started':
                if not is_running_service(result['status']):
                    # 非运行状态时 started 才有效, 设置 action=start
                    action = 'start'
            elif module.params['state'] == 'stopped':
                if is_running_service(result['status']) or is_deactivating_
service(result['status']):
                    action = 'stop'
            else:
                # 其他情况（重启和重新加载）
                if not is_running_service(result['status']):
                    # 如果是非运行的服务, 则直接启动即可
                    action = 'start'
                else:
                    # 将 'restarted/reloaded' 中最后的 'ed' 去掉
                    action = module.params['state'][:-2]
                result['state'] = 'started'

            if action:
                result['changed'] = True
                if not module.check_mode:
                    # 非 check 模式下直接运行相应服务管理的命令
                    (rc, out, err) = module.run_command("%s %s '%s'" %
(systemctl, action, unit))
                    # 处理失败
                    # ...

    # check for chroot
    elif is_chroot(module):
        # 打印警告信息
```

```
                   # ...
         else:
               # 模块执行失败并返回
               # ...

    module.exit_json(**result)


if __name__ == '__main__':
    main()
```

在上面的模块源码中，笔者去掉了部分不常用的参数处理代码并添加了相应注释。整个 systemd 模块的执行流程非常简单，即封装 systemctl 命令去执行。该模块的执行逻辑是先找出 systemctl 命令的完整路径，然后将一些模块选项转换成 systemctl 命令的选项，如 force 参数对应--force。接着对于设置 daemon_reload=yes 和 daemon_reexec=yes 的情况要单独执行相应命令，如有异常，则直接返回模块运行失败的提示信息。然后处理 enabled 参数，对于没有 enabled 的服务，会使用 systemctl enable service_names 命令设置自启动模式，其中 service_name 表示服务名称。最后根据 state 参数对服务运行状态进行控制，运行的相关命令为 systemctl start service_name（或者是 stop、restart、reload）。

3.2.3　Ansible 2 模块的学习建议

对于 Ansible 2 的最新版本，Ansible 官网上有完整的文档和模块说明。而就 Ansible 2.8 而言，也可以在官网上找到其所有模块，如图 3-2 所示。

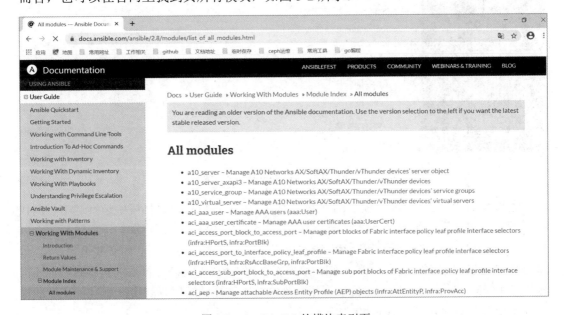

图 3-2　Ansible 2.8 的模块索引页

可以在图 3-2 所示的网页上搜索相应的模块，如搜索 systemd 模块，结果如图 3-3 所示，点进去就可以看到该模块的介绍页了，如图 3-4 所示。其中有完善的文档介绍，也有丰富的示例供使用者参考，如图 3-5 所示。

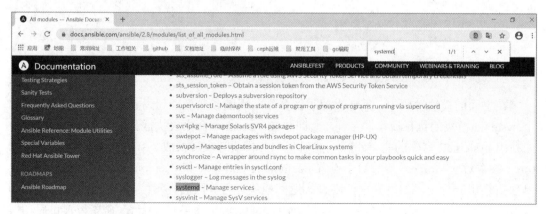

图 3-3　搜索 systemd 模块

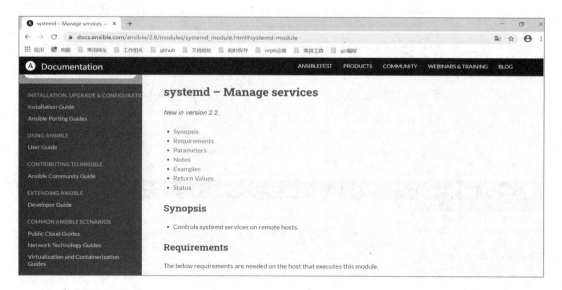

图 3-4　进入 systemd 模块介绍页

通过示例学习了 systemd 模块的基本使用方法后，可以在测试环境中运行这些示例代码并观察输出结果。如果想要学习 systemd 模块的更多用法，掌握模块的运行原理，可以阅读 lib/ansible/modules 目录下对应模块的源码，这是深入学习该模块的必要途径。最后，根据对 Ansible 1 的源码分析经验可知，如果想要深入理解 Ansible 模块的实现过程，还需要阅读 lib/ansible/plugins/action 目录下相应模块的动作插件源码。其实，Ansible 1 中的很多学习经验可以直接用于 Ansible 2 中。

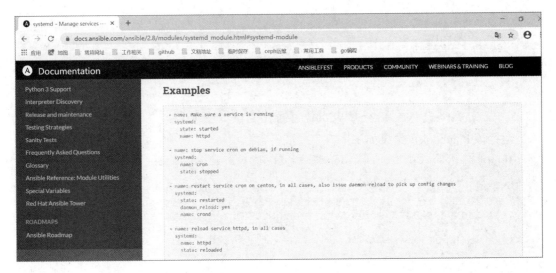

图 3-5　systemd 模块示例

3.3　Ansible 2 中的 Playbook 语法

本节将继续介绍 Ansible 2 中一些基础和高级的 Playbook 语法，其中大部分语法示例均来自于 Ansible 的优秀开源项目 ceph-ansible。在第 6 章中，笔者会详细介绍该项目并上手实践。本节内容正是为学习该项目打基础，因此需要认真学习。

3.3.1　变量

一个完整的 Playbook 项目必定会有大量的配置变量，这使得整个项目的部署更为灵活和强大。例如，设置生成的用户密码、选择启动的服务、控制配置文件生成的参数等。此外，还需要合理放置变量参数的位置，避免相互覆盖并方便修改。下面介绍 Ansible 的Playbook 中常用的设置变量的方式。

（1）在 hosts 文件中设置变量。下面是一个示例 hosts 文件：

```
(ansible2.8.16) [root@master playbook]# cat hosts
[nodes]
ceph-[1:3]

[nodes:vars]
ansible_ssh_pass=@SHENcong19920522
node_var=test_variable_in_hosts
```

接下来编写一个简单的 Playbook 文件，只打印 hosts 中设置的变量，具体内容如下：

```
---
- hosts: all
  gather_facts: false

  tasks:
    - name: debug node_var
      debug:
        msg: "node_var = {{ node_var }}"
```

运行该 Playbook，结果如下：

```
(ansible2.8.16) [root@master playbook]# ansible-playbook -i hosts test_
vars1.yml

PLAY [all]
****************************************************************************

TASK [debug node_var]
****************************************************************************
ok: [ceph-1] => {
    "msg": "node_var = test_variable_in_hosts"
}
ok: [ceph-2] => {
    "msg": "node_var = test_variable_in_hosts"
}
ok: [ceph-3] => {
    "msg": "node_var = test_variable_in_hosts"
}

PLAY RECAP
****************************************************************************

# 忽略汇总结果
# ...
```

（2）在当前目录下新建一个 group_vars 目录并在该目录下新建一个 all.yml 文件，表明这是一个全局的变量文件。其内容如下：

```
(ansible2.8.16) [root@master playbook]# cat group_vars/all.yml
node_var: test_variable_in_group_all_yaml
```

再次执行上面的 Playbook，结果如下：

```
(ansible2.8.16) [root@master playbook]# ansible-playbook -i hosts test_
vars1.yml

PLAY [all]
****************************************************************************

TASK [debug node_var]
****************************************************************************
ok: [ceph-1] => {
    "msg": "node_var = test_variable_in_group_all_yaml"
}
ok: [ceph-2] => {
    "msg": "node_var = test_variable_in_group_all_yaml"
```

```
}
ok: [ceph-3] => {
    "msg": "node_var = test_variable_in_group_all_yaml"
}

PLAY RECAP
***************************************************************************
# 忽略汇总结果
# ...
```

可以看到，group_vars 目录下的对应变量文件会覆盖 hosts 中设置的变量。

（3）将变量写入对应执行的 YAML 文件中

例如，在上面的 Playbook 文件中加上 vars 参数，具体代码如下：

```
(ansible2.8.16) [root@master playbook]# cat test_vars1.yml
---
- hosts: all
  gather_facts: false

  # 设置变量
  vars:
    node_var: test_variable_in_yaml

  tasks:
    - name: debug node_var
      debug:
        msg: "node_var = {{ node_var }}"
```

继续执行该 Playbook，执行结果如下：

```
(ansible2.8.16) [root@master playbook]# ansible-playbook -i hosts test_
vars1.yml

PLAY [all]
***************************************************************************

TASK [debug node_var]
***************************************************************************
ok: [ceph-2] => {
    "msg": "node_var = test_variable_in_yaml"
}
ok: [ceph-1] => {
    "msg": "node_var = test_variable_in_yaml"
}
ok: [ceph-3] => {
    "msg": "node_var = test_variable_in_yaml"
}

PLAY RECAP
***************************************************************************
# 忽略汇总结果
# ...
```

可以看到，YAML 文件中定义的变量值覆盖了 group_all/all.yml 和 hosts 中设置的变量值。

（4）运行时设置变量。这种方式在自动部署的 Playbook 中非常常见。它根据上一个任务的结果并结合 register 语法设置变量及其值，然后在后续的任务中使用该变量。修改上面的 Playbook，添加一个设置变量的任务，具体代码如下：

```
(ansible2.8.16) [root@master playbook]# cat test_vars1.yml
---
- hosts: all
  gather_facts: false

  vars:
    node_var: test_variable_in_yaml

  tasks:
    - name: set node_var
      set_fact:
        node_var: "test_variable_in_runtime"

    - name: debug node_var
      debug:
        msg: "node_var = {{ node_var }}"
```

再次运行该 Playbook，可以看到 node_var 的结果是上一个任务设置的变量值，具体代码如下：

```
(ansible2.8.16) [root@master playbook]# ansible-playbook -i hosts test_
vars1.yml

PLAY [all]
********************************************************************

TASK [set node_var]
********************************************************************
ok: [ceph-1]
ok: [ceph-2]
ok: [ceph-3]

TASK [debug node_var]
********************************************************************
ok: [ceph-1] => {
    "msg": "node_var = test_variable_in_runtime"
}
ok: [ceph-2] => {
    "msg": "node_var = test_variable_in_runtime"
}
ok: [ceph-3] => {
    "msg": "node_var = test_variable_in_runtime"
}

PLAY RECAP
********************************************************************
# 忽略汇总结果
# ...
```

（5）命令行带变量，即通过-e 选项直接传递相应的变量及其值。执行如下 ansible-playbook 命令：

```
(ansible2.8.16) [root@master playbook]# ansible-playbook -i hosts test_
vars1.yml -e "node_var=test_variable_in_console"

PLAY [all]
**************************************************************************

TASK [debug node_var]
**************************************************************************
ok: [ceph-1] => {
    "msg": "node_var = test_variable_in_console"
}
ok: [ceph-2] => {
    "msg": "node_var = test_variable_in_console"
}
ok: [ceph-3] => {
    "msg": "node_var = test_variable_in_console"
}

PLAY RECAP
**************************************************************************
# 忽略汇总结果
# ...
```

可以看到，通过-e 选项设置的变量优先级最高，直接覆盖了 Playbook 中上一个任务设定的变量值。

（6）在角色目录下的 defaults 目录或者 vars 目录下的 main.yml 中设置变量。

这个目录下保存的变量一般是这个角色任务的变量，变量优先级较低，容易被主 playbook 或者 hosts 等文件中设置的变量所覆盖。为了能展示相应的结果，下面新建一个简单的角色并打印相应的变量值，具体操作如下：

```
(ansible2.8.16) [root@master playbook]# mkdir -p roles/test-variable/tasks
(ansible2.8.16) [root@master playbook]# mkdir -p roles/test-variable/
defaults
(ansible2.8.16) [root@master playbook]# mkdir -p roles/test-variable/vars
```

接着准备角色下的 3 个 main.yml 文件，具体内容如下：

```
(ansible2.8.16) [root@master playbook]# cat roles/test-variable/defaults/
main.yml
node_var: test_variable_in_role_defaults
(ansible2.8.16) [root@master playbook]# cat roles/test-variable/vars/
main.yml
node_var: test_variable_in_role_vars
(ansible2.8.16) [root@master playbook]# cat roles/test-variable/tasks/
main.yml
---
- name: debug node_var
```

```
    debug:
      msg: "node_var = {{ node_var }}"
```

调整 Playbook 的主运行文件的内容，具体代码如下：

```
(ansible2.8.16) [root@master playbook]# cat test_vars1.yml
---
- hosts: ceph-1
  gather_facts: false

  vars:
    node_var: test_variable_in_yaml

  tasks:
    - name: set node_var
      set_fact:
        node_var: "test_variable_in_runtime"

    - import_role:
        name: test-variable
```

接着就可以通过不断地注释及运行 Playbook 来查看相关变量的优先级了。具体的运行方式可以参考前面的过程，这里就不再详细介绍了。

经过反复测试，可以得到 Ansible 2 中变量的优先级：角色 default 目录< hosts 文件< group_vars/all.yml < YAML 中的 vars 参数设置的变量<角色 vars 目录<运行时 task 设置的变量<ansible-playbook 命令直接传入的变量及其参数。

3.3.2　条件判断

前面在 2.2 节中介绍 Ansible 1.9.6 的用法时已经介绍过 Ansible 中的条件判断语法了，这在 Ansible 2 中几乎没有任何变化。本节将使用 ceph-ansible 项目中的几个典型写法来辅助读者进一步加强 when 语法的使用。

示例 1：简单的 bool 类型判断，配合 bool 过滤将结果强制转换成 bool 类型。

```
- name: set_fact container_exec_cmd
  set_fact:
    container_exec_cmd: "{{ container_binary }} exec ceph-mon-{{ ansible_
hostname }}"
  when: containerized_deployment | bool
```

示例 2：根据返回结果判断，可以是字典形式，配合 not 对 bool 结果取反。

```
- name: "show ceph status for cluster {{ cluster }}"
  debug:
    msg: "{{ ceph_status.stdout_lines }}"
    delegate_to: "{{ groups[mon_group_name][0] }}"
  run_once: true
  when: not ceph_status.failed
```

示例 3：组合判断。下面第一个任务使用了自定义的模块，对于判断变量是否有定义可以使用 is defined 方式，此外还可以使用==、!=、> 、<、in、not in 等语句进行逻辑

判断。

```
- name: add ceph-mgr systemd service overrides
  config_template:
    src: "ceph-mgr.service.d-overrides.j2"
    dest: "/etc/systemd/system/ceph-mgr@.service.d/ceph-mgr-systemd-
overrides.conf"
    config_overrides: "{{ ceph_mgr_systemd_overrides | default({}) }}"
    config_type: "ini"
  when:
    - ceph_mgr_systemd_overrides is defined
    - ansible_service_mgr == 'systemd'

- import_role:
    name: ceph-osd
    tasks_from: container_options_facts.yml
  when: inventory_hostname in groups.get(osd_group_name, [])
```

示例 4：使用 Python 函数并综合过滤条件进行判断。下面的任务中通过 groups.get() 函数获取 Playbook 中目标主机 mgr_group_name（变量）组下的 hosts 列表，并通过列表的长度判断该任务是否需要执行。

```
- name: fetch ceph mgr keyring
  ceph_key:
    name: "mgr.{{ ansible_hostname }}"
    state: present
    caps:
      mon: allow profile mgr
      osd: allow *
      mds: allow *
    cluster: "{{ cluster }}"
    secret: "{{ (mgr_secret != 'mgr_secret') | ternary(mgr_secret, omit) }}"
    owner: "{{ ceph_uid if containerized_deployment else 'ceph' }}"
    group: "{{ ceph_uid if containerized_deployment else 'ceph' }}"
    mode: "0400"
    dest: "/var/lib/ceph/mgr/{{ cluster }}-{{ ansible_hostname }}/keyring"
  environment:
    CEPH_CONTAINER_IMAGE: "{{ ceph_docker_registry + '/' + ceph_docker_
image + ':' + ceph_docker_image_tag if containerized_deployment else None }}"
    CEPH_CONTAINER_BINARY: "{{ container_binary }}"
  when: groups.get(mgr_group_name, []) | length == 0
```

此外，在编写 Playbook 时，可以对字符串执行一些 Python 操作，例如：

```
- name: set yum package dir path
  set_fact:
    data_dir: "{{ yum_source.split('/')[:-1] | join('/') }}"
```

上面的 yum_source 是一个表示路径的字符串，如/data/test/yum_source。对该字符串调用 split()函数去掉斜线分割的最后一个字符串，对于 yum_source="/data/test/yum_source" 来说，yum_source.split('/')[:-1]的结果就是['', 'data', 'test']，最后使用 join('/')过滤后得到的结果为'/data/test'。这些都是 Python 中的字符串方法，都可以在 when 语句中进行使用。下面是一个简单的 Playbook 文件，内容如下：

```
(ansible2.8.16) [root@master playbook]# cat test_vars2.yml
---
- hosts: ceph-1
  gather_facts: false

  tasks:
    - name: set node_var
      set_fact:
        node_var: "test_variable_in_runtime"

    - name: debug node_var
      debug:
        msg: "node_var = {{ node_var }}"
      when: node_var.endswith('runtime') | bool
```

运行该 Playbook，结果如下：

```
(ansible2.8.16) [root@master playbook]# ansible-playbook -i hosts test_
vars1.yml

PLAY [ceph-1]
*************************************************************************

TASK [set node_var]
*************************************************************************
ok: [ceph-1]

TASK [debug node_var]
*************************************************************************
ok: [ceph-1] => {
    "msg": "node_var = test_variable_in_runtime"
}

PLAY RECAP
*************************************************************************
# 忽略汇总结果
# ...
```

从上面的代码中可以看到，node_var 变量的值为 test_variable_in_runtime，故 node_var.endswith('runtime')的值为 True，因此后面的任务就会执行。接着对 ansible-playbook 使用-e 参数改变 node_var 变量的值，运行结果如下：

```
(ansible2.8.16) [root@master playbook]# ansible-playbook -i hosts test_
vars1.yml -e "node_var=hello"

PLAY [ceph-1]
*************************************************************************

TASK [set node_var]
*************************************************************************
ok: [ceph-1]

TASK [debug node_var]
*************************************************************************
skipping: [ceph-1]
```

```
PLAY RECAP
********************************************************************
# 忽略汇总结果
# ...
```

通过 3.3.1 节的介绍可知，在 ansible-playbook 命令行中设置的参数优先级最高，因此这里 node_var 的结果为 hello，于是 node_var.endswith('runtime') 的结果为 False，任务被跳过不执行。

示例 5：多个表达式结合。下面的任务中接连使用了 and、or、is defined 及 Python 中的字典的 get() 方法，是一个非常典型的参考样例。

```
- name: include deploy_monitors.yml
  include_tasks: deploy_monitors.yml
  when:
    - (mon_socket_stat is defined and mon_socket_stat.get('rc') != 0) or
(ceph_mon_container_stat is defined and ceph_mon_container_stat.get
('stdout_lines', [])|length == 0)
    - not switch_to_containers | default(False) | bool
```

3.3.3　循环语句

Playbook 中的循环语法类似 Python 中的 for 语句，下面是 Ansible 2 中几种典型的循环语法示例。

示例 1：with_items 语法。类似于 Python 中的 for key in data 语句，其中，key 默认为 item 变量。

```
- name: remove data
  file:
    name: "{{ item }}"
    state: absent
  with_items:
    - /etc/grafana/dashboards
    - /etc/grafana/grafana.ini
    - /etc/grafana/provisioning
    - /var/lib/grafana
    - /etc/alertmanager
    - /var/lib/alertmanager
    - /var/lib/prometheus
    - /etc/prometheus
  failed_when: false
```

上面这个示例非常简单，中间多了一个 failed_when 语法，表示任务执行失败时将忽略错误继续执行。为了能更好地展示 with_items 的用法，给出 Playbook 示例如下：

```
(ansible2.8.16) [root@master playbook]# cat test_loop1.yml
- hosts: ceph-1
  gather_facts: false

  vars:
```

```
    test_item_vars:
      - "hello1"
      - "hello2"
      - "hello3"

    test_split_vars: "1#2#3#4"

  tasks:
    - name: test simple for one
      debug:
        msg: "item = {{ item }}"
      with_items:
        - "hello"
        - "world"
        - "xyz"

    - name: test simple for two
      debug:
        msg: "item = {{ item }}"
      with_items: "{{ test_item_vars }}"

    - name: test simple for three
      debug:
        msg: "item = {{ item }}"
      with_items: "{{ test_split_vars.split('#') }}"
```

上面的 Playbook 中准备了 3 个演示 for 循环的任务。第 1 个任务是常规的 with_items 用法，直接在该语句后面添加循环的列表值；第 2 个任务是传给 with_items 一个为列表的变量；第 3 个任务是传给 with_items 一个经过处理后得到的列表值。这 3 种情况都能实现 for 循环功能。以下是该 Playbook 的执行结果：

```
(ansible2.8.16) [root@master playbook]# ansible-playbook -i hosts test_
loop1.yml

PLAY [ceph-1]
********************************************************************************

TASK [test simple for one]
********************************************************************************
ok: [ceph-1] => (item=hello) => {
    "msg": "item = hello"
}
ok: [ceph-1] => (item=world) => {
    "msg": "item = world"
}
ok: [ceph-1] => (item=xyz) => {
    "msg": "item = xyz"
}

TASK [test simple for two]
********************************************************************************
ok: [ceph-1] => (item=hello1) => {
    "msg": "item = hello1"
}
```

```
ok: [ceph-1] => (item=hello2) => {
    "msg": "item = hello2"
}
ok: [ceph-1] => (item=hello3) => {
    "msg": "item = hello3"
}

TASK [test simple for three]
*********************************************************************
ok: [ceph-1] => (item=1) => {
    "msg": "item = 1"
}
ok: [ceph-1] => (item=2) => {
    "msg": "item = 2"
}
ok: [ceph-1] => (item=3) => {
    "msg": "item = 3"
}
ok: [ceph-1] => (item=4) => {
    "msg": "item = 4"
}

PLAY RECAP
*********************************************************************
# 忽略汇总结果
# ...
```

上面的 with_*语法其实使用了 Ansible 中的 lookup 插件，例如 with_items 语法对应 lib/ansible/plugins/lookup 目录下的 items.py。在 Ansible 中有一个更加简单的循环语法——loop，该语法的使用和 with_items 几乎一致。

示例 2：loop 语法，它和 with_items 语法类似。

在前面的 test_loop1.yml 文件中，将所有的 with_items 替换成 loop 后再次运行，其结果和示例 1 是相同的。但 loop 和 with_items 语法还是有一些区别的，具体如下：

- loop 语法不支持字符串变量，而 with_items 语法会将字符串变量作为列表的一个元素。
- 对于以元素为列表的情况，loop 语法并不支持展开列表，会直接将其当作一个整体进行循环；而 with_items 语法则会将该列表展开循环。如果 loop 语法想要支持进一步展开操作，则需要将元素进行 flatten(1)过滤。

下面针对 loop 语法和 with_items 语法的不同，笔者编写了一个测试的 Playbook，具体内容如下：

```
- hosts: ceph-1
  gather_facts: false

  vars:
    test_str: "hello, world"
    test_item_vars:
      - 1
      - [2, 3]
```

```
      - 4
  tasks:
    - name: test with_item for str
      debug:
        msg: "item = {{ item }}"
      with_items: "{{ test_str }}"

    - name: test loop for str
      debug:
        msg: "item = {{ item }}"
      loop: "{{ test_str }}"
      ignore_errors: true

    - name: test simple for one
      debug:
        msg: "item = {{ item }}"
      with_items: "{{ test_item_vars }}"

    - name: test loop
      debug:
        msg: "item = {{ item }}"
      loop: "{{ test_item_vars }}"

    - name: test loop two
      debug:
        msg: "item = {{ item }}"
      loop: "{{ test_item_vars | flatten(1) }}"
```

由于 loop 值为字符串时会报错，因此在该任务中添加了 ignore_errors 标识来忽略本次错误以保证 Playbook 正常运行下去。以下是该 Playbook 的运行结果：

```
(ansible2.8.16) [root@master playbook]# ansible-playbook -i hosts test_
loop2.yml

PLAY [ceph-1]
********************************************************************************

TASK [test with_item for str]
********************************************************************************
ok: [ceph-1] => (item=hello, world) => {
    "msg": "item = hello, world"
}

TASK [test loop for str]
********************************************************************************
fatal: [ceph-1]: FAILED! => {"msg": "Invalid data passed to 'loop', it
requires a list, got this instead: hello, world. Hint: If you passed a
list/dict of just one element, try adding wantlist=True to your lookup
invocation or use q/query instead of lookup."}
...ignoring

TASK [test simple for one]
********************************************************************************
ok: [ceph-1] => (item=1) => {
```

```
        "msg": "item = 1"
}
ok: [ceph-1] => (item=2) => {
    "msg": "item = 2"
}
ok: [ceph-1] => (item=3) => {
    "msg": "item = 3"
}
ok: [ceph-1] => (item=4) => {
    "msg": "item = 4"
}

TASK [test loop]
***********************************************************************
ok: [ceph-1] => (item=1) => {
    "msg": "item = 1"
}
ok: [ceph-1] => (item=[2, 3]) => {
    "msg": "item = [2, 3]"
}
ok: [ceph-1] => (item=4) => {
    "msg": "item = 4"
}

TASK [test loop two]
***********************************************************************
ok: [ceph-1] => (item=1) => {
    "msg": "item = 1"
}
ok: [ceph-1] => (item=2) => {
    "msg": "item = 2"
}
ok: [ceph-1] => (item=3) => {
    "msg": "item = 3"
}
ok: [ceph-1] => (item=4) => {
    "msg": "item = 4"
}

PLAY RECAP
***********************************************************************
# 忽略汇总结果
# ...
```

最后还有一个小问题，如何才能得到循环的位置索引值呢？来看下面这个简单的任务：

```
(ansible2.8.16) [root@master playbook]# cat test_loop3.yml
- hosts: ceph-1
  gather_facts: false

  tasks:
    - name: Count our fruit
      debug:
        msg: "{{ item }} with index {{ my_idx }}"
```

```
        loop:
          - apple
          - banana
          - pear
        loop_control:
          index_var: my_idx
```

这里循环的位置索引值被赋给了 my_index 变量。以下是任务的执行结果：

```
(ansible2.8.16) [root@master playbook]# ansible-playbook -i hosts test_
loop3.yml

PLAY [ceph-1]
***************************************************************************

TASK [Count our fruit]
***************************************************************************
ok: [ceph-1] => (item=apple) => {
    "msg": "apple with index 0"
}
ok: [ceph-1] => (item=banana) => {
    "msg": "banana with index 1"
}
ok: [ceph-1] => (item=pear) => {
    "msg": "pear with index 2"
}

PLAY RECAP
***************************************************************************
# 忽略汇总结果
# ...
```

更多的循环用法可以参考官方文档，网址为 https://docs.ansible.com/ansible/2.8/user_guide/playbooks_loops.html。

3.3.4 等待语句

本节的等待语句主要涉及 until 语法及 wait_for 模块。首先来看下面的典型示例。

```
- name: exit playbook, if can not connect to the cluster
  command: "{{ container_exec_cmd | default('') }} timeout 5 ceph --cluster
{{ cluster }} health"
  register: ceph_health
  until: ceph_health is succeeded
  retries: 5
  delay: 2

- name: container - waiting for clean pgs...
  command: >
    {{ container_binary }} exec ceph-mon-{{ hostvars[groups[mon_group_name]
[0]]['ansible_hostname'] }} ceph --cluster {{ cluster }} -s --format json
  register: ceph_health_post
  until: >
    (((ceph_health_post.stdout | from_json).pgmap.pgs_by_state | length) > 0)
```

```
    and
    (((ceph_health_post.stdout | from_json).pgmap.pgs_by_state | selectattr
('state_name', 'search', '^active\\+clean') | map(attribute='count') |
list | sum) == (ceph_pgs.stdout | from_json).pgmap.num_pgs)
    delegate_to: "{{ groups[mon_group_name][0] }}"
    retries: "{{ health_osd_check_retries }}"
    delay: "{{ health_osd_check_delay }}"
    when: (ceph_pgs.stdout | from_json).pgmap.num_pgs != 0
```

上面是两个使用了 until 语法的任务。第 1 个任务非常简单，就是执行命令并将结果赋给 ceph_health 参数。该任务直到命令执行成功后才结束，否则会多次重试，最大重试次数为 5，重试间隔为 2s。第 2 个任务则较为复杂，不过 until 语法的核心逻辑比较简单，即会等到相应的语句执行成功后才能结束任务，相应的语句表达式中可以使用 and 和 or 等逻辑运算符。为了能更好地理解 until 语法，下面给出一个简单的 Playbook 示例，具体代码如下：

```
(ansible2.8.16) [root@master playbook]# cat test_until.yml
- hosts: ceph-1
  gather_facts: false

  tasks:
    - name: remove test.txt
      file:
        path: /root/test.txt
        state: absent

    # 后台异步执行，5s 后创建/root/test.txt 文件
    - name: create test.txt five seconds later
      shell: sleep 5 && touch /root/test.txt
      async: 1
      poll: 0

    - name: test utils test.txt exiests
      command: ls /root/test.txt
      register: test_file_exist
      until: test_file_exist is succeeded
      retries: 5
      delay: 2
```

上面的 Playbook 首先是移除 ceph-1 节点上的/root/test.txt 文件，确保它不存在；其次是使用 shell 命令创建/root/test.txt 文件，不过采用了异步方式，直接返回，并且创建 test.txt 文件的命令将在 5s 后执行；最后是执行 ls /root/test.txt 命令，一开始该命令会报错，此时 until 语法发挥作用，开始延迟重试，直到第三次重试后发现文件已经创建，此时任务执行完毕。以下是该 Playbook 的运行结果：

```
(ansible2.8.16) [root@master playbook]# ansible-playbook -i hosts test_
until.yml

PLAY [ceph-1]
*******************************************************************************
```

```
TASK [remove test.txt]
*************************************************************************
changed: [ceph-1]

TASK [create test.txt five seconds later]
*************************************************************************
changed: [ceph-1]

TASK [test utils test.txt exiests]
*************************************************************************
FAILED - RETRYING: test utils test.txt exiests (5 retries left).
FAILED - RETRYING: test utils test.txt exiests (4 retries left).
FAILED - RETRYING: test utils test.txt exiests (3 retries left).
changed: [ceph-1]

PLAY RECAP
*************************************************************************
# 忽略汇总结果
# ...
```

接下来是 wait_for 模块及其用法介绍，该模块的功能是等设置条件满足后再执行。其常用参数有以下几个。

- connect_timeout：连接超时时间。
- delay：开始轮询之前等待的秒数。
- host：等待的主机。
- path：指定文件路径，当该文件存在时才能继续。
- port：要轮询的端口号，它和 path 是一对互斥的参数。
- sleep：每次检查间隔的秒数。
- state：选项值有 absent、drained、present、started 和 stopped，默认为 started。对于端口（设置 port 参数）检查而言，started 检查端口是否打开，stopped 检查端口是否关闭，drained 将检查活动的连接情况。对于设置 path 参数而言，absent 和 present 分别用于表示 path 路径的文件被删除或者存在时才执行 wait_for 模块。

有了上面的铺垫之后，下面给出一个示例 Playbook，其内容如下：

```
(ansible2.8.16) [root@master playbook]# cat test_wait_for.yml
- hosts: ceph-1
  gather_facts: false

  handlers:
  - name: restart machine
    shell: sleep 2 && shutdown -r now "Ansible updates triggered"
    async: 1
    poll: 0
    ignore_errors: true

  - name: wait for server to boot
    become: false
    local_action:
      module: wait_for
```

```
    port: 22
    host: "{{ inventory_hostname }}"
    state: started
    delay: 10
    timeout: 500

 - name: remove data
   shell: rm -rf /var/lib/ceph/*

tasks:
- name: is reboot needed
  local_action:
    module: command
      echo requesting reboot
  become: false
  notify:
    - restart machine
    - wait for server to boot
    - remove data
```

上面的任务主要是设置通知回调的 handlers。任务执行完毕后将依次执行以下 3 个回调任务：

（1）异步重启主机。

（2）等待主机重启完毕，正是使用了 wait_for 模块。

（3）重启主机后删除一些数据目录。

注意指定 port 为 22，这是 SSH 监听的端口，通过它来判断服务是否已重启完毕。该 Playbook 的运行效果如下：

```
(ansible2.8.16) [root@master playbook]# ansible-playbook -i hosts test_
wait_for.yml

PLAY [ceph-1]
***********************************************************************

TASK [is reboot needed]
***********************************************************************
changed: [ceph-1 -> localhost]

RUNNING HANDLER [restart machine]
***********************************************************************
changed: [ceph-1]

RUNNING HANDLER [wait for server to boot]
***********************************************************************
ok: [ceph-1 -> localhost]

RUNNING HANDLER [remove data]
***********************************************************************
[WARNING]: Consider using the file module with state=absent rather than
running 'rm'. If you need to use command because file is insufficient you
can add 'warn: false' to this
command task or set 'command_warnings=False' in ansible.cfg to get rid of
this message.
```

```
changed: [ceph-1]

PLAY RECAP
**********************************************************************
# 忽略汇总结果
# ...
```

大约等待十几秒后（虚拟机较快），服务器重启完毕，对应的任务执行完成，然后会继续执行下一个任务。

3.4 本 章 小 结

本章简要介绍了 Ansible 2.8 的一些常用模块的使用并剖析了它们的实现源码，接着又介绍了一些常用的 Playbook 语法并给出了大量 ceph-ansible 中的典型实例，同时针对一些较为复杂的语法进行了实际测试。本章学习起来较为轻松，没有复杂的代码剖析，但是这些内容关系到第 4～6 章的学习，因此非常重要。在第 4 章中介绍 Ansible 2 的源码时会分析这部分 Playbook 语法的实现代码，第 5 章溯源 lineinfile 模块的相关 Bug 时，会以这里分析的 lineinfile 模块源码为重要依据，并且第 6 章中将要介绍的 ceph-ansible 项目中的许多任务原型，都在本章作为典型示例进行了详细介绍。

第 4 章　Ansible 2 核心源码剖析

本章将剖析 Ansible 2 的核心源码。具体的分析流程和 Ansible 1 几乎一致，从 Ansible 的核心源码介绍开始，到 Ansible 模块的运行流程分析以及 Playbook 语法分析，帮助读者逐步深入理解 Ansible 工具的核心逻辑。本次分析的 Ansible 版本为 2.8.16，该版本的测试环境已在第 3 章中搭建完成。

4.1　Ansible 2.8 核心源码初探

本节将从实战角度介绍 Ansible 2.8 中的部分源码内容，通过实际调用和测试来帮助读者理解 Ansible 的源码，为后续剖析 Ansible 2 的模块运行流程和 Playbook 执行做好前期准备。

4.1.1　日志打印

首先介绍在第 3 章中提到的关于使用-v/-vv/-vvv 选项打印调试信息的实现原理。其中涉及的代码主要位于 lib/ansible/utils/display.py 文件中，涉及的类是 Display。以下是 Display 类的实现源码：

```
# 源码位置: lib/ansible/utils/display.py
# ...

class Display(with_metaclass(Singleton, object)):

    def __init__(self, verbosity=0):

        self.columns = None
        self.verbosity = verbosity

        #...

    # 6 个级别的输出方法
    def v(self, msg, host=None):
        return self.verbose(msg, host=host, caplevel=0)

    def vv(self, msg, host=None):
        return self.verbose(msg, host=host, caplevel=1)
```

```
def vvv(self, msg, host=None):
    return self.verbose(msg, host=host, caplevel=2)

def vvvv(self, msg, host=None):
    return self.verbose(msg, host=host, caplevel=3)

def vvvvv(self, msg, host=None):
    return self.verbose(msg, host=host, caplevel=4)

def vvvvvv(self, msg, host=None):
    return self.verbose(msg, host=host, caplevel=5)

# ...

def verbose(self, msg, host=None, caplevel=2):

    to_stderr = C.VERBOSE_TO_STDERR
    # 判断 self.verbosity 与输入级别的大小，判断是否打印
    if self.verbosity > caplevel:
        if host is None:
            self.display(msg, color=C.COLOR_VERBOSE, stderr=to_stderr)
        else:
            self.display("<%s> %s" % (host, msg), color=C.COLOR_VERBOSE,
stderr=to_stderr)

    # ...
```

前面介绍过，Ansible 会根据-vv 选项中 v 出现的次数得到选项值 verbosity。该值正好用来实例化 Display 类。来看下面的操作：

```
>>> from ansible.utils.display import Display
>>> display = Display(2)
>>> display.v('hello, world')
hello, world
>>> display.vv('hello, world')
hello, world
>>> display.vvv('hello, world')
>>> display.vvvv('hello, world')
>>> display.vvvvv('hello, world')
>>>
```

从上面的操作中可以看到，当 Display 对象的 verbosity 值为 2 时，调用对象的 v()和 vv()函数有输出，而更高等级的 vvv()、vvvv()和 vvvvv()函数则没有输出。这个判断逻辑在 verbose()方法中一清二楚。在 Ansible 代码中，会有大量的 v()和多 v 的函数在其中打印调试信息。但实际运行时，只有小于等于 verbosity 等级的 v()函数才能打印出相应的结果，这便是 Ansible 中使用-v 选项控制打印日志的原理。

4.1.2 默认配置

还记得 Ansible 1 中的 constants.py 文件吗？这里定义了 Ansible 中所有的默认配置及

其值，如默认的模块 DEFAULT_MODULE_NAME、默认启动的最大进程数 DEFAULT_FORKS 等。而到了 Ansible 2 中，constants.py 文件依然存在，不过里面的内容已经有了很大变化，虽然有少部分常量定义，但原先的默认模块、默认进程数等并没有在该文件中直接定义，而是通过 YAML 文件进行配置。具体的实现代码如下：

```
# 源码位置：lib/ansible/constants.py
# ...

# 实例化配置管理类
config = ConfigManager()

# 为了能理解 for 循环代码，需要先知道 config.data.get_settings() 的值
for setting in config.data.get_settings():

    value = setting.value
    if setting.origin == 'default' and \
      isinstance(setting.value, string_types) and \
      (setting.value.startswith('{{') and setting.value.endswith('}}')):
        try:
            t = Template(setting.value)
            value = t.render(vars())
            try:
                value = literal_eval(value)
            except ValueError:
                pass
        except Exception:
            pass

        value = ensure_type(value, setting.type)

    # 设置默认配置的值
    set_constant(setting.name, value)

# ...
```

为了能理解上面读取配置的代码，下面先来看下 lib/ansible/config/data.py 文件中的 ConfigData 类，后面会看到该类与这里代码的联系。

```
# 源码位置：lib/ansible/config/data.py
# ...

class ConfigData(object):

    def __init__(self):
        # 全局配置
        self._global_settings = {}
        self._plugins = {}

    def get_setting(self, name, plugin=None):
        # 获取配置值
        setting = None
        if plugin is None:
            setting = self._global_settings.get(name)
```

```
            elif plugin.type in self._plugins and plugin.name in self._plugins
        [plugin.type]:
                setting = self._plugins[plugin.type][plugin.name].get(name)

            return setting

        def get_settings(self, plugin=None):
            # 获取所有的配置值，组成列表
            settings = []
            if plugin is None:
                settings = [self._global_settings[k] for k in self._global_
        settings]
            elif plugin.type in self._plugins and plugin.name in self._plugins
        [plugin.type]:
                settings =
            [self._plugins[plugin.type][plugin.name][k] for k in self._plugins
        [plugin.type][plugin.name]]

            return settings

        def update_setting(self, setting, plugin=None):
            # 更新配置信息
            if plugin is None:
                self._global_settings[setting.name] = setting
            else:
                if plugin.type not in self._plugins:
                    self._plugins[plugin.type] = {}
                if plugin.name not in self._plugins[plugin.type]:
                    self._plugins[plugin.type][plugin.name] = {}
                self._plugins[plugin.type][plugin.name][setting.name] = setting
```

上面的代码比较简单，就是用 self._global_settings 来保存全局的参数，然后查询和更新全局配置，其中如果涉及插件，配置表达式会稍微复杂一些。ConfigData 类的测试示例如下：

```
>>> from ansible.config.data import ConfigData
>>> from ansible.config.manager import Setting
>>> set1 = Setting('test', 'xxx', None, 'string')
>>> set2 = Setting('test2', '22', None, 'int')
>>> data.update_setting(set1)
>>> data.update_setting(set2)
>>> data.get_setting('test')
Setting(name='test', value='xxx', origin=None, type='string')
>>> data.get_settings()
[Setting(name='test2', value='22', origin=None, type='int'), Setting
(name='test', value='xxx', origin=None, type='string')]
```

🔔注意：setting 是 ansible/config/manager.py 中定义的一个具名元组（namedtuple），它有 4 个域，分别是 name、value、origin 及 type。后面会看到，它和配置文件（base.yml）中定义的默认配置字段是一一对应的。

接下来追踪 config= ConfigManager()语句背后的执行过程。ConfigManager 类的实现位

于 lib/ansible/config/manager.py 文件中，具体代码如下：

```python
# 源码位置: lib/ansible/config/manager.py
# ...

class ConfigManager(object):

    DEPRECATED = []
    WARNINGS = set()

    def __init__(self, conf_file=None, defs_file=None):

        self._base_defs = {}
        self._plugins = {}
        self._parsers = {}

        self._config_file = conf_file
        # 1.得到 ConfigData 类的一个实例
        self.data = ConfigData()

        # 2. 读取当前目录下 base.yml 文件中的内容
        self._base_defs =
                self._read_config_yaml_file(defs_file or ('%s/base.yml' %
os.path.dirname(__file__)))

        if self._config_file is None:
            # 3. find_ini_config_file()函数的作用是搜索 ansible.cfg 文件的路径,
            #    搜索顺序: 环境变量、当前目录、属主目录和/etc/ansible/ansible.cfg
            self._config_file = find_ini_config_file(self.WARNINGS)

        if self._config_file:
            if os.path.exists(to_bytes(self._config_file)):
                # 4. 读取配置文件内容, 更新 self._parsers 值
                self._parse_config_file()

        # 5. 将更新配置到 ConfigData 对象中, 也就是 self.data 中
        self.update_config_data()
        try:
            # 似乎没啥作用, 更新另一个 YAML 文件中的内容
            self.update_module_defaults_groups()
        except Exception as e:
            # 打印错误
            self.module_defaults_groups = {}

    # ...
```

通过分析上面的代码，可以总结出 ConfigManager()初始化的过程如下：

（1）实例化 ConfigData 类，赋给 data 属性值，参见编号为 1 的注释。

（2）读取当前文件所在目录下的 base.yml 文件内容，得到 self._base_defs 值，参见编号为 2 的注释。

（3）搜索 Ansible 的配置文件 ansible.cfg，搜索顺序为环境变量、当前目录、属主目录

和/etc/ansible/ansible.cfg，找不到时为 None，参见编号为 3 的注释。

（4）读取配置文件 ansible.cfg 的内容，保存解析的对象到 self._parsers 中，self._parsers 中的 key 为对应的配置文件路径，参见编号为 4 的注释。

（5）更新 base.yml 中的数据到 ConfigData 对象中，也就是 ConfigManager 对象的 data 属性，参见编号为 5 的注释。

下面重点介绍第 4 处和第 5 处的代码，即 _parse_config_file()和 update_module_defaults_groups()方法的具体实现。首先来看 _parse_config_file()方法的实现，具体代码如下：

```python
# 源码位置：lib/ansible/config/manager.py
# ...

class ConfigManager(object):

    # ...

    def _parse_config_file(self, cfile=None):

        if cfile is None:
            cfile = self._config_file

        ftype = get_config_type(cfile)
        if cfile is not None:
            # 是ini文件类型，也就是以.cfg和.ini后缀结尾的文件
            if ftype == 'ini':
                # Python中的configparser模块，专门用于解析配置文件。在Python2
                  中是ConfigParser，在Python3中是configparser
                self._parsers[cfile] = configparser.ConfigParser()
                with open(to_bytes(cfile), 'rb') as f:
                    try:
                        # 读取文件内容，转换成字符串形式
                        cfg_text = to_text(f.read(), errors='surrogate_or_
strict')
                    except UnicodeError as e:
                        # 抛出异常
                        # ...

                    try:
                        if PY3:
                            # 读取文本内容
                            self._parsers[cfile].read_string(cfg_text)
                        else:
                            cfg_file = io.StringIO(cfg_text)
                            self._parsers[cfile].readfp(cfg_file)
                    except configparser.Error as e:
                        raise AnsibleOptionsError("Error reading config file
(%s): %s" % (cfile, to_native(e)))
            # elif ftype == 'yaml':
            #    with open(cfile, 'rb') as config_stream:
            #        self._parsers[cfile] = yaml.safe_load(config_stream)
            else:
```

```
            raise AnsibleOptionsError("Unsupported configuration file
type: %s" % to_native(ftype))

    # ...
```

上面的代码比较简单，只是读取文件信息并保存到 self._parsers[cfile]中。为了能复现上面的操作，先准备一个 ansible.cfg 文件，内容如下：

```
(ansible2.8.16) [root@master ansible-2.8.16]# cat ansible.cfg
[defaults]
library = ./library
action_plugins = plugins/actions
callback_plugins = plugins/callback
roles_path = ./roles
log_path = ansible.log

# 指定开启进程数
forks = 10
host_key_checking = False
gathering = smart
```

接着在虚拟环境中执行如下操作：

```
>>> from ansible.module_utils._text import to_text
>>> from ansible.module_utils.six.moves import configparser
>>> cfile = "/root/ansible-2.8.16/ansible.cfg"
>>> with open(cfile, 'r+') as f:
...     cfg_text = f.read()
...
>>> print(cfg_text)
[defaults]
library = ./library
action_plugins = plugins/actions
callback_plugins = plugins/callback
roles_path = ./roles
log_path = ansible.log

# 指定开启进程数
forks = 10
host_key_checking = False
gathering = smart

>>> c = configparser.ConfigParser()
>>> c.read_string(cfg_text)
>>> c.sections()
['defaults']
>>> c.options("defaults")
['library', 'action_plugins', 'callback_plugins', 'roles_path', 'log_path',
'forks', 'host_key_checking', 'gathering']
>>> c.get("defaults", "library")
'./library'
```

在上述代码中，self._parsers[cfile]的值正是前面演示的 c 变量，配置文件的信息全部保存到了该属性值中。接下来是 update_config_data()方法的实现，具体代码如下：

```
# 源码位置：lib/ansible/config/manager.py
# ...

class ConfigManager(object):

    # ...

    def update_config_data(self, defs=None, configfile=None):
        ''' 更新常量值 '''

        if defs is None:
            defs = self._base_defs

        if configfile is None:
            configfile = self._config_file

        # 处理异常情况
        # ...

        self.data.update_setting(Setting('CONFIG_FILE', configfile, '',
'string'))

        origin = None
        for config in defs:
            # 处理异常情况
            # ...

            try:
                value, origin = self.get_config_value_and_origin(config,
configfile)
            except Exception as e:
                # 处理异常，打印错误信息
                # ...

            # 设置常量
            self.data.update_setting(Setting(config, value, origin, defs
[config].get('type', 'string')))

    # ...
```

在上面的代码中，由于 defs 参数未传入，故会将 self._base_defs 的值赋给它，即 defs 保存了从 base.yml 文件中读取的默认配置。有兴趣的读者可以继续查看 base.yml 中的内容，Ansible 中默认的常量大部分定义在该 YAML 文件中。接下来更新一个 name=CONFIG_FILE 的 setting 值，再接下来更新变量 defs 中的所有值，然后调用 get_config_value_and_origin()方法获取对应配置参数（config）的真实值（value）及其来源（origin）。如果没有在 ansible.cfg 中重新设置，那么对应配置值来源于默认的配置文件，则 origin='default'，否则 origin 为对应的配置文件路径。最后便得到 Setting(config, value, origin, type)并更新到 ConfigData 对象中，即 self.data 中。

根据以上分析可知，update_config_data()方法的核心是调用 get_config_value_and_

origin()找到配置值及其来源，该方法较为复杂，这里不再进行详细描述，但是可以通过代码演示其功能。请看如下操作：

```
>>> from ansible.config.manager import ConfigManager
>>> cfile = '/root/ansible-2.8.16/ansible.cfg'
>>> config = ConfigManager(conf_file=cfile)
>>> config._base_defs
# 结果太长，忽略。其为 base.yml 文件的字典形式
# ...
>>> config._base_defs['DEFAULT_MODULE_NAME']
{'name': 'Default adhoc module', 'default': 'command', 'description':
'Module to use with the ``ansible`` AdHoc command, if none is specified via
``-m``.', 'env': [], 'ini': [{'key': 'module_name', 'section': 'defaults'}]}
```

从上面的代码中可以看到，cm._base_defs 中保存了所有的默认配置信息，例如默认的模块参数 DEFAULT_MODULE_NAME，其值为 command。此外，还有相关描述信息及名称等。再看下面的操作：

```
>>> config.get_config_value_and_origin('DEFAULT_FORKS', cfile)
(10, '/root/ansible-2.8.16/ansible.cfg')
>>> config.get_config_value_and_origin('DEFAULT_MODULE_NAME', cfile)
('command', 'default')
```

由于在 ansible.cfg 中有 forks 参数及其值，它会覆盖默认配置文件中的 DEFAULT_FORKS 值，因此在调用 cm.get_config_value_and_origin('DEFAULT_FORKS', cfile)的结果，origin 会是'cfile'的值。而未被覆盖的参数 DEFAULT_MODULE_NAME 来源于默认的配置文件，因此 origin='default'。

可以看到，一个简单的实例化语句背后竟然有这么多的操作。在实例化 ConfigManager 对象后，self.data 属性值（ConfigData 对象）被 base.yaml 文件中的数据填充。可以通过 config.data.get_settings()语句查看对应的结果。由于默认的内容太长，下面只取前 5 行数据进行展示：

```
>>> config = ConfigManager()
>>> len(config.data.get_settings())
189
>>> config.data.get_settings()[:5]
[Setting(name='CONFIG_FILE', value='/root/ansible-2.8.16/ansible.cfg',
origin='', type='string'), Setting(name='ALLOW_WORLD_READABLE_TMPFILES',
value=False, origin='default', type='boolean'), Setting(name='ANSIBLE_
CONNECTION_PATH', value=None, origin='default', type='path'), Setting
(name='ANSIBLE_COW_SELECTION', value='default', origin='default', type=
'string'), Setting(name='ANSIBLE_COW_WHITELIST', value=['bud-frogs',
'bunny', 'cheese', 'daemon', 'default', 'dragon', 'elephant-in-snake',
'elephant', 'eyes', 'hellokitty', 'kitty', 'luke-koala', 'meow', 'milk',
'moofasa', 'moose', 'ren', 'sheep', 'small', 'stegosaurus', 'stimpy',
'supermilker', 'three-eyes', 'turkey', 'turtle', 'tux', 'udder', 'vader-
koala', 'vader', 'www'], origin='default', type='list')]
```

有了这些知识储备，再回过头去看前面在 constants.py 文件中设置默认配置变量的代码就会十分清晰，这里也不再继续过多描述了。

4.1.3　inventory 文件解析

在 Ansible 2.8 中解析 inventory 文件的代码相比 Ansible 1 有了非常大的变化。在 Ansible 1 中，解析 inventory 文件的所有核心代码全部位于 lib/anisble/inventory 目录下，包括支持外部脚本输入等。然而到了 Ansible 2.8，解析 inventory 文件的代码变成了插件形式，其核心的解析代码位于 lib/ansible/plugins/inventory 目录下。本节将讲解 Ansible 2.8 中关于 inventory 文件的解析过程。首先来看一些辅助的基本函数和类，它们是理解 inventory 代码的前奏。

```
>>> from ansible.parsing.utils.addresses import parse_address
>>> parse_address("192.168.88.204:8888")
('192.168.88.204', 8888)
>>> parse_address("xyz:8888")
('xyz', 8888)
>>> parse_address("xyz[1:3]:8888")
Traceback (most recent call last):
  File "<stdin>", line 1, in <module>
  File "/root/.pyenv/versions/ansible2.8.16/lib/python3.8/site-packages/
ansible/parsing/utils/addresses.py", line 214, in parse_address
    raise AnsibleParserError("Detected range in host but was asked to ignore
ranges")
ansible.errors.AnsibleParserError: Detected range in host but was asked to
ignore ranges
>>> parse_address("xyz[1:3]:8888", allow_ranges=True)
('xyz[1:3]', 8888)
```

上面演示的 parse_address() 函数主要用于解析主机和端口并返回一个二元组。深入 parse_address() 函数源码可以看到，其实现原理只是单纯的正则表达式解析字符串。注意，该函数有一个 allow_ranges 参数，当字符串中包含[]或者各种奇怪的主机名时会抛错。接下来看一个解析文件的类 DataLoader，具体代码如下：

```
# 源码位置: lib/ansible/parsing/dataloader.py
# ...

class DataLoader:

    def __init__(self):

        self._basedir = '.'
        self._FILE_CACHE = dict()
        self._tempfiles = set()
        self._vaults = {}
        self._vault = VaultLib()
        self.set_vault_secrets(None)

    # ...

    def load(self, data, file_name='<string>', show_content=True):
        return from_yaml(data, file_name, show_content, self._vault.secrets)
```

```python
    def load_from_file(self, file_name, cache=True, unsafe=False):
        ''' 从包含 JSON 或者 YAML 语法格式的文件中载入数据 '''

        file_name = self.path_dwim(file_name)
        display.debug("Loading data from %s" % file_name)

        if cache and file_name in self._FILE_CACHE:
            parsed_data = self._FILE_CACHE[file_name]
        else:
            # 获取文件内容
            (b_file_data, show_content) = self._get_file_contents(file_name)
            file_data = to_text(b_file_data, errors='surrogate_or_strict')
            # 导入数据
            parsed_data = self.load(data=file_data, file_name=file_name,
show_content=show_content)
            # 缓存数据
            self._FILE_CACHE[file_name] = parsed_data

        if unsafe:
            # 直接返回，浅拷贝
            return parsed_data
        else:
            # 深度拷贝
            return copy.deepcopy(parsed_data)

    # ...

    def _get_file_contents(self, file_name):
        if not file_name or not isinstance(file_name, (binary_type, text_
type)):
            raise AnsibleParserError("Invalid filename: '%s'" % to_native
(file_name))

        # 得到完整的路径
        b_file_name = to_bytes(self.path_dwim(file_name))
        if not self.path_exists(b_file_name):
            raise AnsibleFileNotFound("Unable to retrieve file contents",
file_name=file_name)

        try:
            # 最基本的读取文件内容的方式
            with open(b_file_name, 'rb') as f:
                data = f.read()
                # 返回文件内容或者加密后的内容
                return self._decrypt_if_vault_data(data, b_file_name)
        except (IOError, OSError) as e:
            # 抛出异常
            # ...

    # ...

    def path_dwim(self, given):
        '''
```

```
        展开给定路径，最终调用的是 unfrackpath()函数处理该路径
        '''

        given = unquote(given)
        # 转换成字符串形式
        given = to_text(given, errors='surrogate_or_strict')

        if given.startswith(to_text(os.path.sep)) or given.startswith(u'~'):
            # 如果 given 是绝对路径（以/开头），或者是以波浪号（~）开头，直接赋给 path
              变量
            path = given
        else:
            # 如果是相对路径，需要加上当前目录组成绝对路径
            basedir = to_text(self._basedir, errors='surrogate_or_strict')
            path = os.path.join(basedir, given)

        return unfrackpath(path, follow=False)

    # ...
```

DataLoader 类的使用非常简单，不需要任何参数即可实例化。下面是 DataLoader 类中几个比较重要的方法：

- load_from_file()方法的作用是加载 YAML 文件内容，得到 JSON 格式的数据。
- load()方法会在 load_from_file()中被调用。
- _get_file_contents()方法是一个比较通用的方法，用于获取输入文件的内容。
- path_dwim()方法将展开~/test.txt、$HOME/xxx.yaml、xxx/test.txt 这样的路径，最终得到完整的绝对路径。

下面在虚拟环境中演示该类的使用，同时也将调用以上方法并复现 load_from_file()方法的实现过程。

```
(ansible2.8.16) [root@master playbook]# ls /root/ansible-2.8.16/playbook/
test_vars2.yml
/root/ansible-2.8.16/playbook/test_vars2.yml
(ansible2.8.16) [root@master playbook]# python
Python 3.8.6 (default, Oct 18 2020, 15:33:08)
[GCC 4.8.5 20150623 (Red Hat 4.8.5-39)] on linux
Type "help", "copyright", "credits" or "license" for more information.
>>> from ansible.parsing.dataloader import DataLoader
>>> from ansible.module_utils._text import to_text
>>> file_name = "/root/ansible-2.8.16/playbook/test_vars2.yml"
>>> d = DataLoader()
>>> (b_file_data, show_content) = d._get_file_contents(file_name)
>>> b_file_data
b'---\n- hosts: ceph-1\n  gather_facts: false\n\n  tasks:\n    - name: set
node_var\n      set_fact:\n        node_var: "test_variable_in_runtime"\n\n
- name: debug node_var\n      debug:\n        msg: "node_var = {{ node_var }}"\n
      when: node_var.endswith(\'runtime\') | bool\n'
>>> file_data = to_text(b_file_data, errors='surrogate_or_strict')
>>> file_data
'---\n- hosts: ceph-1\n  gather_facts: false\n\n  tasks:\n    - name: set
node_var\n      set_fact:\n        node_var: "test_variable_in_runtime"\n\n
```

```
- name: debug node_var\n        debug:\n        msg: "node_var = {{ node_var }}"\n
    when: node_var.endswith(\'runtime\') | bool\n'
>>> print(file_data)
---
- hosts: ceph-1
  gather_facts: false

  tasks:
    - name: set node_var
      set_fact:
        node_var: "test_variable_in_runtime"

    - name: debug node_var
      debug:
        msg: "node_var = {{ node_var }}"
      when: node_var.endswith('runtime') | bool

>>> d.load(data=file_data, file_name=file_name, show_content=show_content)
[{'hosts': 'ceph-1', 'gather_facts': False, 'tasks': [{'name': 'set node_
var', 'set_fact': {'node_var': 'test_variable_in_runtime'}}, {'name':
'debug node_var', 'debug': {'msg': 'node_var = {{ node_var }}'},
'when': "node_var.endswith('runtime') | bool"}]}]
>>> d.load_from_file(file_name)
[{'hosts': 'ceph-1', 'gather_facts': False, 'tasks': [{'name': 'set node_
var', 'set_fact': {'node_var': 'test_variable_in_runtime'}}, {'name':
'debug node_var', 'debug': {'msg': 'node_var = {{ node_var }}'},
'when': "node_var.endswith('runtime') | bool"}]}]
>>> import os
>>> os.getcwd()
'/root/ansible-2.8.16/playbook'
>>> d.path_dwim('text.txt')
'/root/ansible-2.8.16/playbook/text.txt'
>>> d.path_dwim('~/text.txt')
'/root/text.txt'
```

再来看如何解析 inventory 文件相关的类，具体代码如下：

```
# 源码位置：lib/ansible/plugins/inventory/__init__.py
# ...

class BaseInventoryPlugin(AnsiblePlugin):
    """ 解析 Inventory 文件 """

    TYPE = 'generator'
    _sanitize_group_name = staticmethod(to_safe_group_name)

    def __init__(self):

        super(BaseInventoryPlugin, self).__init__()

        self._options = {}
        self.inventory = None
        self.display = display

    def parse(self, inventory, loader, path, cache=True):
```

```python
        # 主要是赋值
        self.loader = loader
        self.inventory = inventory
        self.templar = Templar(loader=loader)

    def verify_file(self, path):
        # 校验文件是否存在，是否可读。正常时返回 True，否则在详情下打印警告信息并返回
          False
        valid = False
        b_path = to_bytes(path, errors='surrogate_or_strict')
        if (os.path.exists(b_path) and os.access(b_path, os.R_OK)):
            valid = True
        else:
            # 忽略打印内容
            self.display.vvv('...')
        return valid

    # ...

    def _read_config_data(self, path):

        config = {}
        try:
            # 调用 self.loader 的 load_from_file() 方法，获取 YAML 文件的相关内容
            config = self.loader.load_from_file(path, cache=False)
        except Exception as e:
            raise AnsibleParserError(to_native(e))

        # 各种异常检测
        # ...

        # 处理 cache 参数
        # ...

        return config

    def _consume_options(self, data):
        for k in self._options:
            if k in data:
                self._options[k] = data.pop(k)

    def _expand_hostpattern(self, hostpattern):
        try:
            # 获取 hosts 文件中的主机形式及端口
            (pattern, port) = parse_address(hostpattern, allow_ranges=True)
        except Exception:
            # 异常时使用原样
            pattern = hostpattern
            port = None

        if detect_range(pattern):
            # 对于出现 ceph-[1:3] 这样包含范围的字符串，需要对其进行展开，得到完整的
              hostnames，如 ceph-[1:3]，经过 expand_hostname_range() 方法展开后
```

```
        得到[ceph-1,ceph-2,ceph-3]
        hostnames = expand_hostname_range(pattern)
    else:
        hostnames = [pattern]

    return (hostnames, port)


class BaseFileInventoryPlugin(BaseInventoryPlugin):
    TYPE = 'storage'

    def __init__(self):
        # 调用父类初始化方法
        super(BaseFileInventoryPlugin, self).__init__()
```

上述代码中定义了类 BaseInventoryPlugin 及其子类 BaseFileInventoryPlugin。前者定义了一些基础方法，后者在其父类基础上几乎没做任何调整。在 Ansible 1.9.6 中，inventory 目录下有 ini.py、script.py 及 dir.py 等文件，这些文件均用于获取 Ansible 的目标主机列表。而在 Ansible 2.8 中，这些代码全部被移动到了 ansible/plugins/inventory 目录下，并且多了很多其他获取主机列表的方式。为简单起见，这里只介绍通过 ini.py 文件获取相应的主机、组及变量信息的方法，该文件位于 ansible/plugins/inventory 目录下。具体代码如下：

```
# 源码位置: lib/ansible/plugins/inventory/ini.py
# ...

class InventoryModule(BaseFileInventoryPlugin):
    # ...

    def __init__(self):

        # 调用父类初始化方法
        super(InventoryModule, self).__init__()

        self.patterns = {}
        self._filename = None

    def parse(self, inventory, loader, path, cache=True):
        # 核心函数，先调用父类的 parse() 方法，只是简单赋值
        super(InventoryModule, self).parse(inventory, loader, path)

        self._filename = path

        try:
            # 如果有 loader 属性，则调用 loder 属性的_get_file_contents() 方法以获
              取文件内容
            if self.loader:
                (b_data, private) = self.loader._get_file_contents(path)
            else:
                # 否则直接读取文件内容
                b_path = to_bytes(path, errors='surrogate_or_strict')
                with open(b_path, 'rb') as fh:
                    b_data = fh.read()
```

```
        try:
            # 批量转换成字符串格式，避免后续多次调用 to_text()进行转换，以提高
            效率
            data = to_text(b_data, errors='surrogate_or_strict').
splitlines()
        except UnicodeError:
            # 处理 GitHub 上的一个 issue: https://github.com/ansible/ansible/
            issues/17593
            # ...

        # 核心处理语句，即调用 _parse()方法
        self._parse(path, data)
    except Exception as e:
        raise AnsibleParserError(e)

    # ...
```

上面定义的 InventoryModule 类继承了 BaseFileInventoryPlugin 类。__init__()方法只是调用父类的初始化函数，同时初始化几个属性值。接着会实现 parse()方法，该方法将解析 inventory 文件，得到组（group）、主机（host）及变量（variable）等信息。从上面的代码中可知，parse()方法中最核心的处理语句用于调用 self._parse()方法解析 inventory 文件的内容。继续追踪_parse()方法的源码，具体如下：

```
# 源码位置: lib/ansible/plugins/inventory/ini.py
# ...

class InventoryModule(BaseFileInventoryPlugin):
    # ...

    def _parse(self, path, lines):

        # 编译准备好的正则表达式，主要设置了 2 个正则表达式，用于匹配 section 和 group
        self._compile_patterns()

        pending_declarations = {}
        groupname = 'ungrouped'
        state = 'hosts'
        self.lineno = 0
        # 遍历 hosts 文件内容，一行一行地处理
        for line in lines:
            # 记住处理的行号
            self.lineno += 1

            line = line.strip()
            # 跳过空行和注释行，
            if not line or line[0] in self._COMMENT_MARKERS:
                continue

            # 检查 hosts 文件中的 section，包含[]这样的形式
            m = self.patterns['section'].match(line)
```

```
        if m:
            # 对于[groupname:vars]，会匹配出 groupname 和 vars，此时 state=vars
            (groupname, state) = m.groups()

            # 将非法的组名进行转换，默认对非法字符使用下划线进行替换
            groupname = to_safe_group_name(groupname)

            # 如果之前的 state 状态没有值，默认本行属于 ansible 的主机
            state = state or 'hosts'
            if state not in ['hosts', 'children', 'vars']:
                # 如果不是这三种状态，直接抛错，例如[xxx:not-valid]，会得到
                  state=not-valid
                # ...

            # 对于关联到组的变量
            if groupname not in self.inventory.groups:
                if state == 'vars' and groupname not in pending_
declarations:
                    # 下面会用到这个字典数据
                    pending_declarations[groupname] =
                          dict(line=self.lineno, state=state, name=
groupname)
                # 在 self.inventory 中添加组
                self.inventory.add_group(groupname)

            if groupname in pending_declarations and state != 'vars':
                if pending_declarations[groupname]['state'] == 'children':
                    self._add_pending_children(groupname, pending_
declarations)
                elif pending_declarations[groupname]['state'] == 'vars':
                    del pending_declarations[groupname]

            continue
        elif line.startswith('[') and line.endswith(']'):
            # 抛出异常
            # ...

        # 处理包含 ansible 主机的普通情况：[groupname]
        if state == 'hosts':
            # 处理这一行内容，得到主机、端口及变量
            hosts, port, variables = self._parse_host_definition(line)
            # 更新变量
            self._populate_host_vars(hosts, variables, groupname, port)

        # 处理[groupname:vars]形式的组变量情况
        elif state == 'vars':
            # 该行为变量
            (k, v) = self._parse_variable_definition(line)
            self.inventory.set_variable(groupname, k, v)

        # 处理[groupname:children]形式的子组情况
        elif state == 'children':
            # 该行为组名，是 groupname 的子组
```

```
            child = self._parse_group_name(line)
            if child not in self.inventory.groups:
                # 更新 pending_declarations
                if child not in pending_declarations:
                    pending_declarations[child] =
                    dict(line=self.lineno, state=state, name=child, parents=
[groupname])
                else:
                    pending_declarations[child]['parents'].append
(groupname)
            else:
                self.inventory.add_child(groupname, child)
        else:
            self._raise_error("Entered unhandled state: %s" % (state))

    for g in pending_declarations:
        # 如果还有状态为 vars 的行，说明其对应的组不存在
        decl = pending_declarations[g]
        if decl['state'] == 'vars':
            # 抛出异常
            # ...
        # 如果还有状态为 children 的行，说明该组对应的父组不存在
        elif decl['state'] == 'children':
            # 抛出异常
            # ...

    # ...

    def _compile_patterns(self):

        self.patterns['section'] = re.compile(
            to_text(r'''^\[
                ([^:\]\s]+)              # 组名
                (?::(\w+))?              # state，可以没有，默认为 hosts
                \]
                \s*                     # 忽略尾部空格
                (?:\#.*)?               # 一行的最后可以加上注释
                $                       #
            ''', errors='surrogate_or_strict'), re.X
        )

        self.patterns['groupname'] = re.compile(
            to_text(r'''^
                ([^:\]\s]+)
                \s*                     # 忽略后面的空格
                (?:\#.*)?               # 匹配该行最后的注释，可有可无
                $                       #
            ''', errors='surrogate_or_strict'), re.X
        )
```

　　上面的代码较为复杂，笔者也做了详细注释，稍后还会在交互模式下来复现上述代码语句以帮助读者更好地理解其解析流程。_compile_patterns()方法中编译了两个正则表达式：

- self.patterns['section']：用于匹配[groupname]、[groupname:state]这样的字符串。
- self.patterns['groupname']：用于匹配组名。

接着对文件内容一行一行地进行匹配，每行的内容会设置两个标签，即所属组（groupname）和本行状态（state）。而本行状态只有三个值：

- hosts：表示本行为主机名。
- vars：表明本行为变量赋值的写法，即 xxx=yyy 这样的形式。
- children：表明本行为子组。

首先在找到[groupname]这样的 section 行后，设置 groupname 和 state 的值，然后继续进行下一行的处理。对于非 section 行，会根据上一个 section 行设置的 groupname 和 state 进行相应的处理。例如，对于 state=vars 的情况，会解析该行的 k=v 形式，然后设置对应组（groupname）变量。其他情况参考相应的 if 分支就可以得到相应的处理逻辑。需要注意 pending_declarations 的属性值，它的作用是先保存[groupname:vars]和[groupname:children]这种形式的 section 的内容，以方便将这些变量以及子组保存到正确的位置，对于找到正确位置的组和变量，会删除对应的 pending_declarations 的 key。扫描并处理完 inventory 中的内容后，如果 pending_declarations 中 key 的值包含 state='vars'或者 state='children'，说明扫描的行所在的 section 指定的组不存在，会直接抛出异常。

上面的 parse()函数中有以下 2 个参数。

- inventory：该参数非常重要，从代码中可以看到，该参数会被赋值给对象的 inventory 属性值，并在后续添加组和子组以及设置变量时使用该属性的一些方法。
- loader：从代码中可以看到，该参数的值也会被赋给对象的 loader 属性值，并用于获取文件的内容。不过，通过代码中的 if-else 语句可以看出，当 self.loader 为空时，会直接使用 open()方法读取文件的内容。因此，从代码的角度来看，可以不用传入 loaders 参数。

此时，需要思考 inventory 应该传入何值，该值需要能管理 inventory 中的数据，支持添加主机、删除组及设置变量等操作。在 Ansible 2 中，为这样的行为专门定义了一个 InventoryData 类，该类的实现代码如下：

```
# 源码位置：lib/ansible/inventory/data.py
# ...

class InventoryData(object):
    '''管理 inventory 数据(主机和组对象)'''

    def __init__(self):

        # 初始化组和主机对象字典
        self.groups = {}
        self.hosts = {}

        self._groups_dict_cache = {}
```

```python
        self.localhost = None

        self.current_source = None

        # 总是创建'all'和'ungrouped'组
        for group in ('all', 'ungrouped'):
            self.add_group(group)
        self.add_child('all', 'ungrouped')

    # ...

    def serialize(self):
        '''序列化操作，就是转成字典'''
        self._groups_dict_cache = None
        data = {
            'groups': self.groups,
            'hosts': self.hosts,
            'local': self.localhost,
            'source': self.current_source,
        }
        return data

    def deserialize(self, data):
        '''反序列化操作，就是从字典中获取数据，设置对象的属性值'''
        self._groups_dict_cache = {}
        self.hosts = data.get('hosts')
        self.groups = data.get('groups')
        self.localhost = data.get('local')
        self.current_source = data.get('source')

    def _create_implicit_localhost(self, pattern):

        # 如果已经设置 self.localhost
        if self.localhost:
            new_host = self.localhost
        else:
            # 生成 Host 对象，pattern 其实就是 Host 对象的 name 属性值
            new_host = Host(pattern)

            new_host.address = "127.0.0.1"
            new_host.implicit = True

            # set localhost defaults
            py_interp = sys.executable
            if not py_interp:
                # 针对 sys.executable 可能没值的情况给定默认值，参见 issue #13585
                py_interp = '/usr/bin/python'
                # 打印警告信息
                # ...

            new_host.set_variable("ansible_python_interpreter", py_interp)
            # 对于 localhost 主机，需要设置 ansible_connection 变量为本地连接方式
            new_host.set_variable("ansible_connection", 'local')
```

```
            # 更新到 self.localhost 的值中
            self.localhost = new_host

        # 返回得到的 Host 对象
        return new_host

    def get_host(self, hostname):
        #根据主机名从 self.hosts 中获取匹配的 Host 对象
        matching_host = self.hosts.get(hostname, None)

        if matching_host is None and hostname in C.LOCALHOST:
            # 如果没有匹配且 hostname 在('127.0.0.1', 'localhost', '::1')中，则
            调用_create_implicit_localhost()方法得到本地的 Host 对象
            matching_host = self._create_implicit_localhost(hostname)

        return matching_host

    def add_group(self, group):
        '''添加组'''

        if group:
            # 处理异常
            # ...
            if group not in self.groups:
                g = Group(group)
                if g.name not in self.groups:
                    self.groups[g.name] = g
                    self._groups_dict_cache = {}
                    display.debug("Added group %s to inventory" % group)
                group = g.name
            else:
                display.debug("group %s already in inventory" % group)
        else:
            raise AnsibleError("Invalid empty/false group name provided: %s"
% group)

        return group

    def remove_group(self, group):
        '''删除组'''

        if group in self.groups:
            del self.groups[group]
            display.debug("Removed group %s from inventory" % group)
            self._groups_dict_cache = {}

        # 除了删除 self.groups 中对应的组外,还要处理对应的 Host 对象中所属组为 group
        的情况
        for host in self.hosts:
            h = self.hosts[host]
            h.remove_group(group)

    def add_host(self, host, group=None, port=None):
```

```python
        '''添加主机'''
        if host:
            # 处理异常
            # ...

            g = None
            if group:
                # 对于传入 group 的情况需要判断，如果组存在，则获取该组，否则抛出异常
                if group in self.groups:
                    g = self.groups[group]
                else:
                    # 传入的组不存在，直接抛出异常
                    raise AnsibleError("Could not find group %s in inventory"
% group)

            if host not in self.hosts:
                # 创建 Host 对象
                h = Host(host, port)
                # 添加到 self.hosts 中
                self.hosts[host] = h
                # 设置变量
                # ...

                if host in C.LOCALHOST:
                    if self.localhost is None:
                        self.localhost = self.hosts[host]
                    # 去掉了部分打印信息
                    # ...
            else:
                # 如果 host 已经存在，直接从 self.hosts 中获取即可
                h = self.hosts[host]

            if g:
                # 如果有组对象，直接添加生成的 Host 对象即可
                g.add_host(h)
                self._groups_dict_cache = {}
        else:
            raise AnsibleError("Invalid empty host name provided: %s" % host)

        return host

    def remove_host(self, host):
        '''移除 host，要注意也要移除相应组下对应的 host'''
        if host.name in self.hosts:
            del self.hosts[host.name]

        for group in self.groups:
            g = self.groups[group]
            g.remove_host(host)

    # ...

    def add_child(self, group, child):
```

```
        ''' 添加主机或者组到父组中 '''

    if group in self.groups:
        # 找到父组 group
        g = self.groups[group]
        if child in self.groups:
            # 如果 child 是一个组名，则给 group 添加一个子组
            g.add_child_group(self.groups[child])
        elif child in self.hosts:
            # 如果 child 是主机，则给组添加一个主机
            g.add_host(self.hosts[child])
        else:
            # 其他情况，抛出异常
            raise AnsibleError("%s is not a known host nor group" % child)
        self._groups_dict_cache = {}
    else:
        raise AnsibleError("%s is not a known group" % group)

# ...
```

上面的代码省略了部分不重要的代码，同时也去掉了一些调试打印语句。Inventory-Data 类中方法众多，不过都非常简单，这里已经给大部分代码做了相应的注释，相信读者理解起来并不困难。为了能更好地理解 InventoryData 类，下面将使用该类完成一些代码演示，具体代码如下：

```
>>> from ansible.inventory.data import InventoryData
>>> inventory_data = InventoryData()
# 添加组 nodes
>>> inventory_data.add_group('nodes')
'nodes'
# 添加组 webservers
>>> inventory_data.add_group('webservers')
'webservers'
# 打印 inventory_data 中的信息
>>> inventory_data.serialize()
{'groups': {'all': all, 'ungrouped': ungrouped, 'nodes': nodes, 'webservers':
webservers}, 'hosts': {}, 'local': None, 'source': None}
# 在组 nodes 下添加一个节点 ceph-1
>>> inventory_data.add_host("ceph-1", 'nodes', 22)
'ceph-1'
# 在组 nodes 下添加一个节点 ceph-2
>>> inventory_data.add_host("ceph-2", 'nodes', 22)
'ceph-2'
# 再次打印 inventory_data 中的信息
>>> inventory_data.serialize()
{'groups': {'all': all, 'ungrouped': ungrouped, 'nodes': nodes, 'webservers':
webservers}, 'hosts': {'ceph-1': ceph-1, 'ceph-2': ceph-2}, 'local': None,
'source': None}
>>> inventory_data.add_host("ceph-3", 'webservers', 22)
'ceph-3'
>>> inventory_data.serialize()
{'groups': {'all': all, 'ungrouped': ungrouped, 'nodes': nodes, 'webservers':
```

```
webservers}, 'hosts': {'ceph-1': ceph-1, 'ceph-2': ceph-2, 'ceph-3':
ceph-3}, 'local': None, 'source': None}
# 将 nodes 组作为 webservers 组的一个子组
>>> inventory_data.add_child('webservers', 'nodes')
>>> inventory_data.serialize()
{'groups': {'all': all, 'ungrouped': ungrouped, 'nodes': nodes, 'webservers':
 webservers}, 'hosts': {'ceph-1': ceph-1, 'ceph-2': ceph-2, 'ceph-3':
ceph-3}, 'local': None, 'source': None}
# 获取 webservers 组下的所有主机
>>> inventory_data.groups['webservers'].hosts
[ceph-3]
# 获取 webservers 组下的所有子组
>>> inventory_data.groups['webservers'].child_groups
[nodes]
>>> inventory_data.groups['nodes'].hosts
[ceph-1, ceph-2]
>>> inventory_data.groups['nodes'].child_groups
[]
```

通过上面的演示，读者可以对 InventoryData 类有一个直观的理解。有了上面的基础后就可以开始解析 inventory 文件的代码了。首先准备一个 inventory 文件，也就是前面经常使用的 hosts 文件，它的内容如下：

```
(ansible2.8.16) [root@master ansible-2.8.16]# cat hosts_test
[center:vars]
ansible_connection=local

[webservers:children]
nodes

[webservers]
ceph-3

[nodes]
ceph-[1:2]

[center]
master

[nodes:vars]
ansible_ssh_user=root
ansible_ssh_pass=@SHENcong19920522

[all:vars]
ansible_ssh_pass=@SHENcong19920522
```

接下来实例化 ini.py 文件中的 InventoryModule 类，并完成如下操作：

```
>>> from ansible.inventory.data import InventoryData
>>> from ansible.plugins.inventory.ini import InventoryModule
>>> inventory_data = InventoryData()
>>> im = InventoryModule()
>>> im.parse(inventory_data, None, '/root/ansible-2.8.16/hosts_test')
>>> im.inventory.serialize()
{'groups': {'all': all, 'ungrouped': ungrouped, 'center': center, 'webservers':
```

```
webservers, 'nodes': nodes}, 'hosts': {'ceph-3': ceph-3, 'ceph-1': ceph-1,
'ceph-2': ceph-2, 'master': master}, 'local': None, 'source': None}
```

上面的代码解析了 hosts_test 文件并得到了相关信息。此外，还可以通过 self.inventory 属性（InventoryData 对象）查看得到的主机、组以及变量等信息。

在插件目录 plugins 下提供的都是各种获取 inventory 信息的方式。那么 Ansible 如何对接这些插件呢？或者说怎么选择使用哪个插件来获取 inventory 信息呢？这里用到的是 InventoryManager 类。该类也是 Ansible 核心代码中调用的类，由它来提供获取 inventory 信息的统一入口，其代码如下：

```python
# 源码位置: lib/ansible/inventory/manager.py
# ...

class InventoryManager(object):

    def __init__(self, loader, sources=None):

        self._loader = loader
        # 得到 InventoryData 对象
        self._inventory = InventoryData()

        # ...

        if sources is None:
            self._sources = []
        elif isinstance(sources, string_types):
            self._sources = [sources]
        else:
            self._sources = sources

        # 最终调用的为 parse_source()方法
        self.parse_sources(cache=True)

    # ...

    def parse_sources(self, cache=False):

        parsed = False
        for source in self._sources:

            if source:
                if ',' not in source:
                    # 展开全路径
                    source = unfrackpath(source, follow=False)
                # 最后调用的是 parse_source()方法
                parse = self.parse_source(source, cache=cache)
                if parse and not parsed:
                    parsed = True

        # 处理结果
        # ...
```

```
    def parse_source(self, source, cache=False):
        parsed = False
        b_source = to_bytes(source)

        if os.path.isdir(b_source):
            # 对于传入的目录而言，变量目录下的文件继续调用 parse_source()方法递归解析
            for i in sorted(os.listdir(b_source)):
                # 判断是否需要忽略该文件的解析
                if IGNORED.search(i):
                    continue
                # 得到遍历文件的全路径
                fullpath = to_text(os.path.join(b_source, i), errors=
'surrogate_or_strict')
                # 递归调用
                parsed_this_one = self.parse_source(fullpath, cache=cache)
                if not parsed:
                    # 设置解析成功，只解析一次
                    parsed = parsed_this_one
        else:
            self._inventory.current_source = source

            failures = []
            # 遍历搜索解析 inventory 的插件，会在默认配置中设置
            for plugin in self._fetch_inventory_plugins():

                # 得到插件名
                plugin_name = to_text(getattr(plugin, '_load_name', getattr
(plugin, '_original_path', '')))

                try:
                    # 调用插件的 verify_file()方法，判断文件是否存在和可读
                    plugin_wants = bool(plugin.verify_file(source))
                except Exception:
                    plugin_wants = False

                if plugin_wants:
                    try:
                        # 调用对应插件的 parse()方法，正是每个插件中 InventoryModule
                            对象的 parse()方法
                        plugin.parse(self._inventory, self._loader, source,
cache=cache)

                        # 处理数据缓存
                        # ...

                        # 设置解析标识为 True
                        parsed = True
                        break
                    except AnsibleParserError as e:
                        # 处理解析异常
                        # ...
```

```
            except Exception as e:
                # 处理其他异常
                # ...
            else:
                display.vvv("%s declined parsing %s as it did not pass it's
verify_file() method"
                                    % (plugin_name, source))
        else:
            # 处理失败的情况并打印一些信息，-vvv 可见
            # ...

        if not parsed:
            if source != '/etc/ansible/hosts' or os.path.exists(source):
                display.warning("Unable to parse %s as an inventory source"
% source)

        # clear up, jic
        self._inventory.current_source = None

        return

    # ...
```

InventoryManager 类中的方法较多，但最核心的方法当属 parse_source()。该方法需要传入 source 参数，即 inventory 文件路径，它也可以是目录地址。目录可以采用递归处理的方式，而文件则遍历默认的插件并依次调用 parse()方法去解析 inventory 文件，最后将解析得到的 InventoryData 对象保存到对象的_inventory 属性中。以下是 self._fetch_inventory_plugins()方法的源码，从中可以看到 Ansible 2 默认的 inventory 解析插件列表。

```
# 源码位置：lib/ansible/inventory/manager.py
# ...

class InventoryManager(object):
    # ...

    def _fetch_inventory_plugins(self):
        display.vvvv('setting up inventory plugins')

        plugins = []
        # 遍历默认配置文件中指定的插件
        for name in C.INVENTORY_ENABLED:
            # inventory_loader 最后加载的是插件中的 InventoryModule 对象
            plugin = inventory_loader.get(name)
            if plugin:
                plugins.append(plugin)
            else:
                display.warning('Failed to load inventory plugin, skipping
```

```
%s' % name)

    # 如果没有指定解析 inventory 的插件，直接抛出异常
    # ...

    return plugins
```

🔔注意：inventory_loader.get(name)语句最终得到的是 plugins/inventory 目录下插件文件中
定义的 InventoryModule 对象。

来看如下语句：

```
>>> from ansible.plugins.loader import inventory_loader
>>> inventory_loader.get('ini')
<ansible.plugins.inventory.ini.InventoryModule object at 0x7fe9621ded60>
>>> inventory_loader.get('host_list')
<ansible.plugins.inventory.host_list.InventoryModule object at 0x7fe9621
eeaf0>
>>>
```

最后在默认配置文件中查看 INVENTORY_ENABLED 的值：

```
# lib/ansible/config/base.yml
# ...
INVENTORY_ENABLED:
  name: Active Inventory plugins
  default: ['host_list', 'script', 'auto', 'yaml', 'ini', 'toml']
  description: List of enabled inventory plugins, it also determines the
order in which they are used.
  env: [{name: ANSIBLE_INVENTORY_ENABLED}]
  ini:
  - {key: enable_plugins, section: inventory}
  type: list
# ...
```

前面介绍的 ini 插件正好也在里面，最后这个流程就基本上可以串起来了。先简单操
作一下 InventoryManager 类：

```
>>> from ansible.inventory.manager import InventoryManager
>>> im = InventoryManager(None, '/root/ansible-2.8.16/hosts_test')
>>> im._inventory.serialize()
{'groups': {'all': all, 'ungrouped': ungrouped, 'center': center, 'webservers':
webservers, 'nodes': nodes}, 'hosts': {'ceph-3': ceph-3, 'ceph-1': ceph-1,
'ceph-2': ceph-2, 'master': master}, 'local': None, 'source': None}
>>> im._inventory.hosts
{'ceph-3': ceph-3, 'ceph-1': ceph-1, 'ceph-2': ceph-2, 'master': master}
>>> im._inventory.groups
{'all': all, 'ungrouped': ungrouped, 'center': center, 'webservers':
webservers, 'nodes': nodes}
```

InventoryManager 类在实例化时会解析传入的 inventory 文件，并将 inventory 信息赋给 _inventory 属性值。接下来给出完整的 inventory 文件的解析流程，如图 4-1 所示。

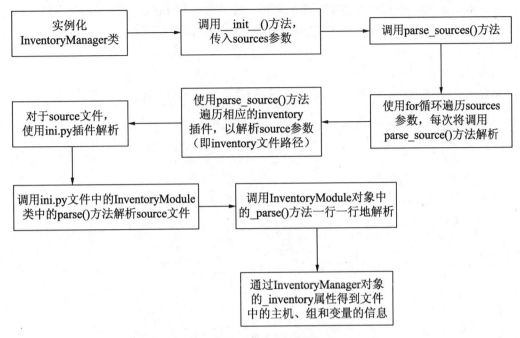

图 4-1　inventory 文件的解析流程

4.1.4　模板渲染

在 Ansible 的 Playbook 文件中通常会设置许多模板变量以提高部署工程的健壮性。在真正执行任务前，Ansible 会将任务中的变量进行渲染，得到最终渲染后的 task。有关模板渲染的代码都在 lib/ansible/template 目录下，内容并不多，只有 5 个代码文件。其中最核心的渲染类定义在 __init__.py 文件中，该类会被 Ansible 大量调用。为了能先看到渲染效果，这里借用 template 目录下定义的 Templar 类完成一个简单的示例，具体操作如下：

```
(ansible2.8.16) [root@master shencong]# python
Python 3.8.6 (default, Oct 18 2020, 15:33:08)
[GCC 4.8.5 20150623 (Red Hat 4.8.5-39)] on linux
Type "help", "copyright", "credits" or "license" for more information.
>>> from ansible.template import Templar
>>> from ansible.parsing.dataloader import DataLoader
>>> loader = DataLoader()
>>> task_vars = {'name': '奇才阁下', 'location': '火星'}
>>> templar = Templar(loader=loader, variables=task_vars)
>>> templar.template("欢迎{{ name }}来到{{ location }}做客")
'欢迎奇才阁下来到火星做客'
```

是不是非常简单？在 Ansible 的核心模块调用以及 Playbook 的执行代码中，对于模板变量的渲染也基本上使用这样的语句。Ansible 对 jinja2 模块进行了二次封装和改造，因此，为了能理解 Ansible 中和模板渲染相关的源码，需要先了解 jinja2 模块的相关用法。以下的诸多示例均来自 jinja2 官网，本节将在虚拟环境中操作这些示例并介绍 Jinja2 中的相关API 用法。

示例 1：最基本的渲染操作

```
# 也可以简写成 from jinja2 import Template，导入的都是同一个类
>>> from jinja2.environment import Template
>>> template = Template('您好，{{ name }}阁下')
>>> template.render(name='奇才')
'您好，奇才阁下'
```

示例 2：渲染模板文件

先准备好一个模板文件 test_template.html，具体内容如下：

```
<p>{{ location }}</p>
{% for animal in animals %}
<li>动物：{{ animal }}</li>
{% endfor %}
```

在虚拟环境的交互模式下输入如下语句：

```
>>> from jinja2 import Environment, FileSystemLoader
>>> env = Environment(loader=FileSystemLoader('./'))
>>> template = env.get_template('test_template.html')
>>> template.render({'location': '广州动物园', 'animals': ['老虎', '狮子',
'大象']})
'<p>广州动物园</p>\n\n<li>动物：老虎</li>\n\n<li>动物：狮子</li>\n\n<li>动
物：大象</li>\n'
```

上面两个操作使用的都是 Template 的 render()方法，该方法是 jinja2 模块的核心渲染方法。通过源码看看该方法的实现：

```
# 源码位置：jinja2/environment.py
# ...

class Template(object):
    # ...

    def render(self, *args, **kwargs):
        vars = dict(*args, **kwargs)
        try:
            return concat(self.root_render_func(self.new_context(vars)))
        except Exception:
            self.environment.handle_exception()

    # ...
```

可以看到，Template.render()方法可以由粒度更小的 Template.root_render_func()方法实现。继续在 Python 交互模式下演示这一现象，具体操作如下：

```
>>> from jinja2.environment import Template
>>> from jinja2.utils import concat
>>> t = Template('您好，{{ name }}阁下')
>>> new_context = t.new_context({'name': '奇才'})
>>> rf = t.root_render_func(new_context)
>>> concat(rf)
'您好，奇才阁下'
```

因此，不用 jinja2 的 render()方法也能实现相关的模板内容渲染工作，而后面会看到上面的渲染语句正是 Templar 对象中 template()方法的最精简版本。

有了上面的基础，就可以开始学习 template 目录下的源码了。首先来看几个 Ansible 封装的 jinja2 类，具体如下：

```
# 源码位置: lib/ansible/template/native_helpers.py
# ...

def ansible_native_concat(nodes):

    head = list(islice(nodes, 2))

    if not head:
        return None

    if len(head) == 1:
        out = head[0]

        if isinstance(out, AnsibleVaultEncryptedUnicode):
            return out.data

        if isinstance(out, StrictUndefined):
            str(out)

        if not isinstance(out, list):
            return out
    else:
        if isinstance(nodes, types.GeneratorType):
            nodes = chain(head, nodes)

        # 核心方法，等同于 jinja2 中的 concat()方法
        out = u''.join([text_type(v) for v in nodes])

    try:
        return literal_eval(out)
    except (ValueError, SyntaxError, MemoryError):
        return out
```

上述函数的内容并不复杂，通过全局搜索可以发现，ansible_native_concat()函数只在 template 目录下的__init__.py 文件中被导入并调用。通过查看 lib/ansible/template/-init-.py 中的源码可知，只有在设置了 DEFAULT_JINJA2_NATIVE 时该函数才会被导入，通过在 base.yml 文件中搜索该默认值可知其为 False，因此默认情况下并不会使用该方法。以下是调用 ansible_native_concat()函数的部分代码：

```
# 源码位置：lib/ansible/template/-init-.py
# ...

if C.DEFAULT_JINJA2_NATIVE:
    try:
        from jinja2.nativetypes import NativeEnvironment as Environment
        # 导入 ansible_native_concat()函数，重命名为 j2_concat
        from ansible.template.native_helpers import ansible_native_concat
as j2_concat
        USE_JINJA2_NATIVE = True
    except ImportError:
        from jinja2 import Environment
        # jinja2 的版本情况可能导致模块导入异常，j2_concat 其实就是 jinja2 中的
          concat()函数
        from jinja2.utils import concat as j2_concat
        from jinja2 import __version__ as j2_version
        display.warning(
            'jinja2_native requires Jinja 2.10 and above. '
            'Version detected: %s. Falling back to default.' % j2_version
        )
else:
    from jinja2 import Environment
    # j2_concat 其实就是 jinja2 中的 concat()函数
    from jinja2.utils import concat as j2_concat

# ...
```

从以上代码中可以看出 ansible_native_concat()函数和 jinja2 模块中的 concat()函数的功能是一致的。而 jinja2 中的 concat()方法只有一条语句，具体代码如下：

```
# 源码位置：jinja2/utils.py
# ...

concat = u"".join

# ...
```

在 Python 交互模式下演示 ansible_native_concat()方法的功能，看它和 jinja2 中的 concat()方法是否一致。具体操作如下：

```
>>> from ansible.template.native_helpers import ansible_native_concat
>>> ansible_native_concat('ceph-1')
'ceph-1'
>>> ansible_native_concat(['ceph-1', 'ceph-2', 'ceph-3'])
'ceph-1ceph-2ceph-3'
>>> from jinja2.utils import concat
>>> concat(['ceph-1', 'ceph-2', 'ceph-3'])
'ceph-1ceph-2ceph-3'
>>> concat('ceph-1')
'ceph-1'
```

最后记住一点，j2_concat 基本上可以认为就是 jinja2 中的 concat()方法。接下来继续介绍 template/safe_eval.py 文件，里面定义了一个 safe_eval()方法。下面是几个关于该方法

的测试示例。

```
>>> from ansible.template.safe_eval import safe_eval
>>> eval('xxx')
Traceback (most recent call last):
  File "<stdin>", line 1, in <module>
  File "<string>", line 1, in <module>
NameError: name 'xxx' is not defined
>>> safe_eval('xxx')
'xxx'
>>> safe_eval('1*2+4*8')
34
>>> eval('1*2+4*8')
34
>>> eval('1*2+4^8')
14
>>> safe_eval('1*2+4^8')
'1*2+4^8'
>>> import os
>>> eval("os.getcwd()")
'/root'
>>> safe_eval("os.getcwd()")
'os.getcwd()'
```

　　从上面的操作中可以看到，Python 内置的 eval()方法支持执行传入的语句，如果输入语句有误，则会直接抛出异常。而对于 safe_eval()来说，则提供了一种安全的执行语句方式，它只支持固定的语句，例如简单的数学表达式（有限操作）等，对于它不支持的语句将直接原样返回文本。继续看 template/template.py 文件中的代码，具体如下：

```
# 源码位置: lib/ansible/template/template.py
# ...

class AnsibleJ2Template(jinja2.environment.Template):
    # 重写 new_context()方法
    def new_context(self, vars=None, shared=False, locals=None):
        if vars is not None:
            if isinstance(vars, dict):
                vars = vars.copy()
                if locals is not None:
                    vars.update(locals)
            else:
                vars = vars.add_locals(locals)
        return self.environment.context_class(self.environment, vars, self.
name, self.blocks)
```

　　上面定义的 AnsibleJ2Template 类继承了 jinja2 中的 Template 类并重写了 new_context() 方法。记住一点：AnsibleJ2Template 就是 jinja2 中的 Template 类。最后还有一个 template/vars.py 文件，其中定义了一个名为 AnsibleJ2Vars 的类，它继承自（如果是 Python 3）collections.abc.Mapping，其内容如下：

```
# 源码位置: lib/ansible/template/vars.py
# ...
```

```
from ansible.module_utils.common._collections_compat import Mapping


__all__ = ['AnsibleJ2Vars']


class AnsibleJ2Vars(Mapping):

    def __init__(self, templar, globals, locals=None, *extras):
        # 初始化必须有 Templar 对象
        self._templar = templar
        self._globals = globals
        self._extras = extras
        # 本地变量
        self._locals = dict()
        if isinstance(locals, dict):
            for key, val in iteritems(locals):
                if val is not missing:
                    if key[:2] == 'l_':
                        self._locals[key[2:]] = val
                    elif key not in ('context', 'environment', 'template'):
                        self._locals[key] = val

    def __contains__(self, k):
        # 判断 key 是否在字典中，这里会检查所有的途径
        if k in self._templar._available_variables:
            return True
        if k in self._locals:
            return True
        for i in self._extras:
            if k in i:
                return True
        if k in self._globals:
            return True
        return False

    def __iter__(self):
        # 迭代魔法函数
        keys = set()
        keys.update(self._templar._available_variables, self._locals, self.
_globals,
*self._extras)
        return iter(keys)

    def __len__(self):
        # 长度魔法函数
        keys = set()
        keys.update(self._templar._available_variables, self._locals, self.
_globals,
*self._extras)
        return len(keys)

    def __getitem__(self, varname):
        # 魔法函数
```

```
        # ...

    def add_locals(self, locals):
        if locals is None:
            return self

        new_locals = self._locals.copy()
        new_locals.update(locals)

        return AnsibleJ2Vars(self._templar, self._globals, locals=new_
locals, *self._extras)
```

可以把 AnsibleJ2Vars 类想象成一个普通的字典形式，只不过该类中重写了一些魔法
函数，这将使得该类区别于普通字典。下面是 Mapping 类和字典的一些简单测试，具体操
作如下：

```
>>> d = {}
>>> type(d)
<class 'dict'>
>>> from ansible.module_utils.common._collections_compat import Mapping
>>> isinstance(d, Mapping)
True
```

有了上面的知识后，下面来操作 AnsibleJ2Vars 类，它和后面即将介绍的 Templar 类关
系密切，但先不过多分析 Templar 类源码。请看如下示例语句：

```
>>> from ansible.template.vars import AnsibleJ2Vars
>>> from ansible.parsing.dataloader import DataLoader
>>> from ansible.template import Templar
>>> loader = DataLoader()
>>> task_vars = {'name': '奇才阁下', 'location': '火星'}
>>> templar = Templar(loader=loader, variables=task_vars)
>>> v = AnsibleJ2Vars(templar, {'global_test': '全局变量值'})
>>> len(v)    #对应调用魔法函数__len__()
3
>>> v['global_test']                  # 对应调用魔法函数 __getitem__()
'全局变量值'
>>> 'global_test' in v                # 对应调用魔法函数 __contains__()
True
>>> 'global_test1' in v
False
>>> v['set_var1'] = '设置变量1'        # 不支持设置 key
Traceback (most recent call last):
  File "<stdin>", line 1, in <module>
TypeError: 'AnsibleJ2Vars' object does not support item assignment
>>> for key in v:                     # 遍历 key，会遍历所有的 key 值
...     print(key, ":", v[key])
...
name : 奇才阁下
location : 火星
global_test : 全局变量值
>>> t1 = v.add_locals({'test_local1': '新的本地变量1'})
>>> len(t1)
```

```
4
>>> for key in t1:
...     print(key, ":", t1[key])
...
name : 奇才阁下
location : 火星
global_test : 全局变量值
test_local1 : 新的本地变量 1
```

通过上面的操作，基本上可以对 AnsibleJ2Vars 类的功能了解得非常清楚。它是一个类似于字典的新型类，和 Templar 对象息息相关。对于该类的对象，可以执行获取长度、判断 key 是否存在、获取 key 对应的值、迭代 key 等操作，而 add_locals()方法会生成一个全新的 AnsibleJ2Vars 对象。

接下来介绍 template 的核心类——Templar，它位于 template/__init__.py 文件中。这里重点解释 templar.template()方法的实现过程，以帮助读者理解 Ansible 是如何封装 jinja2 模块并实现渲染功能的。具体代码如下：

```python
# 源码位置：lib/ansible/templaye/__init__.py
# ...

class AnsibleEnvironment(Environment):

    context_class = AnsibleContext
    template_class = AnsibleJ2Template

    def __init__(self, *args, **kwargs):
        super(AnsibleEnvironment, self).__init__(*args, **kwargs)

        self.filters = JinjaPluginIntercept(self.filters, filter_loader)
        self.tests = JinjaPluginIntercept(self.tests, test_loader)


class Templar:
    # ...

    def __init__(self, loader, shared_loader_obj=None, variables=None):
        variables = {} if variables is None else variables

        self._loader = loader
        self._filters = None
        self._tests = None
        self._available_variables = variables
        self._cached_result = {}

        if loader:
            self._basedir = loader.get_basedir()
        else:
            self._basedir = './'

        if shared_loader_obj:
            self._filter_loader = getattr(shared_loader_obj, 'filter_loader')
            self._test_loader = getattr(shared_loader_obj, 'test_loader')
```

```
        self._lookup_loader = getattr(shared_loader_obj, 'lookup_loader')
    else:
        self._filter_loader = filter_loader
        self._test_loader = test_loader
        self._lookup_loader = lookup_loader

    # ...

    # 可以把它想象成 jinja2 中的 Environment 对象
    self.environment = AnsibleEnvironment(
        trim_blocks=True,
        undefined=AnsibleUndefined,
        extensions=self._get_extensions(),
        finalize=self._finalize,
        loader=FileSystemLoader(self._basedir),
    )

    # ...

    self.SINGLE_VAR = re.compile(r"^%s\s*(\w*)\s*%s$" %
            (self.environment.variable_start_string, self.environment.
variable_end_string))

    # ...

def template(self, variable, convert_bare=False, preserve_trailing_
newlines=True,
        escape_backslashes=True, fail_on_undefined=None, overrides=
None,
        convert_data=True, static_vars=None, cache=True, disable_
lookups=False):

    static_vars = [''] if static_vars is None else static_vars

    # 不渲染不安全的变量，直接返回
    if hasattr(variable, '__UNSAFE__'):
        return variable

    # ...

    try:
        if convert_bare:
            variable = self._convert_bare_variable(variable)

        if isinstance(variable, string_types):
            # 输入的文本为字符串类型
            result = variable
            # 检查是否有 "{{ 变量 }}" 格式的文本，确定该文本是否需要渲染
            if self._contains_vars(variable):
                # 找出单个匹配的结果，也就是 variable="{{ 变量名 }}"这样的情况
                only_one = self.SINGLE_VAR.match(variable)
                if only_one:
                    var_name = only_one.group(1)
                    # 如果匹配单个，只需要直接找到 key 对应的变量值，直接返回即可
```

```
                    if var_name in self._available_variables:
                        resolved_val = self._available_variables[var_name]
                        if isinstance(resolved_val, NON_TEMPLATED_TYPES):
                            return resolved_val
                        elif resolved_val is None:
                            # 如果设置的变量为 None，则返回项目的默认空值
                            return C.DEFAULT_NULL_REPRESENTATION

                # 针对非单个模板变量形式
                sha1_hash = None
                if cache:
                    # 计算变量的 hash 值，得到缓存的唯一 key
                    ~variable_hash = sha1(text_type(variable).encode
('utf-8'))
                    options_hash = sha1(
                        (
                            text_type(preserve_trailing_newlines) +
                            text_type(escape_backslashes) +
                            text_type(fail_on_undefined) +
                            text_type(overrides)
                        ).encode('utf-8')
                    )
                    sha1_hash = variable_hash.hexdigest() + options_
hash.hexdigest()

                if cache and sha1_hash in self._cached_result:
                    # 如果设置了缓存且对应的 key 在字典中，则直接获取缓存的值
                    result = self._cached_result[sha1_hash]
                else:
                    # 其他情况调用 do_template() 方法
                    result = self.do_template(
                        variable,
                        preserve_trailing_newlines=preserve_trailing_
newlines,
                        escape_backslashes=escape_backslashes,
                        fail_on_undefined=fail_on_undefined,
                        overrides=overrides,
                        disable_lookups=disable_lookups,
                    )

                    if not USE_JINJA2_NATIVE:
                        unsafe = hasattr(result, '__UNSAFE__')
                        # 处理一些其他安全场景
                        # ...

                    # 如果需要设置缓存，则将结果保存到字典中
                    if cache:
                        self._cached_result[sha1_hash] = result

            # 返回最后的结果
            return result

        elif isinstance(variable, (list, tuple)):
```

```
                    # 对于输入的 variable 为元组和列表的情况，调用自身处理，传入的值为每
                      个元素
                    return [self.template(
                        v,
                        preserve_trailing_newlines=preserve_trailing_newlines,
                        fail_on_undefined=fail_on_undefined,
                        overrides=overrides,
                        disable_lookups=disable_lookups,
                    ) for v in variable]
                elif isinstance(variable, (dict, Mapping)):
                    # 对于输入 variable 为字典的情况，调用自身处理，传入的值为 key 对应的
                      value
                    d = {}
                    for k in variable.keys():
                        if k not in static_vars:
                            # 反复调用 self.template() 方法实现渲染
                            d[k] = self.template(
                                variable[k],
                                preserve_trailing_newlines=preserve_trailing_
newlines,
                                fail_on_undefined=fail_on_undefined,
                                overrides=overrides,
                                disable_lookups=disable_lookups,
                            )
                        else:
                            d[k] = variable[k]
                    return d
                else:
                    return variable

        except AnsibleFilterError:
            # 异常处理
            # ...
```

上面的代码看着有些复杂，下面会逐个介绍相关语句的功能并复现其执行过程。为了模拟上面的代码，首先做如下操作：

```
>>> from ansible.template import Templar
>>> from ansible.parsing.dataloader import DataLoader
>>> loader = DataLoader()
>>> task_vars = {'name': '奇才阁下', 'location': '火星'}
>>> templar = Templar(loader=loader, variables=task_vars)
```

接着设置待渲染的文本，具体如下：

```
>>> variable = "欢迎{{ name }}来到{{ location }}做客"
```

在 Templar 的初始化中，可以看到 Ansible 实例化了 AnsibleEnvironment 对象，然后得到了一个用于匹配模板变量文本的表达式。具体代码如下：

```
>>> templar.environment.variable_start_string
'{{'
>>> templar.environment.variable_end_string
'}}'
>>> templar.SINGLE_VAR
```

```
re.compile('^{{\\s*(\\w*)\\s*}}$')
```

这个正则表达式专门用于匹配{{ 变量名 }}格式的字符串。来做如下测试：

```
>>> m = templar.SINGLE_VAR.match("{{ name }}")
>>> m
<re.Match object; span=(0, 10), match='{{ name }}'>
>>> m.group(1)
'name'
```

接着在 template()方法中调用_convert_bare_variable()方法，具体如下：

```
>>> templar._convert_bare_variable(variable)
'欢迎{{ name }}来到{{ location }}做客'
```

_convert_bare_variable()方法对当前的变量文本并没有进行任何改动。接着判断变量（variable）的类型，最核心的操作是处理字符串类型，而对于如列表、元组或者字典类型的变量，则遍历所有元素并针对单个元素继续调用 template()方法进行文本渲染。在进入字符串的处理分支后，先使用_contains_vars()方法判断文本中是否包含{{ 变量名 }}形式的字符串。来看该方法的实现，具体代码如下：

```
# 源码位置：lib/ansible/templaye/__init__.py
# ...

class Templar:
    # ...

    def _contains_vars(self, data):
        if isinstance(data, string_types):
            for marker in (self.environment.block_start_string, self.
environment.variable_start_string,
                            self.environment.comment_start_string):
                if marker in data:
                    return True
        return False
```

注意，前面已经演示过 self.environment.block_start_string 和 self.environment.variable_start_string 的值。self.environment.comment_start_string 的值及该函数的测试示例如下：

```
>>> templar.environment.comment_start_string
'{#'
# variable 为前面定义的值，在同一个 console 中
>>> templar._contains_vars(variable)
True
>>> templar._contains_vars('xxxxx')
False
>>> templar._contains_vars('xxxxx{#')
True
>>> templar._contains_vars('xxxxx#}')
False
```

接下来的一大段代码只是匹配字符串是否为{{ 变量名 }}格式，这里匹配的表达式为初始化中指定的 self.SINGLE_VAR 值。如果是这样的格式，处理起来就非常简单，直接获取对应的变量名并在对象的_available_variables 属性中搜索该 key 并返回即可。对于其

他形式，如这里的 variable 文本，则会调用对象的 do_template()方法进行渲染。如果设置了缓存模式，则先要生成唯一的 hash_id 值，然后在缓存中搜索该 key 对应的结果，如果存在，则直接返回。来看如下操作：

```
>>> templar.do_template(variable)
'欢迎奇才阁下来到火星做客'
```

现在来看 do_template()方法，在这里会看到一些非常熟悉的代码，具体如下：

```python
# 源码位置: lib/ansible/templaye/__init__.py
# ...

class Templar:
    # ...

def do_template(self, data, preserve_trailing_newlines=True, escape_
backslashes=True,
fail_on_undefined=None, overrides=None, disable_lookups=False):

        if USE_JINJA2_NATIVE and not isinstance(data, string_types):
            return data

        data_newlines = _count_newlines_from_end(data)

        # ...

        try:
            # 1.第一步得到 environment
            if overrides is None:
                myenv = self.environment.overlay()
            else:
                myenv = self.environment.overlay(overrides)

            # 处理 data 以#jinja2 开头的情况
            # ...

            # 添加 Ansible 常用的过滤器和测试器
            # ...

            if escape_backslashes:
                data = _escape_backslashes(data, myenv)

            try:
                # 2.根据 data 得到 AnsibleJ2Template 对象
                t = myenv.from_string(data)
            except TemplateSyntaxError as e:
                # 抛出异常
                # ...
            except Exception as e:
                # 处理异常情况
                # ...

            t.globals['dict'] = dict
```

```
        # 赋值 t 的 globals 属性
        # ...

        # 3.通过 Templar 对象以及 globals 参数实例化 AnsibleJ2Vars 对象
        jvars = AnsibleJ2Vars(self, t.globals)

        # 4.得到 new_context
        self.cur_context = new_context = t.new_context(jvars, shared=
True)
        # 5.调用 AnsibleJ2Template 对象的 root_render_func()方法
        rf = t.root_render_func(new_context)

        try:
            # 6.得到最终渲染结果
            res = j2_concat(rf)
            if getattr(new_context, 'unsafe', False):
                res = wrap_var(res)
        except TypeError as te:
            # 处理异常
            # ...

        if USE_JINJA2_NATIVE and not isinstance(res, string_types):
            return res

        if preserve_trailing_newlines:
            # 额外处理
            # ...

        return res
    except (UndefinedError, AnsibleUndefinedVariable) as e:
        # 处理异常情况
        # ...

# ...
```

上面整理出了 6 个关键渲染步骤，接着在 Python 的交互模式下演示这 6 个步骤，最终得到渲染结果。具体操作如下：

```
>>> myenv = templar.environment.overlay()               # 对应第 1 处代码
>>> type(myenv)
<class 'ansible.template.AnsibleEnvironment'>
>>> t = myenv.from_string(variable)                      # 对应第 2 处代码
>>> type(t)
<class 'ansible.template.template.AnsibleJ2Template'>
>>> jvars = AnsibleJ2Vars(templar, {})                   # 对应第 3 处代码
>>> new_context = t.new_context(jvars, shared=True)      # 对应第 4 处代码
>>> type(new_context)
<class 'ansible.template.AnsibleContext'>
>>> new_context['name']
'奇才阁下'
>>> new_context['location']
'火星'
>>> rf = t.root_render_func(new_context)                 # 对应第 5 处代码
```

```
>>> from jinja2.utils import concat as j2_concat
>>> j2_concat(rf)                                        # 对应第 6 处代码
'欢迎奇才阁下来到火星做客'
```

下面用纯 jinja2 模块实现这个模板文本的渲染过程，具体操作如下：

```
>>> from jinja2.environment import Template
>>> from jinja2.utils import concat as j2_concat
>>> variable = "欢迎{{ name }}来到{{ location }}做客"
>>> t = Template(variable)
>>> new_context = t.new_context({'name': '奇才阁下', 'location': '火星'})
>>> new_context['name']
'奇才阁下'
>>> new_context['location']
'火星'
>>> rf = t.root_render_func(new_context)
>>> j2_concat(rf)
'欢迎奇才阁下来到火星做客'
```

上述两种渲染代码有什么不同吗？其实 Ansible 中模板引擎的本质就是封装了的 jinja2 模块，例如 AnsibleEnvironment 类对应 jinja2 中的 Environment 类，AnsibleJ2Template 类对应 jinja2 中的 Template 类，最后都使用 jinja2 的 render() 源码中的语句进行模板文本渲染。

4.1.5　插件

Ansible 2 中插件和 Ansible 1 中插件的功能一样，只不过经过长久的迭代后，Ansible 2 的插件模块已经变得庞大而且复杂，除了支持多种功能的插件外，每种功能的插件都有多种实现代码。其中比较重要的插件有通信插件、inventory 文件解析插件（4.1.3 节已介绍过）、动作插件、become 插件、缓存插件、过滤插件、lookup 插件和过滤插件。

所有插件组下的插件都是类似的，都定义着相同的类和部分相同的方法。下面简单介绍一些插件的代码文件。

1. 通信插件

Ansible 2 中的通信插件和 Ansible 1 中详细介绍过的底层通信插件是一样的，只不过这里支持的通信方式更加多样化。下面是 Ansible 2 中的核心通信插件。

- local.py：本地通信。
- paramiko_ssh.py：基于 Paramiko 模块实现的远程通信插件。
- saltstack.py：基于 Saltstack 运维管理工具实现的远程通信插件。
- ssh.py：基于 SSH 命令实现的远程通信插件，是 Ansible 2 中默认使用的通信插件。

这些插件中同样实现了统一的 Connection 类，该类中固定有如下几个方法。

- _connect()：建立连接。

- exec_command()：远程执行 shell 命令。
- put_file()：本地上传文件到远端服务器。
- fetch_file()：从远端服务器下载文件。
- reset()：重置连接。
- close()：关闭连接。

上面几个插件的具体实现细节相比 Ansible 1 变化不大，只是代码略显复杂。

2．策略插件

相比 Ansible 1 而言，策略插件是 Ansible 2 新增的，其内容也超级复杂。从后面的分析中将看到，该插件会将 Ansible 1 中的多进程执行任务移到这里，同时也会在这里调用模块所属的动作插件，具体的细节将会在 4.2 节中详细介绍。

3．动作插件

和 Ansible 1 一样，每个模块都会对应一个动作插件，并且大部分模块会共用一个插件。动作插件中会统一定义一个 ActionModule 类，最核心的就是该类中的 run()方法。通过阅读该方法的代码就能了解该模块的核心运行逻辑，再加上对模块的源码分析，就能完全掌握该模块的运行过程。

4．缓存插件

缓存插件在 Ansible 1.9.6 中已经出现过，Ansible 2 中的缓存插件无论是功能还是实现都和 Ansible 1 类似，这里不再过多描述。

5．become插件

become 插件主要用于实现提权的功能。对于 CentOS 系统而言，使用最多的两个 become 插件为 su.pys 和 sudo.py。前者用于 su 切换用户，后者用于 sudo 提权。sudo.py 插件中的内容如下：

```
# 源码位置：lib/ansible/plugins/become/sudo.py
# ...

from ansible.plugins.become import BecomeBase


class BecomeModule(BecomeBase):

    name = 'sudo'

    # 失败信息
    fail = ('Sorry, try again.',)
    missing = ('Sorry, a password is required to run sudo', 'sudo: a password
is required')
```

```
def build_become_command(self, cmd, shell):
    super(BecomeModule, self).build_become_command(cmd, shell)

    if not cmd:
        return cmd

    # 得到 becomecmd, 默认为 sudo
    becomecmd = self.get_option('become_exe') or self.name

    flags = self.get_option('become_flags') or ''
    prompt = ''
    if self.get_option('become_pass'):
        # 指定提权时输入密码的提示信息
        self.prompt = '[sudo via ansible, key=%s] password:' % self._id
        if flags:
            flags = flags.replace('-n', '')
        prompt = '-p "%s"' % (self.prompt)

    # 获得用户
    user = self.get_option('become_user') or ''
    if user:
        user = '-u %s' % (user)

    # 最后返回组合而成的提权命令
    return ' '.join([becomecmd, flags, prompt, user, self._build_
success_command(cmd, shell)])
```

become 插件中的 BecomeModule 类都实现了 build_become_command()方法，即得到完整的提权命令。

6. lookup插件

lookup 插件在第 3 章介绍 Playbook 语法时提到过，例如 with_items 语法的执行逻辑对应该插件组下的 items.py 插件，with_dict 语法对应 dict.py 插件。这些插件控制着 Playbook 中相应语法的功能，具体调用细节将在后面介绍 Ansible 2 中 Playbook 的运行流程时进行讲解，这里不再过多描述。

4.2　Ansible 2.8 模块运行的核心流程分析

模块的完整运行流程在 Ansible 1.1 和 Ansible 1.9.6 中已经各分析过一次。在 Ansible 2.8 中，Ansible 的核心运行逻辑并没有改变，但是代码变得更加复杂和抽象。本节将一步一步跟踪 Ansible 中 ping 模块的整个执行流程，带领读者梳理 Ansible 2.8 中模块执行的核心源码。

4.2.1 命令入口

第 3 章介绍 Ansible 2 命令行的用法时简单分析了 ansible、ansible-playbook 等命令的入口方法，下面直接分析 ansible 命令入口的源码，具体如下：

```python
# 源码位置: lib/ansible/cli/adhoc.py
# ...

class AdHocCLI(CLI):
    # ...

    def run(self):

        # ...

        # 1. 获取基本对象
        loader, inventory, variable_manager = self._play_prereqs()

        try:
            # 2. 获取匹配的 Ansible 主机列表
            hosts = self.get_host_list(inventory, context.CLIARGS['subset'],
pattern)
        except AnsibleError:
            # 处理异常
            # ...

        # 3. 如果有--list-hosts 选项，则展示匹配的 hosts 主机后退出
        if context.CLIARGS['listhosts']:
            display.display('  hosts (%d):' % len(hosts))
            for host in hosts:
                display.display('    %s' % host)
            return 0

        # 4. 封装任务的 Play 对象
        play_ds = self._play_ds(
                pattern,
                context.CLIARGS['seconds'],
                context.CLIARGS['poll_interval']
            )
        play = Play().load(play_ds, variable_manager=variable_manager,
loader=loader)

        playbook = Playbook(loader)
        playbook._entries.append(play)
        playbook._file_name = '__adhoc_playbook__'

        # ...

        self._tqm = None
        try:
            # 5. 封装 TaskQueueManager 对象
```

```
        self._tqm = TaskQueueManager(
            inventory=inventory,
            variable_manager=variable_manager,
            loader=loader,
            passwords=passwords,
            stdout_callback=cb,
          run_additional_callbacks=C.DEFAULT_LOAD_CALLBACK_PLUGINS,
            run_tree=run_tree,
            forks=context.CLIARGS['forks'],
        )

        # 设置回调
        self._tqm.send_callback('v2_playbook_on_start', playbook)

        # 6. 最核心执行语句，用于运行任务
        result = self._tqm.run(play)

        # 设置回调
        self._tqm.send_callback('v2_playbook_on_stats', self._tqm._stats)
    finally:
        # 清理工作
        # ...

    return result
```

上面代码的执行逻辑非常清楚，笔者整理了 6 处比较重要的代码段，具体分析如下。

第 1 处代码段的含义是获取基本对象。例如之前经常提到的 DataLoader 对象、InventoryManager 对象和 VariableManager 对象，它们三个都是封装的最基本的类，用于导入数据、解析 inventory 文件和管理变量。

第 2 处代码通过 AdHocCLI 对象的 get_host_list()方法获取当前匹配的 Ansible 主机。此时必须传入 inventory 参数和相应的匹配表达式。该方法的实现源码如下：

```
# 源码位置: lib/ansible/cli/__init__.py
# ...

class CLI(with_metaclass(ABCMeta, object)):
    # ...

    @staticmethod
    def get_host_list(inventory, subset, pattern='all'):

        no_hosts = False
        if len(inventory.list_hosts()) == 0:
            # 如果设置了相关警告参数，则调用 display.warning()方法打印警告信息
            # ...
            no_hosts = True

        inventory.subset(subset)

        # 最终调用 InventoryManager 对象的 list_hosts()方法获取匹配的 Ansible 主机
        hosts = inventory.list_hosts(pattern)
        if not hosts and no_hosts is False:
```

```
        # 抛出异常
        # ...

    return hosts
```

通过上面的代码可以看到，get_host_list()方法最终是调用 InventoryManager 对象的 list_hosts()方法并结合 pattern 参数获取匹配的 Ansible 主机，而 InventoryManager 类的详细用法可以参考 4.1.3 节的内容，这里不再过多说明。

第 3 处代码是处理--list-hosts 选项。如果有该选项，则只是展示匹配的 hosts 主机，然后直接退出，否则继续向下执行。

第 4 处代码是封装 Play 对象，接下来作为 TaskQueueManager 对象运行任务（run()方法）的参数。

第 5 处代码是封装 TaskQueueManager 对象，非常重要。

第 6 处代码正是 Ansible 模块执行的核心方法，也是后续追踪模块运行的开始。

到目前为止，ansible 命令的入口代码已经很明确了，为了能进一步掌握 Ansible 2.8 中模块的运行流程，需要深入两个类（Play 和 TaskQueueManager）和一个方法（TaskQueueManager 类中的 run()方法）的学习，这也是 4.2.2 节分析源码的重点。

4.2.2　核心运行模块

本节将重点介绍 Play 类、TaskQueueManager 类以及 TaskQueueManager 类中的 run() 方法的源码，深入分析一个普通 Ansible 模块的运行全流程。

1. Play类

这里 Play 类的含义与 Ansible 1 中一致，它表示一个剧本，而该剧本中会包含多个任务（task）。该类的源码定义如下：

```python
# 源码位置: lib/ansible/playbook/play.py
# ...

class Play(Base, Taggable, Become, CollectionSearch):
    # ...

    def __repr__(self):
        return self.get_name()

    def get_name(self):
        ''' 返回 Play 的名称 '''
        return self.name

    @staticmethod
    def load(data, variable_manager=None, loader=None, vars=None):
        if ('name' not in data or data['name'] is None) and 'hosts' in data:
            if isinstance(data['hosts'], list):
                data['name'] = ','.join(data['hosts'])
```

```
        else:
            data['name'] = data['hosts']
    p = Play()
    if vars:
        p.vars = vars.copy()
    return p.load_data(data, variable_manager=variable_manager, loader=
loader)

    # ...
```

这里需要回顾下 Ansible 1 中关于 Play 类的一些分析经验。当时介绍 Play 类的实例化时提到，需要传入一个非常重要的参数 ds 用于表示剧本的数据信息。现在用前面的代码测试一下，具体操作如下：

```
>>> from ansible.parsing.dataloader import DataLoader
>>> from ansible.inventory.manager import InventoryManager
>>> from ansible.vars.manager import VariableManager
>>> loader = DataLoader()
>>> inventory = InventoryManager(loader=loader, sources="/root/hosts")
>>> variable_manager = VariableManager(loader=loader, inventory=inventory,
version_info=None)
# 可以在原语句后加一个 print()方法打印 play_ds 结果，然后执行[ansible ceph-1 -i
  hosts -m ping]命令来看最后 play_ds 的值，最后用在这里
>>> play_ds = {'name': 'Ansible Ad-Hoc', 'hosts': 'ceph-1', 'gather_facts':
'no', 'tasks': [{'action': {'module': 'ping', 'args': {}}, 'async_val': 0,
'poll': 15}]}
>>> from ansible.playbook.play import Play
>>> play = Play().load(play_ds, variable_manager=variable_manager, loader=
loader)
>>> play._ds
{'name': 'Ansible Ad-Hoc', 'hosts': 'ceph-1', 'gather_facts': 'no', 'tasks':
[{'action': {'module': 'ping', 'args': {}}, 'async_val': 0, 'poll': 15}]}
>>>
```

从上面的操作中可以看到，前面命令入口代码中第 4 处代码的 play_ds 其实就是 Play 对象的数据信息，该 Play 中会包含 hosts 以及 tasks 等属性以用于后续的执行，而这和 Ansible 1 中 Play 的执行过程是一致的，只不过 Ansible 2 中将模块的运行和 Play 的执行统一到了 TaskQueueManager 对象中。接下来对 Play 类中定义的相关方法进行操作以加深印象，具体如下：

```
>>> play.get_roles()
[]
>>> play.get_tasks()
[[TASK: ping]]
>>> play.serialize()
{'hosts': 'ceph-1', 'gather_facts': 'no', 'gather_subset': ['all'],
'gather_timeout': 10, 'fact_path': None, 'vars_files': [], 'vars_prompt':
[], 'roles': [], 'handlers': [], 'pre_tasks': [], 'post_tasks': [], 'tasks':
[BLOCK(uuid=000c295e-8383-2e23-5177-00000000000b) (id=140050401328720)
(parent=None)], 'force_handlers': None, 'max_fail_percentage': None,
'serial': [], 'strategy': 'linear', 'order': None, 'name': 'Ansible Ad-Hoc',
'connection': None, 'port': None, 'remote_user': None, 'vars': {}, 'module_
defaults': None, 'environment': None, 'no_log': None, 'run_once': None,
```

```
'ignore_errors': None, 'ignore_unreachable': None, 'check_mode': None,
'diff': None, 'any_errors_fatal': False, 'debugger': None, 'tags': [],
'become': None, 'become_method': None, 'become_user': None, 'become_flags':
None, 'collections': None, 'uuid': '000c295e-8383-2e23-5177-00000000000a',
'finalized': False, 'squashed': False, 'included_path': None}
>>> play.get_vars()
{}
>>> play.get_handlers()
[]
>>> play.hosts
'ceph-1'
>>> play.tasks
[BLOCK(uuid=000c295e-8383-2e23-5177-00000000000b)(id=140050401328720)
(parent=None)]
```

当然还有载入前置任务（_load_pre_tasks()）、后置任务（_load_post_tasks()）、加载角色（_load_roles()）等操作，都和运行 Playbook 相关，这里就不再演示了。读者只需要掌握 play.get_tasks()的输出即可，后续会根据 hosts 和 tasks 去调用模块执行任务。

2. TaskQueueManager类

再来看整个 Ansible 2 的核心类——TaskQueueManager，该类位于 lib/ansible/executor 目录下。而在 Ansible 1 中，同样功能的核心代码则位于 lib/ansible/runner 目录中，而 Ansible 2 对此目录下的代码进行了全面重构。直接看该类的实现源码：

```python
# 源码位置: lib/ansible/executor/task_queue_manager.py
# ...

class TaskQueueManager:

    # 定义任务的执行状态
    # ...

    def __init__(self, inventory, variable_manager, loader, passwords,
stdout_callback=None, run_additional_callbacks=True, run_tree=False,
forks=None):

        # 设置基本对象
        self._inventory = inventory
        self._variable_manager = variable_manager
        self._loader = loader
        # 回调统计
        self._stats = AggregateStats()
        # 保存 SSH 连接的密码和 sudo 的提取密码
        self.passwords = passwords
        # 设置回调函数
        self._stdout_callback = stdout_callback
        self._run_additional_callbacks = run_additional_callbacks
        self._run_tree = run_tree
        # 指定运行进程数
        self._forks = forks or 5
```

```
        # 回调相关参数
        self._callbacks_loaded = False
        self._callback_plugins = []
        self._start_at_done = False

        # 一些不重要的信息
        # ...

        # 保存和追踪执行失败和无法访问的主机信息
        self._failed_hosts = dict()
        self._unreachable_hosts = dict()

        try:
            # 创建多进程队列
            self._final_q = multiprocessing.Queue()
        except OSError as e:
            # 抛出异常
            # ...

        # A temporary file (opened pre-fork) used by connection
        # plugins for inter-process locking.
        self._connection_lockfile = tempfile.TemporaryFile()

    def _initialize_processes(self, num):
        self._workers = []

        for i in range(num):
            self._workers.append(None)

    # ...
```

上面只给出了 TaskQueueManager 类的初始化方法，内容并不复杂。注意到最后创建了一个多进程队列，它和 Ansible 1 中多进程队列的功能是一样的。下面在前面命令行的基础上手工创建一个 TaskQueueManager 对象，操作如下：

```
>>> from ansible.executor.task_queue_manager import TaskQueueManager
>>> passwords = {'conn_pass': None, 'become_pass': None}
>>> tqm = TaskQueueManager(inventory=inventory, variable_manager=variable_
manager,loader=loader, passwords=passwords)
```

如果这里直接调用 tqm 的 run()方法，会执行 Play 中的 task 任务吗？尝试进行以下操作：

```
>>> result = tqm.run(play)

PLAY [Ansible Ad-Hoc] ********************************************************

TASK [ping] *****************************************************************
fatal: [ceph-1]: FAILED! => {"msg": "the connection plugin '<class 'ansible.
utils.sentinel.Sentinel'>' was not found"}
```

很明显，不能直接这样运行 Play 对象，中间肯定有一些步骤在前面被忽略了。那么产生这个运行错误的原因是什么，又该如何解决这个报错，确保能正确执行 Play 呢？这些问题将在完成 Ansible 2.8 的整个模块运行流程追踪后再解决。

3. run()方法

现在到了最核心的源码追踪部分，这里的过程会非常漫长且涉及多个类和源码文件，请耐心阅读。

```
# 源码位置: lib/ansible/executor/task_queue_manager.py
# ...

class TaskQueueManager:

    # ...

    def run(self, play):

        # 载入回调
        if not self._callbacks_loaded:
            self.load_callbacks()

        # 获取所有变量
        all_vars = self._variable_manager.get_vars(play=play)
        warn_if_reserved(all_vars)
        # 前面介绍过的, 用于模板变量的渲染
        templar = Templar(loader=self._loader, variables=all_vars)

        new_play = play.copy()
        new_play.post_validate(templar)
        new_play.handlers = new_play.compile_roles_handlers() + new_play.
handlers

        # HostVars 对象
        self.hostvars = HostVars(
            inventory=self._inventory,
            variable_manager=self._variable_manager,
            loader=self._loader,
        )

        play_context = PlayContext(new_play, self.passwords, self._connection_
lockfile.fileno())

        # ...

        self.send_callback('v2_playbook_on_play_start', new_play)

        # build the iterator
        iterator = PlayIterator(
            inventory=self._inventory,
            play=new_play,
            play_context=play_context,
            variable_manager=self._variable_manager,
            all_vars=all_vars,
            start_at_done=self._start_at_done,
```

```
    )

    # 调整 worker 进程数
    self._initialize_processes(min(self._forks, iterator.batch_size))

    # 加载具体的策略插件，默认为 linear 插件
    strategy = strategy_loader.get(new_play.strategy, self)

    # 策略插件为空，直接抛出异常
    # ...

    # 标记上一次执行失败的 hosts
    for host_name in self._failed_hosts.keys():
        host = self._inventory.get_host(host_name)
        iterator.mark_host_failed(host)

    self.clear_failed_hosts()

    # ...

    # 最核心的语句，通过策略插件来运行 Play
    play_return = strategy.run(iterator, play_context)

    # 记录执行失败的主机
    # ...

    # 收尾工作
    # ...

    return play_return
```

上面的 run() 方法涉及 PlayContext 类、PlayIterator 类以及默认策略类。其中最重要的当属默认的策略类了，那么 Ansible 2 中默认的策略类究竟是什么呢？加载策略插件的代码如下：

```
# 源码位置：lib/ansible/plugins/loader.py
# ...

strategy_loader = PluginLoader(
    'StrategyModule',
    'ansible.plugins.strategy',
    C.DEFAULT_STRATEGY_PLUGIN_PATH,
    'strategy_plugins',
    required_base_class='StrategyBase',
)
```

此外，在交互模式下执行如下操作可以看到更为直观的结果：

```
>>> play.strategy
'linear'
>>> new_play = play.copy()
>>> new_play.strategy
'linear'
# tqm 就是前面创建的 TaskQueueManager 对象
```

```
>>> strategy_loader.get(new_play.strategy, tqm)
<ansible.plugins.strategy.linear.StrategyModule object at 0x7f78bb54fdf0>
```

通过上面的结果可知，最后得到的 strategy 其实就是 linear.py 插件中的 StrategyModule 对象，然后调用该对象的 run()方法去执行模块任务。该 run()方法有两个参数：PlayIterator 对象和 PlayContext 对象。下面分别学习这两个对象对应类的源码，然后才能去学习 run() 方法的代码。

首先学习 PlayContext 类，其源码实现如下：

```python
# 源码位置: lib/ansible/playbook/play_context.py
# ...

class PlayContext(Base):

    # 定义了众多属性
    # ...

    def __init__(self, play=None, passwords=None, connection_lockfd=None):

        super(PlayContext, self).__init__()

        if passwords is None:
            passwords = {}

        # 设置连接的密码和 sudo 提权密码
        self.password = passwords.get('conn_pass', '')
        self.become_pass = passwords.get('become_pass', '')

        # become 插件
        self._become_plugin = None

        self.prompt = ''
        self.success_key = ''

        # 文件描述符，和加锁相关
        self.connection_lockfd = connection_lockfd

        if context.CLIARGS:
            self.set_attributes_from_cli()

        if play:
            self.set_attributes_from_play(play)

    def set_attributes_from_play(self, play):
        self.force_handlers = play.force_handlers

    def set_attributes_from_cli(self):

        # 设置一些远程连接时的属性，例如私钥文件、SSH 连接的额外参数等
        if context.CLIARGS.get('timeout', False):
            self.timeout = int(context.CLIARGS['timeout'])

        self.private_key_file = context.CLIARGS.get('private_key_file')
```

```python
        self.verbosity = context.CLIARGS.get('verbosity')
        self.ssh_common_args = context.CLIARGS.get('ssh_common_args')
        self.ssh_extra_args = context.CLIARGS.get('ssh_extra_args')
        self.sftp_extra_args = context.CLIARGS.get('sftp_extra_args')
        self.scp_extra_args = context.CLIARGS.get('scp_extra_args')

        self.start_at_task = context.CLIARGS.get('start_at_task', None)

    def set_task_and_variable_override(self, task, variables, templar):
        # ...

    def set_become_plugin(self, plugin):
        # 设置 become 插件
        self._become_plugin = plugin

    def make_become_cmd(self, cmd, executable=None):
        """ 提升命令执行权限的辅助函数 """

        # 提示 deprecated 信息，该函数已经不再使用，可以使用 become 插件代替
        # ...

        if not cmd or not self.become:
            # cmd 为空或者 self.become 为 False 时，直接返回命令
            return cmd

        become_method = self.become_method

        # load/call become plugins here
        plugin = self._become_plugin

        if plugin:
            options = {
                'become_exe': self.become_exe or become_method,
                'become_flags': self.become_flags or '',
                'become_user': self.become_user,
                'become_pass': self.become_pass
            }
            plugin.set_options(direct=options)

            if not executable:
                executable = self.executable

            shell = get_shell_plugin(executable=executable)
            # 使用 become 插件最终获取 sudo 提权的完整 cmd 命令
            cmd = plugin.build_become_command(cmd, shell)
            if self.become_pass:
                # 设置了 self.become_pass 时，则可能在提权中会有输入密码的提示，可由
                  插件设置
                self.prompt = plugin.prompt
        else:
            # 抛出异常
            # ...

        return cmd
```

```python
    def update_vars(self, variables):

        for prop, var_list in C.MAGIC_VARIABLE_MAPPING.items():
            try:
                # 忽略包含 become 字符串的 key
                if 'become' in prop:
                    continue

                # 从该对象中获取对应变量的值
                var_val = getattr(self, prop)
                for var_opt in var_list:
                    if var_opt not in variables and var_val is not None:
                        # 给 variables 添加一个新的 key-value 值
                        variables[var_opt] = var_val
            except AttributeError:
                continue

    def _get_attr_connection(self):
        conn_type = None
        # 确认全局配置中设置的连接属性是否为 smart
        if self._attributes['connection'] == 'smart':
            # 设置为 SSH 方式连接
            conn_type = 'ssh'
            if not check_for_controlpersist(self.ssh_executable) and paramiko
    is not None:
                # SSH 连接方式需要 controlpersist 特性支持，如果不支持且 paramiko
                #    模块存在，则重新设置连接方式为基于 paramiko 模块方式
                conn_type = "paramiko"

        elif self._attributes['connection'] == 'persistent' and paramiko is
    not None:
            # 将 connection 属性设置为 persistent 时，直接设置连接方式为 paramiko
            conn_type = 'paramiko'

        if conn_type:
            # 更新到 connection 属性中
            self.connection = conn_type

        return self._attributes['connection']
```

上述代码中，需要注意在 update_vars()方法中出现的常量 C.MAGIC_VARIABLE_
MAPPING，其内容如下：

```python
# 源码位置：lib/ansible/constants.py
# ...

MAGIC_VARIABLE_MAPPING = dict(

    # base
    connection=('ansible_connection', ),
    module_compression=('ansible_module_compression', ),
    shell=('ansible_shell_type', ),
    executable=('ansible_shell_executable', ),
```

```
# connection common
remote_addr=('ansible_ssh_host', 'ansible_host'),
remote_user=('ansible_ssh_user', 'ansible_user'),
password=('ansible_ssh_pass', 'ansible_password'),
port=('ansible_ssh_port', 'ansible_port'),
pipelining=('ansible_ssh_pipelining', 'ansible_pipelining'),
timeout=('ansible_ssh_timeout', 'ansible_timeout'),
private_key_file=('ansible_ssh_private_key_file', 'ansible_private_
key_file'),

# networking modules
network_os=('ansible_network_os', ),
connection_user=('ansible_connection_user',),

# ssh TODO: remove
ssh_executable=('ansible_ssh_executable', ),
ssh_common_args=('ansible_ssh_common_args', ),
sftp_extra_args=('ansible_sftp_extra_args', ),
scp_extra_args=('ansible_scp_extra_args', ),
ssh_extra_args=('ansible_ssh_extra_args', ),
ssh_transfer_method=('ansible_ssh_transfer_method', ),

# docker TODO: remove
docker_extra_args=('ansible_docker_extra_args', ),

# become
become=('ansible_become', ),
become_method=('ansible_become_method', ),
become_user=('ansible_become_user', ),
become_pass=('ansible_become_password', 'ansible_become_pass'),
become_exe=('ansible_become_exe', ),
become_flags=('ansible_become_flags', ),

# deprecated
sudo=('ansible_sudo', ),
sudo_user=('ansible_sudo_user', ),
sudo_pass=('ansible_sudo_password', 'ansible_sudo_pass'),
sudo_exe=('ansible_sudo_exe', ),
sudo_flags=('ansible_sudo_flags', ),
su=('ansible_su', ),
su_user=('ansible_su_user', ),
su_pass=('ansible_su_password', 'ansible_su_pass'),
su_exe=('ansible_su_exe', ),
su_flags=('ansible_su_flags', ),
)

# ...
```

从这里可以看到在 hosts 甚至 ansible.cfg 中的许多配置选项，如 ansible_connection 等，它们会影响 SSH 远程连接。例如，以什么方式进行 SSH 连接，以什么用户进行远程连接，以及远程提权的用户和密码，等等。接下来将对该类进行实战以帮助读者更好地理解上面的代码，具体操作如下：

```
# 注意，在前面的命令行基础上继续执行
>>> import tempfile
>>> connection_lockfile = tempfile.TemporaryFile()
>>> from ansible.playbook.play_context import PlayContext
# new_paly 和 passwords 就是前面命令行的结果，同一个命令行下不再重复导入
>>> play_context = PlayContext(new_play, passwords, connection_lockfile.
fileno())
```

上面得到了 PlayContext 对象后，来看一下 make_become_cmd()方法的输出。这里通过前面的代码可以看到，想要正确输出提权后的命令，需要完成以下两个动作：

- 加载一个 become 插件。通常情况下，选择 sudo 插件。
- 设置 PlayContext 对象的 become 属性为 True。通过代码可以看到，该属性值为 False 时，将直接返回命令本身，并不进行任何其他操作。

以下是 make_become_cmd()方法的调用示例：

```
>>> from ansible.plugins.loader import become_loader
# 获取对应 become 插件下的 BecomeModule 对象
>>> become_plugin = become_loader.get("sudo")
>>> play_context.set_become_plugin(become_plugin)
# 必须要设置 become 属性为 True，才能得到相应的提权后的命令
>>> play_context.become = True
>>> play_context.make_become_cmd('cat /etc/sudoers')
"sudo   /bin/sh -c 'echo BECOME-SUCCESS-hqjwlyxnnrbupfjntewdeqhxmoloyzbj ;
 cat /etc/sudoers'"
```

再来看 update_vars()方法的操作示例，具体如下：

```
>>> variables = {"x": 123}
>>> play_context.update_vars(variables)
>>> variables
{'x': 123, 'ansible_connection': <class 'ansible.utils.sentinel.Sentinel'>,
'ansible_module_compression': 'ZIP_DEFLATED', 'ansible_shell_executable':
'/bin/sh', 'ansible_ssh_pipelining': False, 'ansible_pipelining': False,
'ansible_ssh_timeout': 10, 'ansible_timeout': 10, 'ansible_ssh_executable':
'ssh'}
```

最后是_get_attr_connection()方法，该方法比较简单，就是选择底层的通信模式。这里的代码主要区分两种模式：

- 基于 SSH 命令进行远程通信。
- 基于 paramiko 模块方式封装远程通信操作。

这两种方式的代码分别对应 plugins/connection 目录下的 ssh.py 和 paramiko_ssh.py 文件，这和 Ansible 1 中是一样的。

```
>>> play_context._get_attr_connection()
<class 'ansible.utils.sentinel.Sentinel'>
>>> play_context._attributes['connection'] = 'ssh'
>>> play_context._get_attr_connection()
'ssh'
```

注意到_get_attr_connection()方法的默认输出并不是前面熟悉的'paramiko'或者'ssh'字符串，而是一个类，而这个类和前面执行 tqm.run(play)的报错信息似乎有些关联。这里先

记住出现的现象，后续深入源码学习时会再次回到这里的代码。

下面继续学习 PlayIterator 类，其源码实现如下：

```
# 源码位置: lib/ansible/executor/play_iterator.py
# ...

class PlayIterator:
    # ...

    def __init__(self, inventory, play, play_context, variable_manager,
all_vars, start_at_done=False):
        # ...

        self._host_states = {}
        start_at_matched = False
        batch = inventory.get_hosts(self._play.hosts, order=self._play.
order)
        self.batch_size = len(batch)
        for host in batch:
            # 得到 HostState 对象并保存
            self._host_states[host.name] = HostState(blocks=self._blocks)
            if play_context.start_at_task is not None and not start_at_done:
                while True:
                    (s, task) = self.get_next_task_for_host(host, peek=True)
                    if s.run_state == self.ITERATING_COMPLETE:
                        break
                    if task.name == play_context.start_at_task or \
                    fnmatch.fnmatch(task.name, play_context.start_at_task) or \
                        task.get_name() == play_context.start_at_task or \
                    fnmatch.fnmatch(task.get_name(), play_context.start_at_
task):
                        start_at_matched = True
                        break
                    else:
                        self.get_next_task_for_host(host)

            # ...

    def get_next_task_for_host(self, host, peek=False):

        s = self.get_host_state(host)

        task = None
        if s.run_state == self.ITERATING_COMPLETE:
            return (s, None)

        (s, task) = self._get_next_task_from_state(s, host=host, peek=peek)

        if not peek:
            self._host_states[host.name] = s

        return (s, task)

    def _get_next_task_from_state(self, state, host, peek, in_child=False):
```

```
            # 代码比较复杂，后续演示即可
            # ...

    # ...
```

PlayIterator 对象中的 get_next_task_for_host()方法会在后面被调用，用于获取下一个执行的主机与任务。由于其对应的_get_next_task_from_state()方法较为复杂，不便于展开描述，因此这里将提前使用后面用到的语句来测试该方法的功能。为了能更好地模拟多主机的情况，将前面手动输入的 play_ds 修改为如下值：

```
play_ds = {'name': 'Ansible Ad-Hoc', 'hosts': 'nodes', 'gather_facts': 'no',
'tasks': [{'action': {'module': 'ping', 'args': {}}, 'async_val': 0, 'poll':
15}]}
```

这样在 Play 中就有 3 个主机任务待执行。和前面一样，生成 inventory、variable_manager、new_play 和 play_context 值，它们会作为参数去实例化 PlayIterator 对象，具体操作如下：

```
# 忽略前面生成的 inventory、variable_manager、new_play 和 play_context 值
# ...
>>> from ansible.executor.play_iterator import PlayIterator
>>> iterator = PlayIterator(inventory=inventory, play=new_play, play_
context=play_context, variable_manager=variable_manager, all_vars={},
start_at_done=False)
```

下面的语句都会在后续源码分析中看到，这里提前借用这些语句演示 PlayIterator 对象的使用。

```
>>> hosts = inventory.get_hosts(iterator._play.hosts, order=iterator.
_play.order)
>>> hosts
[ceph-1, ceph-2, ceph-3]
>>> type(hosts[0])
<class 'ansible.inventory.host.Host'>
```

上面得到的 hosts 为 Host 对象列表，分别代表匹配的 3 个 Ansible 主机。接着调用 PlayIterator 对象的 get_next_task_for_host() 方法反复调用 ceph-1，结果如下：

```
>>> iterator.get_next_task_for_host(hosts[0])
(HostState([BLOCK(uuid=000c295e-8383-cb6e-a711-000000000018)(id=1402421
36800752)(parent=None), BLOCK(uuid=000c295e-8383-cb6e-a711-00000000001b)
(id=140242136851120)(parent=None), BLOCK(uuid=000c295e-8383-cb6e-a711-
00000000000b)(id=140242136851168)(parent=None), BLOCK(uuid=000c295e-8383
-cb6e-a711-00000000001b)(id=140242136851408)(parent=None), BLOCK(uuid=
000c295e-8383-cb6e-a711-00000000001b)(id=140242136851696)(parent=None)]),
TASK: meta (flush_handlers))
>>> iterator.get_next_task_for_host(hosts[0])
(HostState([BLOCK(uuid=000c295e-8383-cb6e-a711-000000000018)(id=1402421
36800752)(parent=None), BLOCK(uuid=000c295e-8383-cb6e-a711-00000000001b)
(id=140242136851120)(parent=None), BLOCK(uuid=000c295e-8383-cb6e-a711-
00000000000b)(id=140242136851168)(parent=None), BLOCK(uuid=000c295e-8383-
cb6e-a711-00000000001b)(id=140242136851408)(parent=None), BLOCK(uuid=
000c295e-8383-cb6e-a711-00000000001b)(id=140242136851696)(parent=None)]),
TASK: ping)
```

```
>>> iterator.get_next_task_for_host(hosts[0])
(HostState([BLOCK(uuid=000c295e-8383-cb6e-a711-000000000018)(id=1402421
36800752)(parent=None), BLOCK(uuid=000c295e-8383-cb6e-a711-00000000001b)
(id=140242136851120)(parent=None), BLOCK(uuid=000c295e-8383-cb6e-a711-
00000000000b)(id=140242136851168)(parent=None), BLOCK(uuid=000c295e-8383-
cb6e-a711-00000000001b)(id=140242136851408)(parent=None), BLOCK(uuid=
000c295e-8383-cb6e-a711-00000000001b)(id=140242136851696)(parent=None)]),
TASK: meta (flush_handlers))
>>> iterator.get_next_task_for_host(hosts[0])
(HostState([BLOCK(uuid=000c295e-8383-cb6e-a711-000000000018)(id=1402421
36800752)(parent=None), BLOCK(uuid=000c295e-8383-cb6e-a711-00000000001b)
(id=140242136851120)(parent=None), BLOCK(uuid=000c295e-8383-cb6e-a711-
00000000000b)(id=140242136851168)(parent=None), BLOCK(uuid=000c295e-8383-
cb6e-a711-00000000001b)(id=140242136851408)(parent=None), BLOCK(uuid=
000c295e-8383-cb6e-a711-00000000001b)(id=140242136851696)(parent=None)]),
TASK: meta (flush_handlers))
>>> iterator.get_next_task_for_host(hosts[0])
(HostState([BLOCK(uuid=000c295e-8383-cb6e-a711-000000000018)(id=1402421
36800752)(parent=None), BLOCK(uuid=000c295e-8383-cb6e-a711-00000000001b)
(id=140242136851120)(parent=None), BLOCK(uuid=000c295e-8383-cb6e-a711-
00000000000b)(id=140242136851168)(parent=None), BLOCK(uuid=000c295e-8383-
cb6e-a711-00000000001b)(id=140242136851408)(parent=None), BLOCK(uuid=
000c295e-8383-cb6e-a711-00000000001b)(id=140242136851696)(parent=None)]),
None)
```

　　peek 参数默认为 False，因此针对主机 ceph-1，每次调用 get_next_task_for_host()方法将得到 ceph-1 主机的多个任务，包括被附加在主任务（执行 ping 模块）前的 meta 任务以及之后的两次 meta 任务。当 ceph-1 没有其他任务时，再次调用该方法，则输出的结果中 task=None。如果设置 peek 为 True，则每次调用的结果都相同，只会停留在当前任务上。了解了这些后，对于理解后续 Ansible 源码已经足够，至于一些代码细节就不在此深入解析了。

　　最后再来看策略插件的 run()方法，该方法位于 linear.py 插件的 StrategyModule 类中。对于该线性的策略插件，源码中有清晰的说明（位于 run()方法中），具体如下：

```
The linear strategy is simple - get the next task and queue it for all hosts,
then wait for the queue to drain before moving on to the next task
```

　　翻译过来就是说线性（linear）策略比较简单，用于获取下一个任务，将所有的 Ansible 目标主机入队，然后等待队列为空，即队列中的所有目标主机执行任务完毕，然后继续下一个任务。这其实和 Ansible 1 中多进程执行任务的过程是一样的，都是通过进程队列来同步执行任务。Ansible 2 中除了这种线性策略外，还支持调试策略（debug.py）、free 策略、host_pinned 策略等。这种插件方式使得 Ansible 2 相比 Ansible 1 功能更强大且扩展性更好。

　　策略插件的 run()方法的源码如下：

```
# 源码位置: lib/ansible/plugins/strategy/linear.py
# ...

class StrategyModule(StrategyBase):
    # ...
```

```python
    def run(self, iterator, play_context):

        # 循环执行每一个任务
        result = self._tqm.RUN_OK
        work_to_do = True
        while work_to_do and not self._tqm._terminated:

            try:
                # 1. 获取执行主机
                hosts_left = self.get_hosts_left(iterator)

                callback_sent = False
                # 用来控制循环执行任务的标识，非常重要
                work_to_do = False

                host_results = []
                # 2. 获取主机与任务
                host_tasks = self._get_next_task_lockstep(hosts_left, iterator)

                # 跳过执行标识
                skip_rest = False
                choose_step = True

                # 忽略任何异常标识
                any_errors_fatal = False

                # 3. 核心代码，用于处理 task
                results = []
                for (host, task) in host_tasks:
                    # 当前任务为空，继续下一个
                    if not task:
                        continue

                    if self._tqm._terminated:
                        break

                    run_once = False
                    # 如果有任务执行，这里便设置为 True
                    work_to_do = True

                    # 获取任务的动作。如果是模块任务，action 表示模块名
                    try:
                        action = action_loader.get(task.action, class_only=
True)
                    except KeyError:
                        action = None

                    # 如果该任务所属的角色已经被执行过，则不需要再运行
                    # ...

                    if task.action == 'meta':
                        # 处理默认的 meta 任务
                    else:
```

```
# 处理 step
# ...

# 获取任务变量
task_vars = self._variable_manager.get_vars(
        play=iterator._play, host=host, task=task)
self.add_tqm_variables(task_vars, play=iterator._play)
# 得到 Templar 对象
templar = Templar(loader=self._loader, variables=
task_vars)

# 得到渲染后的 run_once 值
run_once = templar.template(task.run_once) or action
and getattr(action,
        'BYPASS_HOST_LOOP', False)

# 检查是否设置 any_errors_fatal 标识，如果设置为 True，一旦
  任何一个节点的该任务执行异常将中止后续任务的执行
if (task.any_errors_fatal or run_once) and not task.
ignore_errors:
    any_errors_fatal = True

# Playbook 第一次执行时，执行 v2_playbook_on_task_start()
  回调函数
if not callback_sent:
    saved_name = task.name
    try:
        # 这里说明 task 的 name 属性可以写成模板形式
        task.name = to_text(templar.template(task.name,
                fail_on_undefined=False), nonstring=
'empty')
    except Exception:
        # 打印调试信息
        # ...

    # 执行任务开始时的回调函数
    self._tqm.send_callback(
            'v2_playbook_on_task_start', task, is_conditional=
False)
    task.name = saved_name
    # 设置为 True，则下次执行 task 时不会再执行上面的回调函数了
    callback_sent = True

# 记录执行的 hosts
self._blocked_hosts[host.get_name()] = True
# 核心代码
self._queue_task(host, task, task_vars, play_context)
del task_vars

# 如果前面设置了只执行一次，则直接跳出 host 循环
if run_once:
    break
```

```
                     # 4. 非常核心的一个处理任务结果队列的方法
                     results += self._process_pending_results(
                         iterator, max_passes=max(1, int(len(self._tqm._workers)
    * 0.1)))

                     # 如果上面的结果为跳过，则直接返回继续执行下一个主机或者任务组
                     if skip_rest:
                         continue

                     # 5. 等到所有主机执行任务完毕
                     if self._pending_results > 0:
                         results += self._wait_on_pending_results(iterator)

                     # 所有主机执行该任务的结果列表
                     host_results.extend(results)

                     # 6. 保存当前活跃的持久连接
                     self.update_active_connections(results)

                     # 处理后续的结果
                     # ...

              except (IOError, EOFError) as e:
                  # most likely an abort, return failed
                  return self._tqm.RUN_UNKNOWN_ERROR

         return super(StrategyModule, self).run(iterator, play_context,
    result)
```

这个 run()方法是整个 Ansible 2 任务执行的核心代码。笔者去掉了所有 display.debug()
语句并对大部分重点语句进行了相应的注释说明。run()方法主要是通过 while 循环来不停
地执行任务队列，直到没有任何任务为止。对于 while 循环中的代码，下面将重点分析上
面标注的 6 处代码段。

第 1 处代码调用 get_hosts_left()方法获取该任务执行的目标主机。StrategyModule 类中
并没有该方法，而是在继承的父类中进行了定义与实现，具体如下：

```
# 源码位置: lib/ansible/plugins/strategy/__init__.py
# ...

class StrategyBase:

    # ...

    def get_hosts_left(self, iterator):
        '''返回本次迭代中可用的 hosts 列表，除去了不可达主机'''

        hosts_left = []
        for host in self._inventory.get_hosts(iterator._play.hosts, order=
    iterator._play.order):
            if host.name not in self._tqm._unreachable_hosts:
                hosts_left.append(host)
        return hosts_left
```

```
# ...
```

该方法的实现比较简单，只是获取将执行本次 task 的主机列表。

第 2 处代码是根据目标主机与待执行的任务，得到一个主机任务列表 host_tasks，其实现源码如下：

```
# 源码位置: lib/ansible/plugins/strategy/linear.py
# ...
class StrategyModule(StrategyBase):
    # ...

    def _get_next_task_lockstep(self, hosts, iterator):
        # ...

        host_tasks = {}
        for host in hosts:
            host_tasks[host.name] = iterator.get_next_task_for_host(host,
peek=True)

        num_setups = 0
        num_tasks = 0
        num_rescue = 0
        num_always = 0

        # 返回(host, state_task)形式的二元组列表
        host_tasks_to_run = [(host, state_task)
                             for host, state_task in iteritems(host_tasks)
                             if state_task and state_task[1]]

        if host_tasks_to_run:
            try:
                lowest_cur_block = min(
                    (iterator.get_active_state(s).cur_block for h, (s, t) in
host_tasks_to_run
                     if s.run_state != PlayIterator.ITERATING_COMPLETE))
            except ValueError:
                lowest_cur_block = None
        else:
            lowest_cur_block = None

        # 遍历 host_tasks_to_run，k 就是前面的 host，v 就是 state_task
        for (k, v) in host_tasks_to_run:
            (s, t) = v

            s = iterator.get_active_state(s)
            if s.cur_block > lowest_cur_block:
                continue

            if s.run_state == PlayIterator.ITERATING_SETUP:
                num_setups += 1
            elif s.run_state == PlayIterator.ITERATING_TASKS:
                # 如果是模块任务，一般会执行这里的代码
```

```
                num_tasks += 1
            elif s.run_state == PlayIterator.ITERATING_RESCUE:
                num_rescue += 1
            elif s.run_state == PlayIterator.ITERATING_ALWAYS:
                num_always += 1

        # 注释掉自行添加的 print()语句，后面会在测试中用到
        # print("num_setups = %d, num_tasks = %d,
                num_rescue = %d, num_always = %d" %
                (num_setups, num_tasks, num_rescue, num_always))

        def _advance_selected_hosts(hosts, cur_block, cur_state):
            rvals = []
            for host in hosts:
                host_state_task = host_tasks.get(host.name)
                if host_state_task is None:
                    continue
                (s, t) = host_state_task
                s = iterator.get_active_state(s)
                if t is None:
                    continue
                if s.run_state == cur_state and s.cur_block == cur_block:
                    new_t = iterator.get_next_task_for_host(host)
                    rvals.append((host, t))
                else:
                    rvals.append((host, noop_task))
            # 最后返回的就是(host, task)形式的二元组列表
            # print("rvals={}".format(rvals))
            return rvals

        # ...
        if num_tasks:
            return _advance_selected_hosts(hosts, lowest_cur_block,
PlayIterator.ITERATING_TASKS)

        # ...

        return [(host, None) for host in hosts]

    # ...
```

这段代码可能不好理解，但是对于 Python 代码而言，调试起来是非常方便的。这里笔者分别在上面的代码中添加了多个 print()语句，打印了运行时的变量结果。接下来在虚拟环境中测试 ping 模块的执行，具体操作如下：

```
(ansible2.8.16) [root@master ansible-2.8.16]# ansible ceph-1 -i hosts -m
ping
num_setups = 0, num_tasks = 1, num_rescue = 0, num_always = 0
rvals=[(ceph-1, TASK: meta (flush_handlers))]
num_setups = 0, num_tasks = 1, num_rescue = 0, num_always = 0
rvals=[(ceph-1, TASK: ping)]
ceph-1 | SUCCESS => {
    "ansible_facts": {
        "discovered_interpreter_python": "/usr/bin/python"
```

```
    },
    "changed": false,
    "ping": "pong"
}
num_setups = 0, num_tasks = 1, num_rescue = 0, num_always = 0
rvals=[(ceph-1, TASK: meta (flush_handlers))]
num_setups = 0, num_tasks = 1, num_rescue = 0, num_always = 0
rvals=[(ceph-1, TASK: meta (flush_handlers))]
num_setups = 0, num_tasks = 0, num_rescue = 0, num_always = 0
```

从上面的输出可以看到，一个简单的 ping 模块的执行同样会在之前和之后加上一些 meta 类型的任务，这和早期 Ansible 1 的模块运行方式类似。

第 3 处代码非常关键。该处的 for 循环主要遍历前面得到的(host, task)形式的二元组列表，对于 action 则分两种情况处理，即 meta 和其他动作（action）。特别是对后者的处理，先是获取 task 的变量，处理 run_once、any_errors_fatal 等参数，然后对 Playbook 的任务设置相应的回调函数，由参数 callback_sent 控制，会在 Playbook 的第一次任务开始时执行，后续执行时 callback_sent 已经被设置为 True，最后将任务加入任务队列中。加入任务队列的语句如下：

```
# 加入任务队列的参数有待执行的主机、任务、任务变量以及一些远程连接的相关参数
self._queue_task(host, task, task_vars, play_context)
```

第 4 处代码调用了策略插件中一个非常重要的方法——_process_pending_results()。该方法用于处理结果队列，其实现代码如下：

```
# 源码位置: lib/ansible/plugins/strategy/__init__.py
# ...

class StrategyBase:

    # ...

    @debug_closure
    def _process_pending_results(self, iterator, one_pass=False, max_
passes=None):
        # ...

        cur_pass = 0
        while True:
            try:
                self._results_lock.acquire()
                # 如果结果队列中有数据，则从左弹出一个任务的结果
                task_result = self._results.popleft()
            except IndexError:
                break
            finally:
                self._results_lock.release()

            # 对 task_result 中的信息进行提取，例如执行的主机、任务等
            original_host = get_original_host(task_result._host)
            queue_cache_entry = (original_host.name, task_result._task)
            found_task = self._queued_task_cache.get(queue_cache_entry)
```

```
['task']
        original_task = found_task.copy(exclude_parent=True, exclude_
tasks=True)
        original_task._parent = found_task._parent
        original_task.from_attrs(task_result._task_fields)

        task_result._host = original_host
        task_result._task = original_task

        # 处理一些回调情况，例如重试回调、失败和跳过的回调等
        if '_ansible_retry' in task_result._result:
            self._tqm.send_callback('v2_runner_retry', task_result)
            continue
        elif '_ansible_item_result' in task_result._result:
            if task_result.is_failed() or task_result.is_unreachable():
                self._tqm.send_callback('v2_runner_item_on_failed', task_
result)
            elif task_result.is_skipped():
                self._tqm.send_callback('v2_runner_item_on_skipped', task_
result)
            else:
                if 'diff' in task_result._result:
                    if self._diff or getattr(original_task, 'diff', False):
                        self._tqm.send_callback('v2_on_file_diff', task_
result)
                self._tqm.send_callback('v2_runner_item_on_ok', task_
result)
            continue

        if original_task.register:
            # 处理 task 中的 register 语法，注册变量
            # ...

        role_ran = False
        if task_result.is_failed():
            # 处理任务执行失败的情况
            # ...
        elif task_result.is_unreachable():
            # 处理主机不可达的情况
            # ...
        elif task_result.is_skipped():
            # 处理跳过的情况
            # ...
        else:
            role_ran = True

            # 考虑原 task 中有循环的情况，例如使用 with_items 等循环语法
            if original_task.loop:
                result_items = task_result._result.get('results', [])
            else:
                result_items = [task_result._result]

            for result_item in result_items:
                # 循环处理结果
```

```
        # ...

        # 处理需要比较文件不同的情况
        if 'diff' in task_result._result:
            if self._diff or getattr(original_task, 'diff', False):
                self._tqm.send_callback('v2_on_file_diff', task_result)

        if not isinstance(original_task, TaskInclude):
            # 处理其他情况
            # ...

        # 最后，回调 v2_runner_on_ok() 方法
        self._tqm.send_callback('v2_runner_on_ok', task_result)

    # 处理中的任务数减 1
    self._pending_results -= 1
    if original_host.name in self._blocked_hosts:
        # 从 _blocked_hosts 队列中移除当前处理完毕的任务
        del self._blocked_hosts[original_host.name]

    if original_task._role is not None and role_ran:
        # 处理角色任务情况
        # ...

    ret_results.append(task_result)

    if one_pass or max_passes is not None and (cur_pass + 1) >=
max_passes:
        break

    cur_pass += 1

    # 返回最后的结果
    return ret_results
```

_process_pending_results() 方法的逻辑非常清晰，即从结果队列中弹出一个任务的执行结果，然后对该结果提取信息并进行回调操作，最后还会将结果追加到类变量 ret_results 中并返回。cur_pass 变量用于控制该方法一次处理多少个 result。在这里的调用中，传入的 max_passes 参数为：

```
max(1, int(len(self._tqm._workers) * 0.1))
```

而 self._tqm._workers 队列的大小就是启动进程数，默认为 5，因此这里计算的结果为：

```
max_passes = max(1, int(0.5)) = 1
```

第 5 处代码中的 self._pending_results 参数表示当前正在执行的任务数。该值初始化时置为 0，在调用 self._queue_task() 方法将任务添加到任务队列时会将该值加 1。_queue_task() 方法的实现代码如下：

```
# 源码位置：lib/ansible/plugins/strategy/__init__.py
# ...
```

```
class StrategyBase:

    # ...

    def _queue_task(self, host, task, task_vars, play_context):

        # ...

        try:
            # ...

            self._pending_results += 1
        except (EOFError, IOError, AssertionError) as e:
            # 发生异常，大概率是终止进程
            display.debug("got an error while queuing: %s" % e)
            return
        display.debug("exiting _queue_task() for %s/%s" % (host.name,
task.action))

        # ...
```

此外，self._wait_on_pending_results()方法表示等待解析结果队列中的结果完成。该方法依旧是调用_process_pending_results()方法来实现，具体代码如下：

```
# 源码位置：lib/ansible/plugins/strategy/__init__.py
# ...

class StrategyBase:

    # ...

    def _wait_on_handler_results(self, iterator, handler, notified_hosts):

        ret_results = []
        handler_results = 0

        display.debug("waiting for handler results...")
        while (self._pending_results > 0 and
               handler_results < len(notified_hosts) and
               not self._tqm._terminated):

            if self._tqm.has_dead_workers():
                raise AnsibleError("A worker was found in a dead state")

            # 调用_process_pending_results()方法
            # 每处理完一个结果，self._pending_results 值会减 1
            results = self._process_pending_results(iterator)
            # 将结果加入 ret_results 中
            ret_results.extend(results)
            # 回调结果
            handler_results += len([
                r._host for r in results if r._host in notified_hosts and
                r.task_name == handler.name])
            # 如果还有任务没有完成，则休眠一段时间后再次执行
```

```
        if self._pending_results > 0:
            time.sleep(C.DEFAULT_INTERNAL_POLL_INTERVAL)

    display.debug("no more pending handlers, returning what we have")

    return ret_results

    # ...
```

_wait_on_handler_results()方法的逻辑也非常简单，即循环调用_process_pending_results()方法直到任务结果全部处理完毕（或者是一些其他情况）。

第 6 处代码调用了 update_active_connections()方法，该方法是 Ansible 针对 SSH 连接做的一些优化措施，代码比较简单，只是保存连接的 Socket 套接字信息，具体如下：

```
# 源码位置: lib/ansible/plugins/strategy/__init__.py
# ...

class StrategyBase:

    # ...

    def update_active_connections(self, results):
        ''' 更新当前活跃连接数 '''
        for r in results:
            if 'args' in r._task_fields:
                # 保存 socket_path
                socket_path = r._task_fields['args'].get('_ansible_socket')
                if socket_path:
                    if r._host not in self._active_connections:
                        self._active_connections[r._host] = socket_path

    # ...
```

在分析完 linear.py 插件中 StrategyModule 类的 run()方法后，会有一个明显的疑惑，即整个 linear.py 插件中的 run()代码都看完了，但 Ansible 2 核心的模块执行过程是什么呢？类似 Ansible 1 中合并成完整的模块源码，然后上传到远端主机去执行的代码在哪里呢？

在前面的分析中提到了一个非常重要的动作：将任务加入任务队列中。之后就是从结果队列中获取任务执行的结果并进行处理后才返回。于是真正的模块核心动作需要从这个添加任务的方法中开始，而该方法是在策略插件的基类中定义的，具体代码如下：

```
# 源码位置: lib/ansible/plugins/strategy/__init__.py
# ...

class StrategyBase:

    # ...

    def _queue_task(self, host, task, task_vars, play_context):
        ''' 处理任务队列，发送到 worker 进程处理 '''

        if task.action not in action_write_locks.action_write_locks:
```

```
        # 获取锁
        action_write_locks.action_write_locks[task.action] = Lock()

    try:

        shared_loader_obj = SharedPluginLoaderObj()

        queued = False
        starting_worker = self._cur_worker
        while True:
            worker_prc = self._workers[self._cur_worker]
            if worker_prc is None or not worker_prc.is_alive():
                self._queued_task_cache[(host.name, task._uuid)] = {
                    'host': host,
                    'task': task,
                    'task_vars': task_vars,
                    'play_context': play_context
                }

                # WorkerProcess 对象
                worker_prc = WorkerProcess(self._final_q, task_vars,
host, task,
                    play_context, self._loader, self._variable_manager,
shared_loader_obj)
                self._workers[self._cur_worker] = worker_prc
                # 触发回调
                self._tqm.send_callback('v2_runner_on_start', host, task)
                # worker 进程启动
                worker_prc.start()
                queued = True
            # 当前 worker 进程数加 1
            self._cur_worker += 1
            # 超过了则从 0 开始
            if self._cur_worker >= len(self._workers):
                self._cur_worker = 0
            if queued:
                break
            elif self._cur_worker == starting_worker:
                time.sleep(0.0001)

        # 记录正在执行的任务数，也是将要返回的结果数
        self._pending_results += 1
    except (EOFError, IOError, AssertionError) as e:
        # 异常，需要直接返回
        return
    display.debug("exiting _queue_task() for %s/%s" % (host.name,
task.action))
```

为了理解上面的函数，需要了解一些前置的基础。例如一开始 Ansible 会对某些模块先创建锁，下面先来看看不需要创建锁的模块有哪些，具体操作如下：

```
>>> from ansible.executor import action_write_locks
>>> action_write_locks.action_write_locks
{None: <Lock(owner=None)>, 'pkgin': <Lock(owner=None)>, 'swupd': <Lock
```

```
(owner=None)>, 'installp': <Lock(owner=None)>, 'macports': <Lock
(owner=None)>, 'pacman': <Lock(owner=None)>, 'apt': <Lock(owner=None)>,
'opkg': <Lock(owner=None)>, 'slurp': <Lock(owner=None)>, 'pkg5': <Lock
(owner=None)>, 'urpmi': <Lock(owner=None)>, 'swdepot': <Lock(owner=None)>,
'portage': <Lock(owner=None)>, 'setup': <Lock(owner=None)>, 'atomic_
container': <Lock(owner=None)>, 'zypper': <Lock(owner=None)>, 'pkgng':
<Lock(owner=None)>, 'dnf': <Lock(owner=None)>, 'apk': <Lock(owner=None)>,
'yum': <Lock(owner=None)>, 'copy': <Lock(owner=None)>, 'file': <Lock
(owner=None)>, 'xbps': <Lock(owner=None)>, 'sorcery': <Lock(owner=None)>,
'stat': <Lock(owner=None)>, 'homebrew': <Lock(owner=None)>, 'svr4pkg':
<Lock(owner=None)>}
```

接着是 self._cur_worker 和 self._workers 两个属性，它们都可以从 __init__.py 方法中得到初始值，具体如下：

```python
# 源码位置: lib/ansible/plugins/strategy/__init__.py
# ...

class StrategyBase:

    def __init__(self, tqm):
        self._tqm = tqm
        self._inventory = tqm.get_inventory()
        self._workers = tqm.get_workers()
        self._variable_manager = tqm.get_variable_manager()
        self._loader = tqm.get_loader()
        # 后面会介绍
        self._final_q = tqm._final_q

        # ...

        # internal counters
        self._pending_results = 0
        self._cur_worker = 0

        # ...

    # ...
```

tqm 参数是一个 TaskQueueManager 对象，其_workers 属性之前也提到过，具体如下：

```python
# 源码位置: lib/ansible/executor/task_queue_manager.py
# ...

class TaskQueueManager:
    # ...

    def __init__(...)
        # ...

        # 指定运行进程数
        self._forks = forks or 5

        # ...
```

```
    def _initialize_processes(self, num):
        self._workers = []

        for i in range(num):
            self._workers.append(None)

    # ...

    def run(self, play):

        # ...

        self._initialize_processes(min(self._forks, iterator.batch_size))

        # ...


    # ...


    def get_workers(self):
        return self._workers[:]

    # ...
```

看到上面的简化代码后，明白 self._workers 的含义了吗？就是初始化了一个大小为 min(self._forks, iterator.batch_size)、元素全为 None 的列表。而 StrategyBase 中的 _cur_worker 属性是 _workers 属性的索引。从 _queue_task() 方法中可以看到 _workers 列表中保存的都是 WorkerProcess 对象，而这个对象正是 Ansible 中 task 执行进程的逻辑体。通过 worker_prc. start() 方法的代码就可以立马猜到 WorkerProcess 类肯定继承了 Python 中的进程类。 WorkerProcess 类的源码如下：

```
# 源码位置: lib/ansible/executor/process/worker.py
# ...

class WorkerProcess(multiprocessing.Process):

    def __init__(self, final_q, task_vars, host, task, play_context, loader,
                 variable_manager, shared_loader_obj):

        super(WorkerProcess, self).__init__()
        self._final_q = final_q
        self._task_vars = task_vars
        self._host = host
        self._task = task
        self._play_context = play_context
        self._loader = loader
        self._variable_manager = variable_manager
        self._shared_loader_obj = shared_loader_obj

        self._loader._tempfiles = set()

    # ...
```

果然，这里的 WorkerProcess 类正是继承了 Python 中的 multiprocessing.Process。此外，从前面的 StrategyBase 类的 __init__.py 文件中可以看到，该类的 _final_q 属性其实就是 TaskQueueManager 对象中的 _final_q 值，对应的代码如下：

```python
# 源码位置: lib/ansible/executor/task_queue_manager.py
# ...

class TaskQueueManager:
    # ...

    def __init__(...):
        # ...

        try:
            self._final_q = multiprocessing.Queue()
        except OSError as e:
            # 抛出异常
            # ...

        self._connection_lockfile = tempfile.TemporaryFile()

    # ...
```

可以看到，StrategyBase 类中的 _final_q 属性其实就是 multiprocessing 模块中的 Queue 对象，而这个值在实例化 WorkerProcess 类时又会传给其 _final_q 属性，这个队列最终用于同步多进程操作返回的结果。了解了上述这些基础后，就可以完整总结出 _queue_task() 方法的执行逻辑了，如图 4-2 所示。

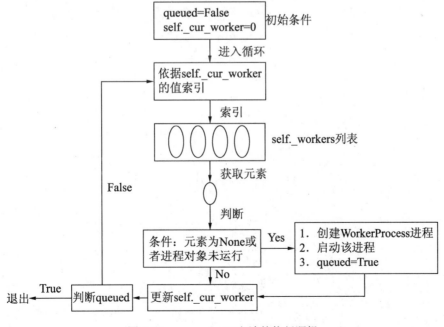

图 4-2　_queue_task() 方法的执行逻辑

理解了_queue_task()方法的逻辑后，可知处理任务的逻辑被封装在 WorkerProcess 类中。于是，追踪的目标开始转向 WorkerProcess 类，其实现代码如下：

```python
# 源码位置: lib/ansible/executor/process/worker.py
# ...

class WorkerProcess(multiprocessing.Process):

    # ...

    def run(self):

        try:
            return self._run()
        except BaseException as e:
            self._hard_exit(e)

    def _run(self):

        try:
            # 实例化 TaskExecutor 类并调用对象的 run() 方法，获取 task 执行结果
            executor_result = TaskExecutor(
                self._host,
                self._task,
                self._task_vars,
                self._play_context,
                self._new_stdin,
                self._loader,
                self._shared_loader_obj,
                self._final_q
            ).run()

            # 对返回结果进行处理并封装成 TaskResult 对象
            self._host.vars = dict()
            self._host.groups = []
            task_result = TaskResult(
                self._host.name,
                self._task._uuid,
                executor_result,
                task_fields=self._task.dump_attrs(),
            )

            # 将 task 的运行结果加入结果队列中
            self._final_q.put(task_result)

        except AnsibleConnectionFailure:
            # 连接异常处理
            self._host.vars = dict()
            self._host.groups = []
            # 封装 TaskResult 对象
            task_result = TaskResult(
                self._host.name,
                self._task._uuid,
                dict(unreachable=True),
```

```
                    task_fields=self._task.dump_attrs(),
                )
                # 放入结果队列
                self._final_q.put(task_result, block=False)

        except Exception as e:
            # 其余异常处理
            if not isinstance(e, (IOError, EOFError, KeyboardInterrupt,
SystemExit)) or
                isinstance(e, TemplateNotFound):
                try:
                    self._host.vars = dict()
                    self._host.groups = []
                    # 封装 TaskResult 对象
                    task_result = TaskResult(
                        self._host.name,
                        self._task._uuid,
                        dict(failed=True, exception=to_text(traceback.format_
exc()), stdout=''),
                        task_fields=self._task.dump_attrs(),
                    )
                    # 放入结果队列
                    self._final_q.put(task_result, block=False)
                except Exception:
                    # 打印警告信息
                    # ...
            finally:
                self._clean_up()

        display.debug("WORKER PROCESS EXITING")
```

　　了解 multiprocessing.Process 类的读者应该知道,该类创建的进程对象启动(调用 start()
方法)后会运行 run()方法。因此对于 task 的核心执行逻辑只需要关注这里的 run()方法,
且 run()方法最终调用的又是_run()方法,于是继续分析_run()方法的实现源码。从上面的
源码可知,最终 Ansible 中 task 的核心处理过程最终又到了 TaskExecutor 类中。在
WorkerProcess 对象的_run()方法中,先是通过对象属性实例化 TaskExecutor 类,然后调用
该对象的 run()方法并得到 task 的运行结果,最后将结果封装成 TaskResult 对象后放到结
果队列中。如果遇到异常,则得到异常结果并放入结果队列中。

　　从上面的分析来看,最后 task 运行的核心逻辑全部在 TaskExecutor 对象的 run()方法
中。继续追踪该 run()方法的实现源码,具体如下:

```
# 源码位置: lib/ansible/executor/task_executor.py
# ...

class TaskExecutor:

    # ...

    def run(self):

        try:
```

```
    try:
        items = self._get_loop_items()
    except AnsibleUndefinedVariable as e:
        # 保存异常，后续会用到
        items = None
        self._loop_eval_error = e

    if items is not None:
        # 处理包含 with_items 语法的 for 循环的 task
        if len(items) > 0:
            # self._run_loop()方法最终也是一个一个调用 self._execute()方
              法获取执行结果
            item_results = self._run_loop(items)

            # ...
        else:
            res = dict(changed=False, skipped=True,
                skipped_reason='No items in the list', results=[])
    else:
        res = self._execute()

    if 'changed' not in res:
        res['changed'] = False

    def _clean_res(res, errors='surrogate_or_strict'):
        # 整理结果，转换成 JSON 形式
        # ...

    # 进一步整理结果
    res = _clean_res(res)
    return res
except AnsibleError as e:
    # 返回异常结果
    return dict(failed=True, msg=wrap_var(to_text(e, nonstring=
'simplerepr')),
        _ansible_no_log=self._play_context.no_log)
except Exception as e:
    # 返回异常结果
    return dict(failed=True, msg='Unexpected failure during module
execution.',
        exception=to_text(traceback.format_exc()),
            stdout='', _ansible_no_log=self._play_context.no_log)
finally:
    try:
        # 关闭连接
        self._connection.close()
    except AttributeError:
        pass
    except Exception as e:
        display.debug(u"error closing connection: %s" % to_text(e))
```

这部分代码的逻辑和 Ansible 1 中类似，先是对 task 中存在 for 循环的情况（with_items 等语法）进行检测，然后分两种情况进行处理，但是二者最终落到一个最小 task 上时都是调用 self._execute()方法来获取该 task 的执行结果，具体如下：

```python
# 源码位置: lib/ansible/executor/task_executor.py
# ...

class TaskExecutor:

    # ...

    def _run_loop(self, items):

        # ...

        no_log = False
        items_len = len(items)
        for item_index, item in enumerate(items):

            # ...

            # 最终还是调用 self._execute()方法获取最小粒度 task 的执行结果
            res = self._execute(variables=task_vars)

            # ...

    # ...
```

再次将目光转移到 self._execute()方法上。这部分的内容非常长，请耐心阅读。

```python
# 源码位置: lib/ansible/executor/task_executor.py
# ...

class TaskExecutor:

    # ...

    def _execute(self, variables=None):
        '''核心任务执行方法'''
        # ...

        # 1. 更新 self._play_context 中的属性值，尤其是 connection 属性
        context_validation_error = None
        try:
            # print("run before:{}".format(self._play_context._get_attr_
connection()))
            self._play_context = self._play_context.set_task_and_variable_
override(
                        task=self._task, variables=variables, templar=
templar)
            # print("run after:{}".format(self._play_context._get_attr_
connection()))

            self._play_context.post_validate(templar=templar)

            if not self._play_context.remote_addr:
                self._play_context.remote_addr = self._host.address
```

```
                    self._play_context.update_vars(variables)

            except AnsibleError as e:
                context_validation_error = e

            # 处理一些和 Playbook 相关的参数，如条件判断、include、include_tasks 等
              语法
            # ...

            # 2. 处理底层 SSH 连接参数
            if (not self._connection or
                    not getattr(self._connection, 'connected', False) or
                    self._play_context.remote_addr != self._connection._play_
context.remote_addr):
                self._connection = self._get_connection(variables=variables,
templar=templar)
                # 笔者添加的测试打印语句
                # print("connection:{}".format(self._connection))
                # self._connection._connect()
                # print(self._connection.exec_command('hostname'))
            else:
                self._connection._play_context = self._play_context

            self._set_connection_options(variables, templar)

            # 3. 最核心的地方，获取该模块处理的 _handler 属性值
            self._handler = self._get_action_handler(connection=self._connection,
templar=templar)

            # ...

        result = None
        # 没有 until 语法时，retries=1，下面的 for 循环只会执行 1 次
        for attempt in xrange(1, retries + 1):
            # 考虑重试机制，每次调用 self._handler.run() 方法执行 task 并获取结果
            try:
                # 4. 最后调用 self._handler 的 run() 方法执行 task 并返回结果
                result = self._handler.run(task_vars=variables)
            except AnsibleActionSkip as e:
                return dict(skipped=True, msg=to_text(e))
            except AnsibleActionFail as e:
                return dict(failed=True, msg=to_text(e))
            except AnsibleConnectionFailure as e:
                return dict(unreachable=True, msg=to_text(e))
            display.debug("handler run complete")

            # 处理本次任务执行后的返回结果
            # ...

            else:
                if retries > 1:
                    result['attempts'] = retries - 1
                    result['failed'] = True
```

```
        # ...

        return result
```

这里对_execute()方法做了大量简化。上述代码中有 4 处需要详细说明，具体分析如下。

第 1 处代码主要是更新 self._play_context 中的属性值，这部分属性值和 SSH 连接参数相关。在介绍 PlayContext 对象并演示其使用时，发现初始的 PlayContext 对象的相关属性值全为<class 'ansible.utils.sentinel.Sentinel'>这样的值。其中最关键的是 connection 属性值，它关系到 Ansible 底层 SSH 连接远程主机的方式。笔者在上面的代码中标注了 2 个 print() 函数，在虚拟环境中添加这两处 print()函数，然后简单执行 ping 模块命令，具体如下：

```
(ansible2.8.16) [root@master ~]# ansible ceph-1 -i hosts  -m ping
run before:<class 'ansible.utils.sentinel.Sentinel'>
run after:ssh
ceph-1 | SUCCESS => {
    "ansible_facts": {
        "discovered_interpreter_python": "/usr/bin/python"
    },
    "changed": false,
    "ping": "pong"
}
```

可以看到，在经过 PlayContext 对象的 set_task_and_variable_override()方法处理后，相应的 SSH 参数全部会使用正确的值以保证后续的 SSH 连接正常。

第 2 处代码主要是获取与远端的 SSH 连接，最终得到 self._connection 值，该值为对应连接插件中的 Connection 对象。在得到该对象后，可以通过该对象的_connect()方法与远端主机建立连接，然后就可以使用 exec_command()方法远程执行命令。笔者同样给出了几行测试语句，将其写到虚拟环境下的 Ansible 源码的对应位置上，然后运行执行 ping 模块的命令，具体如下：

```
(ansible2.8.16) [root@master ~]# ansible ceph-1 -i hosts -m ping
run before:<class 'ansible.utils.sentinel.Sentinel'>
run after:ssh
connection:<ansible.plugins.connection.ssh.Connection object at 0x7f2f29
6b0ac0>
(0, b'ceph-1\r\n', b'Shared connection to ceph-1 closed.\r\n')
ceph-1 | SUCCESS => {
    "ansible_facts": {
        "discovered_interpreter_python": "/usr/bin/python"
    },
    "changed": false,
    "ping": "pong"
}
```

可以看到，上面的 exec_command()方法成功执行了 hostname 命令。

第 3 处和第 4 处代码非常重要，获取对应 action 处理的 handler，然后调用 run()方法。先来看获取 handler 的方法，具体如下：

```
# 源码位置：lib/ansible/executor/task_executor.py
# ...

class TaskExecutor:

    # ...

    def _get_action_handler(self, connection, templar):

        # 模块前缀，按照_分割模块名称
        module_prefix = self._task.action.split('_')[0]

        collections = self._task.collections

        # 核心代码，获取动作插件名
        if self._shared_loader_obj.action_loader.has_plugin(self._task.
action, collection_list=collections):
            handler_name = self._task.action
        elif all((module_prefix in C.NETWORK_GROUP_MODULES,
                module_prefix in self._shared_loader_obj.action_loader)):
            handler_name = module_prefix
        else:
            handler_name = 'normal'
            collections = None

        # 加载模块对应的动作插件
        handler = self._shared_loader_obj.action_loader.get(
            handler_name,
            task=self._task,
            connection=connection,
            play_context=self._play_context,
            loader=self._loader,
            templar=templar,
            shared_loader_obj=self._shared_loader_obj,
            collection_list=collections
        )

        if not handler:
            raise AnsibleError("the handler '%s' was not found" % handler_
name)

        return handler

    # ...
```

从上面的代码中可以看到一个非常熟悉的身影——action 插件。这里同样是根据 task 中的 action 去获取对应的 action 插件，如果能找到对应 action 名称的插件，则 handler_name 就等于对应的插件名，没有对应插件的则统一为 normal。如果是 ping 模块，这里得到的 handler_name = 'normal'，最终得到的 handler 为 normal.py 下的 ActionModule 对象。最终调用的 self._handler.run() 语句即为对应 action 插件下 ActionModule 对象中的 run() 方法。下面继续追踪 normal.py 文件中的 run() 方法，具体代码如下：

```
# 源码位置: lib/ansible/plugins/action/normal.py
# ...

class ActionModule(ActionBase):

    def run(self, tmp=None, task_vars=None):

        self._supports_check_mode = True
        self._supports_async = True

        # 父类的 run() 方法并没有做太多动作
        result = super(ActionModule, self).run(tmp, task_vars)
        del tmp

        if not result.get('skipped'):

            if result.get('invocation', {}).get('module_args'):
                del result['invocation']['module_args']

            wrap_async = self._task.async_val and not self._connection.
has_native_async

            #调用父类的 self._execute_module() 方法获取模块执行的最终结果
            result = merge_hash(result,
                    self._execute_module(task_vars=task_vars, wrap_async=
wrap_async))

            if self._task.action == 'setup':
                result['_ansible_verbose_override'] = True

        if not wrap_async:
            # 清除临时文件
            self._remove_tmp_path(self._connection._shell.tmpdir)

        return result
```

normal 插件下的 run() 方法也非常简单，首先调用了父类的 run() 方法。以下是该方法的实现源码：

```
# 源码位置: lib/ansible/plugins/action/__init__.py
# ...

class ActionBase(with_metaclass(ABCMeta, object)):
    # ...

    @abstractmethod
    def run(self, tmp=None, task_vars=None):

        result = {}

        # 提示不再需要 tmp 参数，可删除
        # ...

        # 处理一些异常情况
```

```
        # ...

        if self._VALID_ARGS:
            task_opts = frozenset(self._task.args.keys())
            bad_opts = task_opts.difference(self._VALID_ARGS)
            if bad_opts:
                raise AnsibleActionFail(
                    'Invalid options for %s: %s' % (self._task.action, ','.
join(list(bad_opts))))

        if self._connection._shell.tmpdir is None and self._early_needs_
tmp_path():
            # 在远端服务器上创建临时文件
            self._make_tmp_path()

        return result
```

从上面的代码中可以看到，ActionBase 类中的 run() 方法并没有做任何有用的事情。通过代码可以看到，在调用该 run() 方法后 result 的结果为{}。如果设置了需要临时目录且当前临时目录为 None，则会调用 _make_tmp_path() 方法在连接的远端主机上创建一个临时目录，用于后续上传模块脚本等。如果有兴趣，可以看一看这个 action 插件中 __init__.py 文件的内容，会看到一些非常熟悉的方法，这些方法都在 Ansible 1 的源码分析中详细介绍过，例如 _low_level_execute_command()、_execute_module()、_transfer_file() 以及 _transfer_data() 等方法。这里的很多方法都是基于远程连接下对远端主机封装的各种操作，例如上面几个远程执行命令、远程传输文件和数据，还有远程修改文件权限（_remote_chmod()）、修改属组（_remote_chown()）等待。有了前面分析 Ansible 1 的基础后再看这些代码都不难，这里就不再继续追踪这些方法的实现细节了。继续回到前面 ActionModule 类中的 run() 方法，排除了调用父类的 run() 方法获取结果后，继续看后面的代码，可以看到后面又调用了 self._execute_module() 方法得到最终的 result。这个方法同样在 Ansible 1 中出现过，和这里的功能一致但是实现方法有所不同。_execute_module() 方法同样是在父类中实现，具体代码如下：

```
# 源码位置：lib/ansible/plugins/action/__init__.py
# ...

class ActionBase(with_metaclass(ABCMeta, object)):
    # ...

    def _execute_module(self, module_name=None, module_args=None, tmp=None,
task_vars=None, persist_files=False, delete_remote_tmp=None, wrap_async=
False):
        # tmp 和 delete_remote_tmp 参数不再使用，进行提示
        # ...

        tmpdir = self._connection._shell.tmpdir

        if not self._is_pipelining_enabled("new", wrap_async) and tmpdir is
None:
```

```
    # 创建临时目录
    self._make_tmp_path()
    # 获取生成的临时目录
    tmpdir = self._connection._shell.tmpdir

if task_vars is None:
    task_vars = dict()

# 获取模块名以及模块参数
if module_name is None:
    module_name = self._task.action
if module_args is None:
    module_args = self._task.args

self._update_module_args(module_name, module_args, task_vars)

# ...

# 这里和 Ansible 1 一样，最终得到模块代码的内容，即 module_data，module_style
  一般为 new
(module_style, shebang, module_data, module_path) =
        self._configure_module(module_name=module_name, module_args=
module_args, task_vars=task_vars)

# ...

self._used_interpreter = shebang
remote_module_path = None

if not self._is_pipelining_enabled(module_style, wrap_async):
    # we might need remote tmp dir
    if tmpdir is None:
        self._make_tmp_path()
        tmpdir = self._connection._shell.tmpdir

    # 获取当前执行模块的脚本名称，例如对于 ping 模块，得到的就是 ping.py
    remote_module_filename = self._connection._shell.get_remote_
filename(module_path)
    # 完整模块文件的路径：临时目录/AnsiballZ_ping.py
    remote_module_path = self._connection._shell.join_path(
                tmpdir, 'AnsiballZ_%s' % remote_module_filename)

# ...

if remote_module_path or module_style != 'new':
    if module_style == 'binary':
        self._transfer_file(module_path, remote_module_path)
    else:
        # 最终调用这里的代码将 module_data 内容写到模块的临时文件中
        self._transfer_data(remote_module_path, module_data)
    # 针对 module_style 的值进行处理，ping 等模块执行时 module_style='new'
    # ...
```

```
        if wrap_async and not self._connection.always_pipeline_modules:
            # 处理异步执行任务
            # ...

        else:

            if self._is_pipelining_enabled(module_style):
                in_data = module_data
            else:
                cmd = remote_module_path

            # 最终通过 shell 插件获取执行完整模块脚本的命令
            cmd = self._connection._shell.build_module_command(
                environment_string, shebang, cmd, arg_path=args_file_path).
strip()

        # 修正传送到远端主机上的文件权限
        if remote_files:
            # 移除 None 和空的情况
            remote_files = [x for x in remote_files if x]
            self._fixup_perms2(remote_files, self._get_remote_user())

        # 最后的远程调用，和 Ansible 1 中的含义一致。远程执行 cmd 命令
        res = self._low_level_execute_command(cmd, sudoable=sudoable, in_
data=in_data)

        # 解析执行模块脚本命令返回的结果
        data = self._parse_returned_data(res)

        # 对结果进行一系列加工处理
        # ...

        return data
```

在上面的_execute_module()方法中，终于看到了和 Ansible 1 类似的执行模块的基本过程：首先通过对象的_configure_module()方法得到完整的模块源码（合并模块本身的代码与公共代码段），然后调用_transfer_data()方法将得到的完整模块源码写入指定的临时文件中，接着生成相应的执行该脚本的命令，最后调用_low_level_execute_command()方法远程执行该命令，得到脚本输出结果。至此，Ansible 2 的模块运行流程分析完毕，至于细节问题，就需要读者自行去钻研了。

4．遗留问题解决

前面其实能看出一些问题，报错的原因如下：

```
the connection plugin '<class 'ansible.utils.sentinel.Sentinel'>' was not
found,
```

正常连接时该连接的值应该是 smart、ssh 或 paramiko 中的一个，默认是 smart，并且会优先使用 SSH 命令方式进行远程连接操作。为了找出代码的问题所在，先准备好下面的一段代码，将命令行操作集合到文件中。

```
# 代码位置: /root/test_play_run.py

from ansible.parsing.dataloader import DataLoader
from ansible.inventory.manager import InventoryManager
from ansible.vars.manager import VariableManager

loader = DataLoader()
inventory = InventoryManager(loader=loader, sources="/root/hosts")
variable_manager = VariableManager(loader=loader, inventory=inventory,
version_info=None)
play_ds = {'name': 'Ansible Ad-Hoc', 'hosts': 'ceph-1', 'gather_facts':
'no', 'tasks': [{'action': {'module': 'ping', 'args': {}}, 'async_val': 0,
'poll': 15}]}
from ansible.playbook.play import Play
play = Play().load(play_ds, variable_manager=variable_manager, loader=
loader)
p = Play()
# print(p.serialize())
from ansible.executor.task_queue_manager import TaskQueueManager
passwords = {'conn_pass': None, 'become_pass': None}
tqm = TaskQueueManager(
        inventory=inventory,
        variable_manager=variable_manager,
        loader=loader,
        passwords=passwords,
        forks=5
    )
result = tqm.run(play)
```

通过前面对 Ansible 2.8 的源码分析可知，上述简化的代码一定漏掉了一些属性配置，才会导致其无法得到和执行 ansible 命令一样的结果，而上述代码中的 loader、inventory 变量和源码中的赋值语句基本一致，并没有太大区别。此外，variable_manager 又是通过 loader 和 inventory 实例化得到的，而且其实现源码并没有其他特殊的变量参与，所以 variable_manager 变量的嫌疑也不大。最大的可能就是这里的 play 变量，一个很明显的原因就是 play_ds 值的来源，笔者通过在 Ansible 源码中添加 print() 函数打印了运行 ping 命令时的 play_ds 值，然后直接用到了这里，而源码中的写法如下：

```
# 源码位置: lib/ansible/cli/adhoc.py
# ...

play_ds = self._play_ds(pattern, context.CLIARGS['seconds'], context.
CLIARGS['poll_interval'])
```

源码中的一些参数值是通过 context 参数获取的，那么这个 context 参数是什么呢？打印出来就知道了。

```
# 源码位置: lib/ansible/cli/adhoc.py
# ...

class AdHocCLI(CLI):

    # ...
```

```
def run(self):

    super(AdHocCLI, self).run()

    # 在对应的虚拟环境中添加 print()方法
    print(context.CLIARGS)

    # ...
```

只需要在对应的虚拟环境中的 Ansible 模块目录下添加上面的 print()语句，即可打印 context.CLIARGS 的值，运行结果如下：

```
(ansible2.8.16) [root@master ~]# ansible ceph-1 -i hosts -m ping
ImmutableDict({'verbosity': 0, 'become': False, 'become_method': 'sudo',
'become_user': None, 'become_ask_pass': False, 'inventory': ('/root/hosts',),
'listhosts': None, 'subset': None, 'poll_interval': 15, 'seconds': 0,
'one_line': None, 'tree': None, 'ask_pass': False, 'private_key_file':
None, 'remote_user': None, 'connection': 'smart', 'timeout': 10, 'ssh_
common_args': '', 'sftp_extra_args': '', 'scp_extra_args': '', 'ssh_extra_
args': '', 'check': False, 'syntax': None, 'diff': False, 'extra_vars': (),
'ask_vault_pass': False, 'vault_password_files': (), 'vault_ids': (),
'forks': 5, 'module_path': None, 'basedir': None, 'module_args': '',
'module_name': 'ping', 'args': ('ceph-1',)})
ceph-1 | SUCCESS => {
    "ansible_facts": {
        "discovered_interpreter_python": "/usr/bin/python"
    },
    "changed": false,
    "ping": "pong"
}
```

看到 context.CLIARGS 的结果了吗？它是一个类字典的对象，其中有一个 key 为 connection，对应的 value 正好为 smart。而前面简化的手工操作中并没有对这个 context 的操作，所以会漏掉一些关键信息，从而导致后续执行不成功。另外，自行得到的 Play 对象和直接用 Ansible 得到的 Play 对象一致吗？在 Play 对象中有一个 serialize()方法，它可以打印代表这个 Play 对象的一些重要属性值。笔者在前面的 Python 代码中注释掉了这行打印序列化对象的操作，现在放开并执行，结果如下：

```
(ansible2.8.16) [root@master ~]# python test_play_run.py
{'hosts': None, 'gather_facts': None, 'gather_subset': ['all'], 'gather_
timeout': 10, 'fact_path': None, 'vars_files': [], 'vars_prompt': [],
'roles': [], 'handlers': [], 'pre_tasks': [], 'post_tasks': [], 'tasks':
[], 'force_handlers': None, 'max_fail_percentage': None, 'serial': [],
'strategy': 'linear', 'order': None, 'name': '', 'connection': None, 'port':
None, 'remote_user': None, 'vars': {}, 'module_defaults': None, 'environment':
None, 'no_log': None, 'run_once': None, 'ignore_errors': None, 'ignore_
unreachable': None, 'check_mode': None, 'diff': None, 'any_errors_fatal':
False, 'debugger': None, 'tags': [], 'become': None, 'become_method': None,
'become_user': None, 'become_flags': None, 'collections': None, 'uuid':
'000c295e-8383-19ea-5bb4-00000000000d', 'finalized': False, 'squashed':
False, 'included_path': None}
```

```
PLAY [Ansible Ad-Hoc] ************************************************
state: TASK: Gathering Facts

TASK [ping] *********************************************************
fatal: [ceph-1]: FAILED! => {"msg": "the connection plugin '<class 'ansible.
utils.sentinel.Sentinel'>' was not found"}
```

在脚本执行的结果中可以看到,得到的 Play 对象的 connection 属性值为 None。于是修改虚拟环境下的 Ansible 源码,同样打印出一开始得到的 Play 对象的序列化结果。具体修改如下:

```python
# 源码位置: lib/ansible/cli/adhoc.py
# ...

class AdHocCLI(CLI):

    # ...

    def run(self):

        super(AdHocCLI, self).run()

        # print(context.CLIARGS)

        # ...

        play_ds = self._play_ds(pattern, context.CLIARGS['seconds'], context.
CLIARGS['poll_interval'])
        play = Play().load(play_ds, variable_manager=variable_manager,
loader=loader)
        # 新增一行注释
        print('ansible:', play.serialize())

        # ...
```

再次在虚拟环境下执行 Ansible 中的 ping 模块(记得去掉前一次打印的 print() 语句),结果如下:

```
(ansible2.8.16) [root@master ~]# ansible ceph-1 -i hosts -m ping
ansible: {'hosts': 'ceph-1', 'gather_facts': 'no', 'gather_subset': ['all'],
'gather_timeout': 10, 'fact_path': None, 'vars_files': [], 'vars_prompt':
[], 'roles': [], 'handlers': [], 'pre_tasks': [], 'post_tasks': [], 'tasks':
[BLOCK(uuid=000c295e-8383-8d99-5892-00000000000b)(id=140123251675600)
(parent=None)], 'force_handlers': None, 'max_fail_percentage': None,
'serial': [], 'strategy': 'linear', 'order': None, 'name': 'Ansible Ad-Hoc',
'connection': 'smart', 'port': None, 'remote_user': None, 'vars': {},
'module_defaults': None, 'environment': None, 'no_log': None, 'run_once':
None, 'ignore_errors': None, 'ignore_unreachable': None, 'check_mode':
False, 'diff': False, 'any_errors_fatal': False, 'debugger': None, 'tags':
[], 'become': False, 'become_method': 'sudo', 'become_user': None, '
become_flags': None, 'collections': None, 'uuid': '000c295e-8383-8d99-5892
-00000000000a', 'finalized': False, 'squashed': False, 'included_path':
None}
state: TASK: Gathering Facts
ceph-1 | SUCCESS => {
    "ansible_facts": {
```

```
        "discovered_interpreter_python": "/usr/bin/python"
    },
    "changed": false,
    "ping": "pong"
}
```

可以看到，这里 Play 中有很多属性都与前面脚本中得到的 Play 不一致。其中报错的 connection 属性在这里也不一样，脚本执行的结果中为 None，而命令行执行的结果中为 smart。那么简单地在脚本中将得到的 play 变量的 connection 属性值修改为 smart，是不是就不会报错了？

```
# 代码位置：/root/test_play_run.py

from ansible.parsing.dataloader import DataLoader
from ansible.inventory.manager import InventoryManager
from ansible.vars.manager import VariableManager

loader = DataLoader()
inventory = InventoryManager(loader=loader, sources="/root/hosts")
variable_manager = VariableManager(loader=loader, inventory=inventory,
version_info=None)
# play_ds = {'name': 'Ansible Ad-Hoc', 'hosts': 'ceph-1', 'gather_facts':
'no', 'tasks': [{'action': {'module': 'ping', 'args': {}}, 'async_val': 0,
'poll': 15}]}
play_ds = {'name': 'Ansible Ad-Hoc', 'hosts': 'ceph-1', 'gather_facts':
'no', 'tasks': [{'action': {'module': 'debug', 'args': {'msg': 'hello,
world'}}, 'async_val': 0, 'poll': 15}]}
from ansible.playbook.play import Play
play = Play().load(play_ds, variable_manager=variable_manager, loader=
loader)
p = Play()
# print(p.serialize())
play.connection = 'smart'                      # --> 最终定位到问题

from ansible.executor.task_queue_manager import TaskQueueManager
passwords = {'conn_pass': None, 'become_pass': None}
tqm = TaskQueueManager(
        inventory=inventory,
        variable_manager=variable_manager,
        loader=loader,
        passwords=passwords,
        forks=5
    )
result = tqm.run(play)
```

为了查看方便，笔者换成了执行 debug 模块并打印了一个简单的 "hello, world"，最后再次运行该脚本，结果如下：

```
(ansible2.8.16) [root@master ~]# python test_play_run.py

PLAY [Ansible Ad-Hoc] ***********************************************************
state: TASK: Gathering Facts

TASK [debug] ********************************************************************
```

```
ok: [ceph-1] => {
    "msg": "hello, world"
}
```

脚本成功执行，总算是跳过了之前的报错。但是这里显示的是 Playbook 的执行结果，并不是前面看到的命令行输出结果。那么是哪里出了问题呢？其实只需要对比下源码中实例化 TaskQueueManager 类的语句即可，具体如下：

```
# 源码位置: lib/ansible/cli/adhoc.py
# ...

class AdHocCLI(CLI):
    # ...

    def run(self):

        # ...

        # 设置回调的地方
        if self.callback:
            cb = self.callback
        elif context.CLIARGS['one_line']:
            cb = 'oneline'
        elif C.DEFAULT_LOAD_CALLBACK_PLUGINS and
                C.DEFAULT_STDOUT_CALLBACK != 'default':
            cb = C.DEFAULT_STDOUT_CALLBACK
        else:
            # 一般的默认值
            cb = 'minimal'

        try:
            # 封装 TaskQueueManager 对象
            self._tqm = TaskQueueManager(
                inventory=inventory,
                variable_manager=variable_manager,
                loader=loader,
                passwords=passwords,
                # 设置输出回调的地方
                stdout_callback=cb,
                run_additional_callbacks=C.DEFAULT_LOAD_CALLBACK_PLUGINS,
                run_tree=run_tree,
                forks=context.CLIARGS['forks'],
            )

            # ...

        finally:
            # 清理工作
            # ...

        return result
```

从上面的代码中可以看到，源码封装 TaskQueueManager 对象后多了一个输出回调参数 cb。从 ansible 命令行的代码中来看，该值为 minimal。下面在测试代码中加上这个参数，具体如下：

```
tqm = TaskQueueManager(
        inventory=inventory,
        variable_manager=variable_manager,
        loader=loader,
        passwords=passwords,
        stdout_callback='minimal',
        forks=5
    )
```

再次执行该脚本，便可以看到和通过控制台输入 ansible 命令一样的结果了，具体如下：

```
(ansible2.8.16) [root@master ~]#python test_play_run.py
ceph-1 | SUCCESS => {
    "msg": "hello, world"
}
```

另外，为了弄清楚上面出现的 connection 属性未被赋值的原因，需要继续看 Play 类的实现，看是哪一步借助了哪些额外信息去得到完整的 Play 对象。首先来看 Play 类的源码，具体如下：

```
# 源码位置：lib/ansible/playbook/play.py
# ...

class Play(Base, Taggable, Become, CollectionSearch):

    # ...
```

前面也介绍过这个 Play 类，但只是简单使用了下。这里单从 Play 类的源码中看不到任何有效信息，下面继续追踪其父类 Base，具体代码如下：

```
# 源码位置：lib/ansible/playbook/base.py
# ...

class Base(FieldAttributeBase):

    # ...

    # connection/transport，会从 context 中取默认值
    _connection = FieldAttribute(isa='string', default=context.cliargs_
deferred_get('connection'))
    _port = FieldAttribute(isa='int')
    _remote_user = FieldAttribute(isa='string', default=context.cliargs_
deferred_get('remote_user'))

    # ...

    _check_mode = FieldAttribute(isa='bool', default=context.cliargs_
deferred_get('check'))
    _diff = FieldAttribute(isa='bool', default=context.cliargs_deferred_
get('diff'))

    # ...
```

看到这里终于知道为什么 ansible 命令中得到的 Play 对象的 connections 属性值为 smart，而手工编写的代码得到的 Play 对象的 connection 值却为 None 了。这是因为源码中的 context 模块可以通过 cliargs_deferred_get() 从 context.CLIARGS 中获取相关 key 对应的 value 值并将其作为对象属性的默认值。

关于 context 的研究这里不再继续展开，笔者给出如下几处代码，有兴趣的读者可以继续学习 context.CLIARGS 的初始化过程。

```python
# 源码位置: lib/ansible/cli/__init__.py
# ...

class CLI(with_metaclass(ABCMeta, object)):

    # ...

    def parse(self):

        self.init_parser()
        options, args = self.parser.parse_args(self.args[1:])
        options, args = self.post_process_args(options, args)
        options.args = args
        context._init_global_context(options)

    # ...
```

在 Ansible 的命令行中会调用 CLI 类中的 parse() 方法去根据 options 的值初始化 context 模块中的 CLIARGS 的值。再来看 context 模块中的源码：

```python
# 源码位置: lib/ansible/context.py
# ...

CLIARGS = CLIArgs({})

def _init_global_context(cli_args):

    global CLIARGS
    # 根据命令行的输入初始化 CLIARGS 值
    CLIARGS = GlobalCLIArgs.from_options(cli_args)


def cliargs_deferred_get(key, default=None, shallowcopy=False):

    def inner():
        value = CLIARGS.get(key, default=default)
        if not shallowcopy:
            return value
        elif is_sequence(value):
            return value[:]
        elif isinstance(value, (Mapping, Set)):
            return value.copy()
        return value
    return inner
```

在 context 模块中，CLIARGS 其实就是一个类字典的对象，而 _init_global_context()

方法则是通过命令行中的参数来初始化这个全局的 CLIARGS，cliargs_deferred_get()方法则是从全局的 CLIARGS 中获取相应 key 对应的值。

4.2.3 流程总结

经过漫长的源码追踪后，我们终于找到了最后的合成模块源码以及传输模块源码内容到远端并执行的代码语句。虽然还有很多细节没有厘清，例如 Ansible 2 中完整的模块代码合成过程、远程连接的细节等，但这些细节和 Ansible 1 大体相同，所以没有必要重复介绍。最后，笔者用两张图来简单总结上面的源码追踪过程，图 4-3 展示了 ansible 命令的简要执行过程，图 4-4 细粒度地展示了默认策略插件的执行流程。

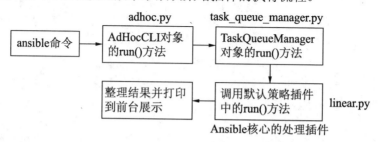

图 4-3　ansible 命令的简要执行过程

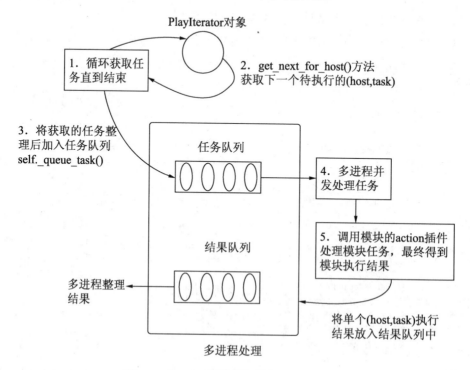

图 4-4　Ansible 中默认策略插件的执行流程

4.3　Ansible 2.8 中 Playbook 的执行流程追踪

本节中将带领读者继续探索 Ansible 2.8 中 Playbook 的执行流程以及部分 Playbook 语法背后的逻辑。

4.3.1　Playbook 命令执行入口

ansible-playbook 命令的源码文件只有 ansible 一行内容,表示该命令的内容和 ansible 命令文件内容一致。根据前面对 ansible 命令的分析可知,执行 ansible-playbook 命令最终会调用 PlaybookCLI 类中的 run()方法。该类的定义位于 lib/ansible/cli/playbook.py 文件中,源码内容如下:

```
# 源码位置: lib/ansible/cli/playbook.py
# ...

class PlaybookCLI(CLI):

    def init_parser(self):

        # 解析命令行参数
        # ...

    def post_process_args(self, options, args):

        # 继续处理选项
        # ...

    def run(self):

        super(PlaybookCLI, self).run()

        sshpass = None
        becomepass = None
        passwords = {}

        b_playbook_dirs = []
        # 处理异常
        # ...

        # 设置 Playbook 文件的路径
        set_collection_playbook_paths(b_playbook_dirs)

        # 从 context 中获取信息
        if not (context.CLIARGS['listhosts'] or context.CLIARGS
['listtasks'] or
                context.CLIARGS['listtags'] or context.CLIARGS['syntax']):
```

```
            (sshpass, becomepass) = self.ask_passwords()
            passwords = {'conn_pass': sshpass, 'become_pass': becomepass}

        # 创建基础对象
        loader, inventory, variable_manager = self._play_prereqs()

        # 获取主机列表
        CLI.get_host_list(inventory, context.CLIARGS['subset'])

        # 清除 facts 的缓存数据
        # ...

        # 生成 PlaybookExecutor 对象
        pbex = PlaybookExecutor(playbooks=context.CLIARGS['args'],
inventory=inventory,
                            variable_manager=variable_manager, loader=
loader,
                            passwords=passwords)

        # 运行 Playbook
        results = pbex.run()

        # 展示最终统计结果
        # ...
```

和 ansible 命令源码一样，ansible-playbook 命令的入口方法较为简单，核心语句就两行：生成 PlaybookExecutor 对象，然后调用该对象的 run() 方法获取 Playbook 的执行结果。需要注意实例化 PlaybookExecutor 类时的 playbooks 参数，它的值为 context.CLIARGS['args']，其含义为针对 ansible-playbook 命令，除去开始的 ansible-playbook 选项及选项值外，命令的剩余部分组成的元组形式便是 context.CLIARGS['args']。例如下面的 ansible-playbook 命令写法去掉 ansible-playbook 命令，去掉 -i 选项以及其值 hosts，最后剩 playbook/vars1.yml 和 playbook/vars2.yml，它们组成的元组便是 playbook 参数的值。

```
[root@master ansible-2.8.16]# ansible-playbook -i hosts playbook/vars1.yml
playbook/vars2.yml
```

最后实例化时 playbook 参数的值为('playbook/vars1.yml', 'playbook/vars2.yml')，这表示多个 playbook 集，ansible-playbook 命令会依次执行元组中的 Playbook 文件。

4.3.2 核心代码追踪

通过 4.3.1 节的分析，可知 Ansible 中 Playbook 的核心执行代码在 executor 目录下的 playbook_executor.py 文件中，这里定义了最关键的 PlaybookExecutor 类，其源码如下：

```
# 源码位置: lib/ansible/executor/playbook_executor.py
# ...

class PlaybookExecutor:
```

```
    def __init__(self, playbooks, inventory, variable_manager, loader,
passwords):
        self._playbooks = playbooks
        self._inventory = inventory
        self._variable_manager = variable_manager
        self._loader = loader
        self.passwords = passwords
        self._unreachable_hosts = dict()

        if context.CLIARGS.get('listhosts') or context.CLIARGS.get('listtasks')
or \
                context.CLIARGS.get('listtags') or context.CLIARGS.get('syntax'):
            # 如果只是展示相关数据并不要求执行，不用创建 TaskQueueManager 对象
            self._tqm = None
        else:
            # 创建 TaskQueueManager 对象并赋值给 self._tqm
            self._tqm = TaskQueueManager(
                inventory=inventory,
                variable_manager=variable_manager,
                loader=loader,
                passwords=self.passwords,
                forks=context.CLIARGS.get('forks'),
            )

        # 检查 SSH 优化功能
        check_for_controlpersist(C.ANSIBLE_SSH_EXECUTABLE)

    def run():
        # 核心，运行 Playbook 的方法
        # ...


    def _get_serialized_batches(self, play):
        # 获取待执行的一批主机列表
        # ...

    def _generate_retry_inventory(self, retry_path, replay_hosts):
        # 生成 Playbook 的 retry 文件，例如对于 site.yml 文件，调用该方法会生成 site.
          retry 文件
        try:
            makedirs_safe(os.path.dirname(retry_path))
            with open(retry_path, 'w') as fd:
                for x in replay_hosts:
                    fd.write("%s\n" % x)
        except Exception as e:
            # 打印警告信息
            # ...
            return False

        return True
```

上面的代码省略了部分内容较多的方法，只是简单地浏览了 PlaybookExecutor 类的内

容。可以看到该核心类只有 4 个简单的方法：

（1）__init__()方法。初始化方法。其中最关键的要属_tqm 属性，在没有--list-tasks 等展示选项时，该属性的值为 TaskQueueManager 对象。前面已经分析过，该对象用于执行 Ansible 中的 tasks，对于 Playbook 的执行也是一样，最终都会将 Playbook 文件拆解成一个个 task 来依次执行。

（2）run()方法。Playbook 文件执行的核心方法，后面会详细追踪其实现过程。

（3）_get_serialized_batches()方法。该方法用于控制每次执行 task 的主机列表，其相关参数则是剧本中的 serial。serial 参数在 Ansible 1 中介绍过，当 serial 为 1 时，每个任务主机都是串行执行。例如本次任务的 ansible 目标主机为 ceph-1~ceph-3，如果设置了 serial=2，则该方法的输出为[[ceph-1, ceph-2], [ceph-3]]，即第一次批量执行的主机为[ceph-1, ceph-2]，第二次批量执行的主机为 ceph-3。以下是关于 serial 参数的一个示例：

```
[root@master ansible-2.8.16]# cat playbook/test_serial.yml
---
- hosts: nodes
  gather_facts: false
  serial: 2

  tasks:
    - name: test ping
      ping:
        data: hello
```

这里设置了并行执行的主机数为 2。该 Playbook 的执行效果如下：

```
[root@master ansible-2.8.16]# ansible-playbook -i hosts playbook/test_
serial.yml

PLAY [nodes] ***************************************************

TASK [test ping] **********************************************
ok: [ceph-2]
ok: [ceph-1]

PLAY [nodes] ***************************************************

TASK [test ping] **********************************************
ok: [ceph-3]

PLAY RECAP ****************************************************
# 忽略统计结果
# ...
```

（4）_generate_retry_inventory()方法。该方法用于生成 Playbook 执行失败后的 retry 文件。例如对于 site.yml 文件，如果该 Playbook 执行失败，则会在对应目录下生成一个 site.retry 文件。生成该 retry 文件名的逻辑可以从 run()方法中找到。

了解了 PlaybookExecutor 类的全貌后，下面便可以专注 run()方法的源码解读了。

```
# 忽略统计结果
```

```
# ...
# 源码位置：lib/ansible/executor/playbook_executor.py
# ...

class PlaybookExecutor:

    # ...

    def run(self):

        result = 0
        entrylist = []
        entry = {}
        try:
            # 预先加载 connection、shell 和 become 的所有插件
            list(connection_loader.all(class_only=True))
            list(shell_loader.all(class_only=True))
            list(become_loader.all(class_only=True))

            # 遍历执行 Playbook
            for playbook_path in self._playbooks:
                pb = Playbook.load(playbook_path,
                        variable_manager=self._variable_manager, loader=
self._loader)

                if self._tqm is None:             # 只有列举功能
                    entry = {'playbook': playbook_path}
                    entry['plays'] = []
                else:
                    # 确保 tqm 要先加载回调
                    self._tqm.load_callbacks()
                    self._tqm.send_callback('v2_playbook_on_start', pb)

                i = 1
                # 获取 Playbook 中的剧本，其实 Playbook 表示剧本集
                plays = pb.get_plays()

                # 开始遍历剧本执行
                for play in plays:
                    # 设置剧本执行的基目录
                    if play._included_path is not None:
                        self._loader.set_basedir(play._included_path)
                    else:
                        self._loader.set_basedir(pb._basedir)

                    # 清除在 inventory 中应用的所有过滤器
                    self._inventory.remove_restriction()

                    # 获取该 Play 中的所有变量
                    all_vars = self._variable_manager.get_vars(play=play)
                    templar = Templar(loader=self._loader, variables=all_vars)
                    # 设置 Play 的 vars_prompt 属性，该属性用于实现人机交互，从界面输
                        入变量
```

```
                    setattr(play, 'vars_prompt', templar.template(play.vars_
prompt))

                    if play.vars_prompt:
                        for var in play.vars_prompt:
                            # 处理人机交互，最终调用 display.do_var_prompt()方法
                            # ...

                    # 处理模板变量，渲染得到最终的变量值
                    all_vars = self._variable_manager.get_vars(play=play)
                    templar = Templar(loader=self._loader, variables=all_
vars)

                    play.post_validate(templar)

                    if context.CLIARGS['syntax']:
                        continue

                    if self._tqm is None:
                        # 这里只作展示
                        entry['plays'].append(play)
                    else:
                        # 更新不可达主机信息
                        self._tqm._unreachable_hosts.update(self._unreachable_
hosts)

                        # 上一次失败和不可达主机数目
                        previously_failed = len(self._tqm._failed_hosts)
                        previously_unreachable = len(self._tqm._unreachable_
hosts)

                        break_play = False
                        # 获取本次批量执行的主机列表
                        batches = self._get_serialized_batches(play)
                        if len(batches) == 0:
                            # 当没有待执行的主机时，设置相应的回调方法
                            # 本次 Play 开始执行并已经结束
                            self._tqm.send_callback('v2_playbook_on_play_start',
play)
                            self._tqm.send_callback('v2_playbook_on_no_hosts_
matched')

                        for batch in batches:
                            # 限制 inventory 中的目标 hosts，这次是分批执行
                            self._inventory.restrict_to_hosts(batch)
                            # 最核心的语句，即调用 TaskQueueManager 对象的 run()方
                              法去执行
                            result = self._tqm.run(play=play)

                            # 处理本次 Play 执行结果
                            # ...

                        if break_play:
                            break
```

```
                i = i + 1                     # 记录剧本编号
        finally:
            if self._tqm is not None:
                self._tqm.cleanup()
            if self._loader:
                self._loader.cleanup_all_tmp_files()

        # ...

        return result
```

从上面的代码中可以简要地看到整个 Playbook 的执行过程，对于单个 Playbook 文件，最终会按照 Play 的粒度去依次执行。而对于 Play 的执行，最终也是调用 TaskQueueManager 类中的 run() 方法。图 4-5 展示了执行 Playbook 中单个 Play 的完整流程。

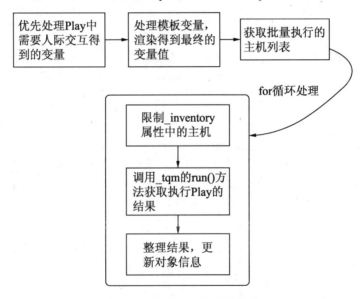

图 4-5　单个 Play 的执行流程

源码追踪到这里就很清楚了，Ansible 的命令行运行以及 Playbook 的解析与执行最终调用的都是同一套核心代码，只是上层的封装和解析不同。在了解了 Playbook 的执行过程后，读者可以思考下 Ansible 是如何解析 Playbook 文件并从中得到 Play 列表的。从上面的代码来看，其实就只有两行代码，具体如下：

```
pb = Playbook.load(playbook_path, variable_manager=self._variable_manager,
loader=self._loader)
plays = pb.get_plays()
```

这里使用了 Playbook 类来解析 Playbook 文件。参考上面的语句，可以在 Python 的交互式模式下进行测试。请看下面的示例：

```
[root@master playbook]# cat /root/ansible-2.8.16/playbook/test_playbook.yml
---
- hosts: nodes
  gather_facts: false

  tasks:
    - name: get name variable
      shell: hostname
      register: name_out

    - name: debug hostname
      debug:
        msg: "hello {{ name_out.stdout }}"
- hosts: center
  gather_facts: false

  tasks:
    - name: ping test
      ping:
        data: hello, center
```

可以看到上面定义了一个 Playbook，该 Playbook 中有两个 Play。第一个 Play 的目标
主机为 nodes 组内的所有主机，第二个 Play 为 center 组内的主机。下面手工解析该 Playbook，
具体操作如下：

```
>>> from ansible.playbook import Playbook
>>> from ansible.parsing.dataloader import DataLoader
>>> from ansible.inventory.manager import InventoryManager
>>> from ansible.vars.manager import VariableManager
>>> loader = DataLoader()
>>> inventory = InventoryManager(loader=loader, sources="/root/hosts")
>>> variable_manager = VariableManager(loader=loader, inventory=
inventory, version_info=None)
>>> pb = Playbook.load('/root/ansible-2.8.16/playbook/test_playbook.yml',
variable_manager=variable_manager, loader=loader)
>>> pb.get_plays()
[nodes, center]
>>> plays = pb.get_plays()
>>> plays[0].serialize()
{'hosts': 'nodes', 'gather_facts': False, 'gather_subset': ['all'], 'gather_
timeout': 10, 'fact_path': None, 'vars_files': [], 'vars_prompt': [],
'roles': [], 'handlers': [], 'pre_tasks': [], 'post_tasks': [], 'tasks':
[BLOCK(uuid=000c295e-8383-0c39-b4d7-00000000000a)(id=139940503071040)
(parent=None)], 'force_handlers': None, 'max_fail_percentage': None, 'serial':
[], 'strategy': 'linear', 'order': None, 'name': 'nodes', 'connection':
None, 'port': None, 'remote_user': None, 'vars': {}, 'module_defaults':
None, 'environment': None, 'no_log': None, 'run_once': None, 'ignore_
errors': None, 'ignore_unreachable': None, 'check_mode': None, 'diff':
None, 'any_errors_fatal': False, 'debugger': None, 'tags': [], 'become':
None, 'become_method': None, 'become_user': None, 'become_flags': None,
'collections': None, 'uuid': '000c295e-8383-0c39-b4d7-000000000009',
'finalized': False, 'squashed': False, 'included_path': None}
>>> plays[0].get_tasks()
[[TASK: get name variable, TASK: debug hostname]]
```

```
>>> plays[1].get_tasks()
[[TASK: ping test]]
```

4.3.3　Playbook 中部分语法的源码解析

本节将简单介绍几个 Ansible 的 Playbook 语法的实现代码，进一步加深对 Playbook 文件的相应语法的理解。

1. register语法

register 语法就是指定保存本次 task 执行结果的变量。在 Ansible l 和 Ansible 2 中都介绍过该语法的使用，非常简单且常用。该语法的处理代码在 Ansible 源码中有多处，每处的写法都大致类似，具体如下：

```
# 源码位置：lib/ansible/executor/task_executor.py

# 第 1 处代码
if self._task.register:
    vars_copy[self._task.register] = wrap_var(result)

# 第 2 处代码
if self._task.register:
    variables[self._task.register] = wrap_var(result)
```

上面的代码也能说明 register 的功能，正是将本次的结果放到全局变量中，对应的 key 正是 register 语法指定的值。通过这样的语句即可知 register 注册的变量是什么类型，它有哪些字段，代表什么含义。关于返回结果的字段有哪些，直接通过源码一查即知。

注意：可以在 VSCode 中搜索 task.register 关键字以获取处理 register 语法的所有代码位置，如图 4-6 所示。

图 4-6　全局搜索 register 语法

2．delegate_to语法

delegate_to 语法在第 2 章中已经介绍过。现在修改之前的示例，给出有 run_once 语法和没有 run_once 语法的两个 task，具体信息如下：

```
(ansible2.8.16) [root@master playbook]# cat test_delegate_to.yml
---
- hosts: nodes
  gather_facts: false

  tasks:
    - name: test delegate and run once
      shell: hostname
      register: name_out
      delegate_to: master
      run_once: yes

    - name: debug hostname
      debug:
        msg: "{{ name_out.stdout }}"

    - name: test delegate
      shell: ls /tmp/{{ inventory_hostname }}
      register: file_out
      delegate_to: master

    - name: debug file
      debug:
        msg: "{{ file_out.stdout }}"
```

第一个任务和之前一样，委托到 master 节点执行 hostname 命令获取主机名，由于执行的结果一致，所以可以使用 run_once 语法控制只执行一次。此外，在 master 主机的/tmp目录下新建 ceph-1、ceph-2 和 ceph-3 目录，每个目录下分别有一个文件。现在需要通过委托的方式获取对应主机名目录下的文件名，对每个主机而言目录都是不一样的，所以委托操作必须每个都执行，不能使用 run_once 语法，具体如下：

```
(ansible2.8.16) [root@master playbook]# ls /tmp/ceph-*
/tmp/ceph-1:
1.txt

/tmp/ceph-2:
2.txt

/tmp/ceph-3:
3.txt
```

最后执行该 Playbook，结果如下：

```
(ansible2.8.16) [root@master playbook]# ansible-playbook -i hosts test_
delegate_to.yml

PLAY [nodes] *********************************************************
```

```
TASK [test delegate and run once] ***********************************
changed: [ceph-1 -> master]

TASK [debug hostname] ***********************************************
ok: [ceph-1] => {
    "msg": "master"
}
ok: [ceph-3] => {
    "msg": "master"
}
ok: [ceph-2] => {
    "msg": "master"
}
[WARNING]: Removed restricted key from module data: ansible_facts =
{'discovered_ ... thon'}


TASK [test delegate] ***********************************************
changed: [ceph-1 -> master]
changed: [ceph-3 -> master]
changed: [ceph-2 -> master]

TASK [debug file] *************************************************
ok: [ceph-1] => {
    "msg": "1.txt"
}
ok: [ceph-2] => {
    "msg": "2.txt"
}
ok: [ceph-3] => {
    "msg": "3.txt"
}

PLAY RECAP ********************************************************
# 忽略统计结果
# ...
```

现在是不是对 delegate_to 语法有了一点了解了？再来看 Ansible 中对于 delegate_to 语法的处理，具体如下：

```
# 源码位置：lib/ansible/executor/task_executor.py
# ...

class TaskExecutor:

    # ...

    def _execute(self, variables=None):

        # ...

        try:
            self._play_context = self._play_context.set_task_and_variable_
override(
                            task=self._task, variables=variables, templar=
```

```
templar)
        # ...
    except AnsibleError as e:
        context_validation_error = e

    # ...
```

同样还是在 TaskExecutor 类的_execute()方法中调用 PlayContext 对象的 set_task_and_variable_override()方法，重新处理远程连接的变量，特别是涉及委托的语法。set_task_and_variable_override()方法的源码如下：

```
# 源码位置: lib/ansible/playbook/play_context.py
# ...

class PlayContext(Base):

    # ...

    def set_task_and_variable_override(self, task, variables, templar):

        # 复制当前的 PlayContext 对象
        new_info = self.copy()

        # ...

        if task.delegate_to is not None:
            # 得到委托主机
            delegated_host_name = templar.template(task.delegate_to)
            # 获取委托变量，例如设置的委托主机的远程连接方式，如 SSH 登录的账号、密码等
            delegated_vars = variables.get('ansible_delegated_vars', dict()).
get(
                                delegated_host_name, dict())

            # 获取委托主机的 SSH 连接方式
            delegated_transport = C.DEFAULT_TRANSPORT
            for transport_var in C.MAGIC_VARIABLE_MAPPING.get('connection'):
                if transport_var in delegated_vars:
                    delegated_transport = delegated_vars[transport_var]
                    break

            for address_var in ('ansible_%s_host' % delegated_transport,)
                    + C.MAGIC_VARIABLE_MAPPING.get('remote_addr'):
                # 如果前面已经有了该信息，直接退出 for 循环
                if address_var in delegated_vars:
                    break
            else:
                # for 循环结束后，设置委托信息的 ansible_host 值
                delegated_vars['ansible_host'] = delegated_host_name

            for port_var in ('ansible_%s_port' % delegated_transport,)
                    + C.MAGIC_VARIABLE_MAPPING.get('port'):
                # 如果前面已经有了该信息，直接退出 for 循环
                if port_var in delegated_vars:
```

```
                    break
            else:
                # for 循环结束后，设置委托信息的 ansible_port 值
                if delegated_transport == 'winrm':
                    delegated_vars['ansible_port'] = 5986
                else:
                    delegated_vars['ansible_port'] = C.DEFAULT_REMOTE_PORT

            for user_var in ('ansible_%s_user' % delegated_transport,)
                    + C.MAGIC_VARIABLE_MAPPING.get('remote_user'):
                if user_var in delegated_vars and delegated_vars[user_var]:
                    break
            else:
                # for 循环结束后，设置委托信息的 ansible_user 信息
                delegated_vars['ansible_user'] = task.remote_user or self.
remote_user
        else:
            # ...

        attrs_considered = []
        for (attr, variable_names) in iteritems(C.MAGIC_VARIABLE_MAPPING):
            for variable_name in variable_names:
                if attr in attrs_considered:
                    continue
                if task.delegate_to is not None:
                    # 核心设置，如果任务设置了委托，则重新更新 PlayContext 对象中的
                      信息
                    if isinstance(delegated_vars, dict) and variable_name in
                    delegated_vars:
                        # new_info 为该 PlayContext 对象的一个副本，重新更新主机以及
                          连接信息
                        setattr(new_info, attr, delegated_vars[variable_name])
                        attrs_considered.append(attr)
                elif variable_name in variables:
                    # ...

        # ...

        if len(delegated_vars) > 0:
            # 进一步处理对应目标主机的远程 SSH 连接方式，设置到 new_info 中
            for connection_type in C.MAGIC_VARIABLE_MAPPING.get('connection'):
                if connection_type in delegated_vars:
                    break
            else:
                # for 循环退出后会再次执行这里
                remote_addr_local = new_info.remote_addr in C.LOCALHOST
                inv_hostname_local = delegated_vars.get('inventory_hostname')
in C.LOCALHOST
                if remote_addr_local and inv_hostname_local:
                    setattr(new_info, 'connection', 'local')
                elif getattr(new_info, 'connection', None) == 'local' and
                    (not remote_addr_local or not inv_hostname_local):
```

```
                    setattr(new_info, 'connection', C.DEFAULT_TRANSPORT)
        # ...
```

上面的代码基本可以确定 delegate_to 语法的功能，首先是从 variables 中获取 key 为 ansible_delegated_vars 的值并赋给 delegated_vars，这是保存委托主机信息的变量。接着在该字典变量中保存 delegated_vars 中关于 SSH 登录委托主机的信息，包括主机地址、端口、远程登录用户以及使用的底层通信模式。接着是重新更新 PlayContext 对象中关于目标主机的信息，这样后续执行对应 task 时会远程登录到委托主机上，执行方式并不改变，委托语法只是改变了远程执行的主机。至于 run_once 语法的控制，出现在策略插件的 run()方法中，具体如下：

```python
# 源码位置: lib/ansible/plugins/strategy/linear.py
# ...

class StrategyModule(StrategyBase):

    # ...

    def run(self, iterator, play_context):

        # ...
        while work_to_do and not self._tqm._terminated:
            # ...
            try:
                # ...

                for (host, task) in host_tasks:

                    # ...
                    run_once = False

                    # ...

                    if task.action == 'meta':
                        # ...
                    else:
                        # ...

                        run_once = templar.template(task.run_once) or action and
                                getattr(action, 'BYPASS_HOST_LOOP', False)

                        # ...
                        # 将任务添加到进程队列中
                        self._queue_task(host, task, task_vars, play_context)
                        del task_vars

                    if run_once:
                        break

                # 获取结果队列中的数据
                # ...
```

```
    # ...
    except (IOError, EOFError) as e:
        return self._tqm.RUN_UNKNOWN_ERROR

return super(StrategyModule, self).run(iterator, play_context, result)
```

笔者大幅简化了 linear.py 插件中 run()方法的代码，只保留了和 run_once 语法相关的代码。从上面的代码中可以清楚的看出 run_once 语法功能，for 循环语句将对应主机和 task 添加到进程的任务队列中，如果设置了 run_once 语法，那么上述代码会将第一个主机加入到任务队列后设置 run_once=True，接着直接跳出 for 循环（上述 break 语句），从而该任务只会被第一个主机执行一次。

3. until 语法

until 语法在第 3 章介绍过，现在来学习该语法的实现源码，然后与现象进行对比，进一步加深对该语法的理解。该语法的实现同样是在 task_executor.py 文件中，具体如下：

```
# 源码位置: lib/ansible/executor/task_executor.py
# ...

class TaskExecutor:

    # ...

    def _execute(self, variables=None):

        # ...

        # 对于设置了 until 语法的 task，需要获取重试参数以及延迟时间，均有默认值
        if self._task.until:
            retries = self._task.retries
            if retries is None:
                retries = 3
            elif retries <= 0:
                retries = 1
            else:
                retries += 1
        else:
            retries = 1

        # 获取重试延迟时间
        delay = self._task.delay
        if delay < 0:
            delay = 1

        # ...

        result = None
        for attempt in xrange(1, retries + 1):

            # ...
```

```
            if self._task.until:
                result['attempts'] = attempt

            # ...

            if retries > 1:
                cond = Conditional(loader=self._loader)
                cond.when = self._task.until
                # 检查 until 条件是否符合
                if cond.evaluate_conditional(templar, vars_copy):
                    # 如果符合 until 条件，直接退出循环
                    break
                else:
                    # 如果不符合条件，继续尝试
                    if attempt < retries:
                        # 设置相关参数
                        result['_ansible_retry'] = True
                        result['retries'] = retries
                        # 重新将任务添加到队列中
                        self._final_q.put(TaskResult(self._host.name, self.
_task._uuid,
                            result, task_fields=self._task.dump_attrs()), block=
False)

                        # 休息 delay 秒
                        time.sleep(delay)
                        # 再次获取 self._handler 值用于下次执行
                        self._handler = self._get_action_handler(
                            connection=self._connection, templar=templar)
        else:
            if retries > 1:
                result['attempts'] = retries - 1
                result['failed'] = True
```

还记得 execute()方法吗？前面 TaskExecutor 对象的 run()方法中的核心语句就是调用 execute()方法。以上代码即为 execute()方法中处理 until 语法的代码段，其处理逻辑如下：

（1）对于设置了 until 语法的 task，需要获取重试参数以及延迟时间，均有默认值（默认重试次数为 3 次，默认重试延时 1s）。

（2）使用 for 循环进行重试。如果没有设置 until 语法，则重试次数为 1。

（3）在解析结果的同时，也会解析 until 条件。如果 until 条件不满足，则将任务添加到任务队列中并休眠 delay 秒后再次获取对应动作模块的_handler 属性值。

（4）直到 until 条件成功或者是达到重试次数，退出循环，结束该任务的执行。

好了，对于 Playbook 的语法解析就到这里。以上对三个基础语法的介绍只是为了起到一个抛砖引玉的作用，关于 Playbook 语法还有很多值得探索的地方，限于篇幅不再一一展开介绍。第 5 章将进入 Ansible 的高级源码学习部分——版本的迭代分析和部分 Bug 溯源。

4.4　本 章 小 结

本章详细分析了 Ansible 2.8 中的一些核心源码。首先从基础的功能模块介绍起，例如显示类（Display）、默认配置设置、inventory 解析、模板渲染等；接着是跟踪 Ansible 2.8 中模块的运行流程，这里的代码与 Ansible 1 相比，除了更加复杂外，很多功能都已经插件化；最后跟踪了 Playbook 的整个运行过程，这里的代码同样是异常复杂，笔者也只能去掉众多细枝末节，带领读者掌握 Playbook 执行中最核心的代码。

第 5 章　Ansible 源码的演进跟踪

本章将探索 Ansible 源码的演进之路并复现部分 Bug，主要包括以下两部分内容：
- Ansible 1.1 到 Ansible 1.2 的完整演进过程，同时也会根据下一个版本修复的代码找出上一个版本的 Bug，从而帮助读者进一步理解和熟悉 Ansible 1 的源码。
- Ansible 2.8.0 到 Ansible 2.8.16 的部分 Bug 溯源及相关 Bug 修复过程分析。

5.1　Ansible 1.1 到 Ansible 1.2 的演进之路

由于 Ansible 的源码中没有完整的 GitHub 分支管理，而且源码中没有记录版本的新增特性、Bug 修复及安全问题修复等方面的信息，因此只能借助文件比较工具来比较两个版本的源码改动情况。笔者使用的是试用版的 Beyond Compare 工具，读者也可以自行搜索开源免费的文件比较器或者评估版的 Beyond Compare 工具完成本节内容的学习。

本节对比的两个版本是 Ansible 1.1 和 Ansible 1.2。首先下载这两个版本的源码包并使用文件比较器打开两者的根目录，然后选择比较全部文件内容，对比结果如图 5-1 所示。

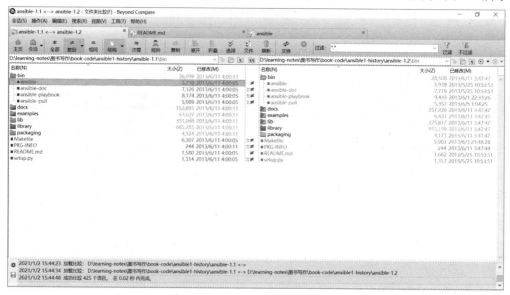

图 5-1　Ansible 1.1 和 Ansible 1.2 源文件对比

　　下面将完整对比两个版本软件的大部分的文件内容，找出升级特性和部分 Bug 并进行修复。

　　首先对比两个 Ansible 版本的 README.md 文件，对比结果如图 5-2 所示。该文件类似于产品说明。通过对比可以看出，Ansible 1.1 属于个人产品，而 Ansible 1.2 及其之后的版本都有了正式的组织名称。这一点从开发者的邮箱变更及 Ansible 的官网地址可以看出。

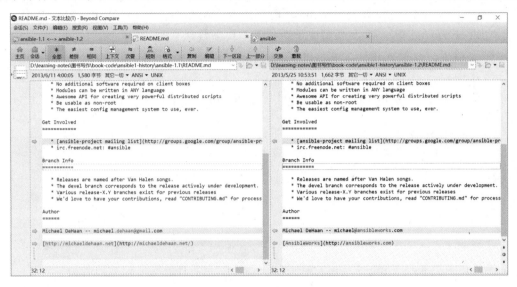

图 5-2　两个 Ansible 版本的说明文件对比

　　接下来对比两个 Ansible 版本的命令文件，结果如图 5-3 所示。

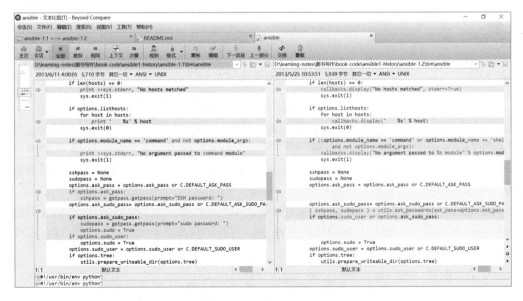

图 5-3　两个 Ansible 版本的命令文件对比

从上面的代码对比中可以看到，Ansible 1.2 的 ansible 文件相比 Ansible 1.1 有如下变化：

- 移除了大部分的 print 语句，改为调用 callbacks.py 文件中的 display()函数，而该函数也封装了 print 语句。除此之外，display()函数还支持日志文件打印、日志加锁等功能。
- 在 Ansible 1.1 中只检查 command 模块是否传入模块参数，而在 Ansible 1.2 中则检查 command 模块和 shell 模块是否传入模块参数。
- 调整了交互模式代码，将 getpass.getpass()语句封装到了 utils.ask_passwords()方法中。

接着对比两个 Ansible 版本中放置模块代码的 library 目录，对比结果如图 5-4 所示。

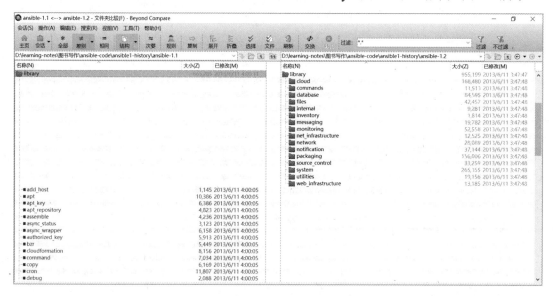

图 5-4 两个版本的 library 目录对比

可以看到，从 Ansible 1.2 开始，对模块进行了归类，将它们按照功能进行了划分，如处理文件的 files 目录、执行命令的 commands 目录、与数据库相关的 database 目录、与网络相关的 network 目录等。例如，前面经常用到的 copy 模块和 file 模块就位于 files 目录下，command 模块和 shell 模块等位于 commands 目录下。随着模块的增多，这种模块分类的方式势在必行。

下面以 copy 模块和 lineinfile 模块为例来说明两个版本中对应模块的改动部分。首先对比 copy 模块，由于模块文件的路径已改变，需要新打开一个文件比较器窗口，因此采用文件比较模式。对比结果如图 5-5 所示。

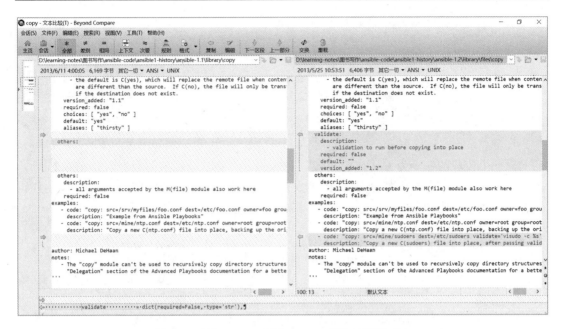

图 5-5　两个 Ansible 版本中 copy 模块的源码对比

通过对比结果可以看到，copy 模块在 Ansible 1.2 版本中新增了一个 validate 参数用于校验。该参数功能的实现比较简单，具体代码如下：

```
# 代码位置：Ansible 1.2 的 copy 模块
# ...

if validate:
    (rc,out,err) = module.run_command(validate % src)
    if rc != 0:
        module.fail_json(msg="failed to validate: rc:%s error:%s" % (rc,err))
```

上面这段代码非常容易理解，假设 validate 的值为 ls %s，copy 模块中上传的文件路径为 src=/root/hosts，此时校验的命令为 ls /root/hosts。如果该命令执行失败，则 copy 模块执行失败。下面是一个简单的示例：

```
(ansible1.2) [root@master ~]# ansible ceph-1 -i hosts -m copy -a "src=hosts
dest=/tmp/hosts validate='xxx %s'"
ceph-1 | FAILED >> {
    "failed": true,
    "msg": "failed to validate: rc:127 error:/bin/sh: xxx: command not
found\n"
}

(ansible1.2) [root@master ~]# ansible ceph-1 -i hosts -m copy -a "src=hosts
dest=/tmp/hosts validate='ls %s'"
ceph-1 | success >> {
    "changed": true,
    "dest": "/tmp/hosts",
    "gid": 0,
```

```
        "group": "root",
        "md5sum": "3767870164ea11b7900f9c82db538df4",
        "mode": "0644",
        "owner": "root",
        "size": 174,
        "src": "/root/.ansible/tmp/ansible-1609518771.39-58784853774143/source",
        "state": "file",
        "uid": 0
}
```

第一个校验的命令为 xxx /root/hosts，校验失败；第二个校验的命令为 ls /root/hosts，校验成功。此外，除了新增的 validate 参数外，copy 模块将 Ansible 1.1 中移动文件的代码进行了替换，对比如下：

```
# Ansible 1.1 中 copy 模块的部分代码片段
dest_tmp = "%s.%s.%s.tmp" % (dest,os.getpid(),time.time())
shutil.copyfile(src, dest_tmp)
module.atomic_replace(dest_tmp, dest)

# Ansible 1.2 中 copy 模块的部分代码片段，和上面的功能一致
module.atomic_move(src, dest)
```

这是由于 Ansible 中的公共代码块（module_common.py）已经发生了变化，Ansible 1.2 中的原子移动方法足以替换 Ansible 1.1 中的三行代码。

下面对比 lineinfile 模块，该模块的改动也不大，没有增加新的参数，只是对原有的代码进行了优化，对比结果如图 5-6 所示。

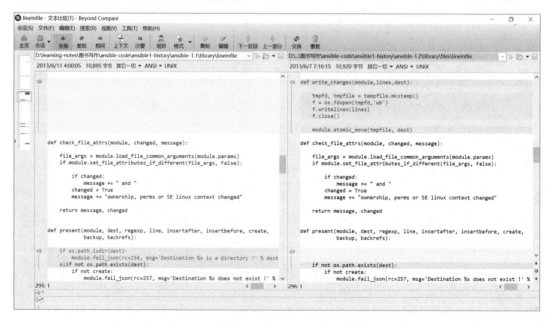

图 5-6　两个版本的 lineinfile 模块源码对比

　　由于无法完整显示对比结果，图 5-6 中只截取了两处不同的地方。在 Ansible 1.1 中，对 dest 是否为目录的判断在 present()和 absent()方法中重复出现，于是 Ansible 1.2 将 dest 是否为目录的判断直接挪到了 main()方法中，具体如下：

```
# Ansible 1.2 的 lineinfile 模块源码
# ...

def main():

    # ...

    if os.path.isdir(dest):
        module.fail_json(rc=256, msg='Destination %s is a directory !' % dest)

    # ...
```

　　此外，下面三行代码在 present()和 absent()方法中也是重复的。

```
f = open(dest, 'wb')
f.writelines(lines)
f.close()
```

　　以上代码只是简单地将 lines 内容写入 dest 文件中，在 Ansible 1.2 中使用函数实现了这个功能，具体代码如下：

```
# Ansible 1.2 中的 lineinfile 模块源码

def write_changes(module,lines,dest):

    tmpfd, tmpfile = tempfile.mkstemp()
    f = os.fdopen(tmpfd,'wb')
    f.writelines(lines)
    f.close()

    module.atomic_move(tmpfile, dest)
```

　　为了避免同时操作 dest 文件，先将内容写入临时文件中，然后再调用公共模块中的原子移动方法将临时文件变成 dest 文件。

　　接下来对比 lib/ansible 目录下的源码。首先对比公共函数模块 module_common.py。对比结果如图 5-7 所示。在 Ansible 1.2 中，该文件主要用于对一些函数进行优化，如加上 try...catch 异常检测机制等。其中变动最大的是移除了公共模块中的 atomic_replace()方法，使用 atomic_move()方法代替，这会直接影响调用 atomic_replace()方法的模块，如 copy 模块。

　　接着对比 constants.py 文件，该文件的变化不大，只是在 Ansible 1.2 中新增了几个常量，如图 5-8 所示。

图 5-7　两个版本的 module_common.py 文件对比

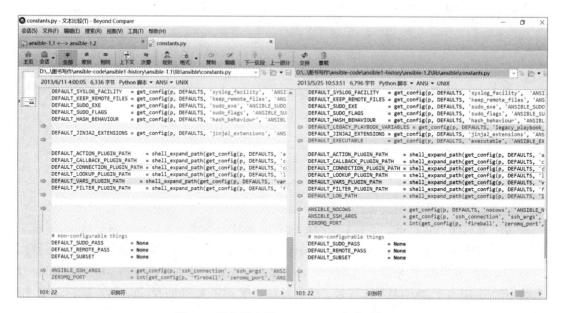

图 5-8　两个版本的 constants.py 文件对比

callback.py 文件在 Ansible 1.2 中的改动较多，主要体现在以下几个方面：

- 实现了一个 display() 方法，该方法替换了 Ansible 1.1 中的所有 print 语句。
- 新增了关于日志锁的代码。

- 使模块执行结果可以显示除了黑色以外的其他颜色。

下面重点介绍第 3 个改动。读者可以在终端上测试多个版本的 Ansible 模块执行命令，会发现在 Ansible 1.1 中，输出结果是单一的颜色，即黑色，而在 Ansible 1.2 中，输出结果是有颜色的。但是在 Ansible 1.1 中的输出结果中不都是黑色，加上 -vv 选项后将输出 ansible 命令的详细执行过程，这些详细信息的字体颜色为蓝色。此外，ansible-playbook 命令执行的输出均有正常颜色（如有改变的为黄色、无改变的为绿色，错误的为红色等），仅仅是执行 ansible 命令时没有输出对应的颜色，Ansible 1.1 中打印模块输出的结果调用的代码如下：

```python
# 源码包：Ansible 1.1
# 源码位置：lib/ansible/callbacks.py
# ...

def host_report_msg(hostname, module_name, result, oneline):
    ''' 针对单个节点整理 JSON 结果 '''

    failed = utils.is_failed(result)
    msg = ''
if module_name in [ 'command', 'shell', 'raw' ] and
        'ansible_job_id' not in result and result.get('parsed',True) != False:
    if not failed:
        msg = command_generic_msg(hostname, result, oneline, 'success')
    else:
        msg = command_generic_msg(hostname, result, oneline, 'FAILED')
else:
    if not failed:
        msg = regular_generic_msg(hostname, result, oneline, 'success')
    else:
        msg = regular_generic_msg(hostname, result, oneline, 'FAILED')

    # msg 就是模块最后的输出结果
    return msg

# ...

class CliRunnerCallbacks(DefaultRunnerCallbacks):
    # ...

    def _on_any(self, host, result):
        result2 = result.copy()
        result2.pop('invocation', None)
        # 这里就是最后打印文本的地方
        print host_report_msg(host, self.options.module_name, result2,
self.options.one_line)
        # ...
```

Ansible 1.1 中模块的运行结果最终由 host_report_msg()函数获取，打印则由 CliRunner-Callbacks 类中的_on_any()方法完成。上面的代码中直接使用了 print 语句打印结果，因此效果为黑色。在 Ansible 1.2 中对这两处代码进行了调整，具体如下：

```python
# 源码包：Ansible 1.2
# 源码位置：lib/ansible/callbacks.py
# ...

def host_report_msg(hostname, module_name, result, oneline):

    failed = utils.is_failed(result)
    msg = ('', None)
    if module_name in [ 'command', 'shell', 'raw' ] and
            'ansible_job_id' not in result and result.get('parsed',True) != False:
        if not failed:
            msg = (command_generic_msg(hostname, result, oneline, 'success'),
'green')
        else:
            msg = (command_generic_msg(hostname, result, oneline, 'FAILED'),
'red')
    else:
        if not failed:
            msg = (regular_generic_msg(hostname, result, oneline, 'success'),
'green')
        else:
            msg = (regular_generic_msg(hostname, result, oneline, 'FAILED'),
'red')
    return msg

# ...

class CliRunnerCallbacks(DefaultRunnerCallbacks):
    # ...

    def _on_any(self, host, result):
        result2 = result.copy()
        result2.pop('invocation', None)
        # 得到结果主题和颜色
        (msg, color) = host_report_msg(host, self.options.module_name,
result2, self.options.one_line)
        # 调用 display()函数，用 color 颜色输出结果
        display(msg, color=color)
        # ...
```

在上面的代码中，host_report_msg()函数的主体内容不变，但是对每种情况的结果手工添加了一个 color 值，形成了一个二元组(msg, color)。另外，在 CliRunnerCallbacks 类的方法中不再使用 print 打印模块的执行结果，而是使用 display()函数。该函数支持输入颜色

参数以实现不同颜色的打印效果，从而改变 Ansible 1.1 中命令行输出颜色单一的问题，这也算实现了一个新功能。

　　继续对比 lib/ansible/callback_plugins 目录下的 noop.py 文件，该文件只修正了两个方法名，目的是使函数名变得统一，属于简单优化，对比结果如图 5-9 所示。

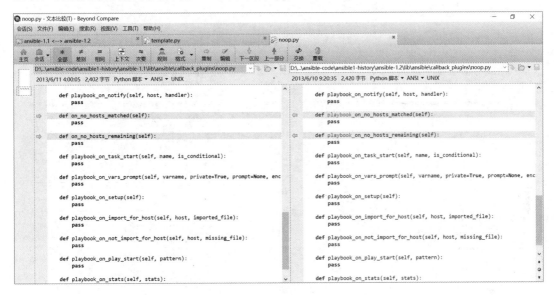

图 5-9　noop.py 文件对比

　　下面接着对比分析 inventory 目录，重点分析 __init__.py 和 ini.py 这两个文件。对于 __init__.py 文件，笔者找出了三处优化，具体如下：

　　（1）将 type() 方法判断换成了 isinstance() 方法判断。示例如下：

```
type(host_list) in [ str, unicode ]   -> isinstance(host_list, basestring)
type(restriction) != list         -> not isinstance(restriction, list)
```

　　（2）替换部分表达式，例如：

```
x.find(":") != -1              -> ":" in x
```

　　（3）优化了 Inventory 类中的部分方法，包括_gen_host() 方法、subset() 方法及 basedir() 方法，如图 5-10 所示。

　　以上 3 个方法的优化主要是考虑到一些特殊的情况，如调用_gen_host() 方法根据主机名获取对应的 Host 对象时，不会区分 localhost 和 127.0.0.1。

　　另外，在 Ansible 1.2 中新增了_vars_plugins 和_playbook_basedir 属性，同时还新增了 playbook_basedir() 和 set_playbook_basedir() 方法。

　　接下来对比 ini.py 文件。由第 1 章的分析可知，该文件中的代码用于解析 hosts 文件，得到相应的目标主机、组及变量信息。该文件在 Ansible 1.2 中的改动并不大，对比结果如图 5-11 所示。

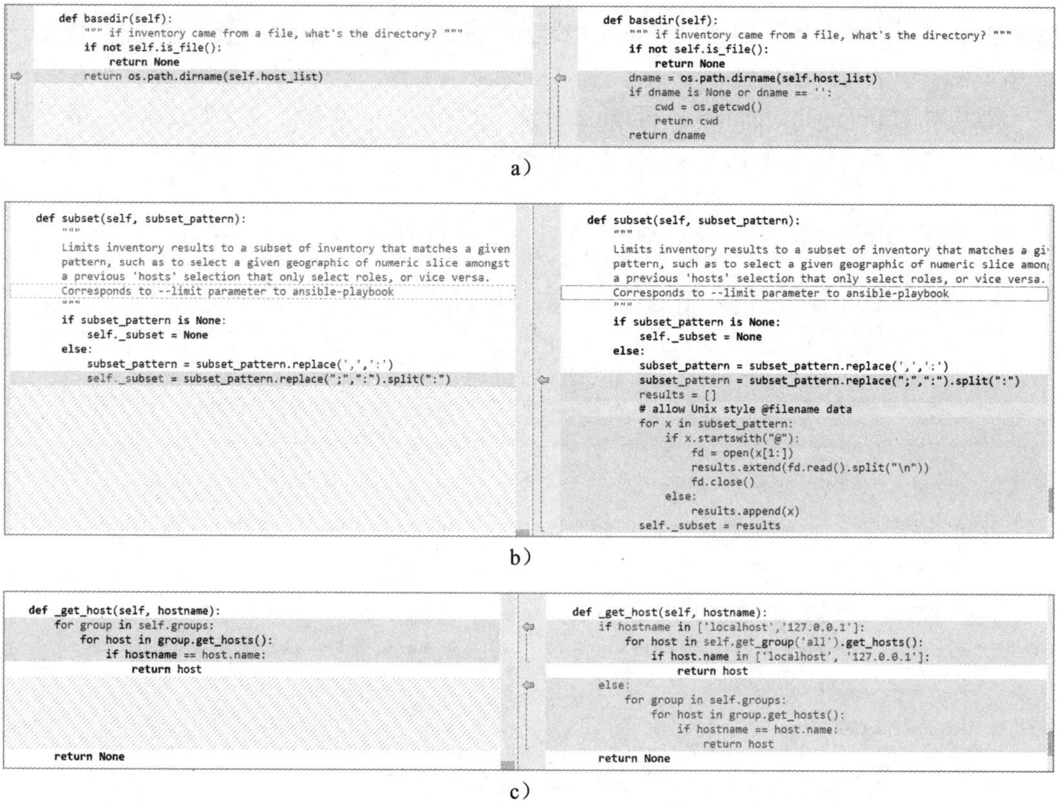

图 5-10　Inventory 类代码优化

图 5-11　两个版本中的 ini.py 文件对比

在 Ansible 1.2 中，对 ini.py 文件的改动是为了解决两个问题：

（1）对于组定义，允许在组标题后面添加注释，但是这部分处理漏掉了一处代码，后面测试时会看到这个遗漏之处并进行修正。

（2）对于设置组变量的部分，如果该组不存在，则需要创建新组并将其加入 all 的子组中，而在 Ansible 1.1 中只会创建新组，但是并不会将其加入 all 的子组中。

首先来看第一个问题，先准备一个带注释的 hosts 文件，具体如下：

```
(ansible1.1) [root@master ansible-1.1-test]# cat test_hosts
[nodes]                                # 用于测试
ceph-[1:3]

[not-exist-group:vars]
test=123

[nodes:vars]
ansible_ssh_user=root
ansible_ssh_pass=@SHENcong19920522
ansible_ssh_connection=ssh
```

接着进入 Ansible 1.1 的虚拟环境，然后导入 InventoryParser 类并使用该类解析 hosts 文件。为了使数组对象的显示更为清晰，需要给 Group 类添加一个__repr__()方法，具体代码如下：

```
# 源码包: ansible 1.1 和 ansible 1.2
# 源码位置: lib/ansible/inventory/group.py
# ...

class Group(object):

    # ...

    # 新增一个魔法函数，打印 Group 对象时显示 Group 的名称
    def __repr__(self):
        return self.name

    # ...
```

以上改动需要在 Ansible 1.1 和 Ansible 1.2 的虚拟环境中进行。接下来进入 Ansible 1.1 虚拟环境的交互模式执行以下动作：

```
(ansible1.1) [root@master ansible-1.1-test]# python
Python 2.7.18 (default, Oct 13 2020, 23:55:15)
[GCC 4.8.5 20150623 (Red Hat 4.8.5-39)] on linux2
Type "help", "copyright", "credits" or "license" for more information.
>>> from ansible.inventory.ini import InventoryParser
>>> parser = InventoryParser('/root/ansible-1.1-test/test_hosts')
>>> parser.groups
{'ungrouped': ungrouped, 'nodes # \xe7\x94\xa8\xe4\xba\x8e\xe6\xb5\x8b\
xe8\xaf\x95': nodes # 用于测试, 'all': all, 'not-exist-group': not-exist-
                       group, 'nodes': nodes}
```

```
>>> parser.groups['all'].child_groups
[ungrouped, nodes  # 用于测试]
```

以上结果反映了 Ansible 1.1 中存在的两个问题：一是将组中的注释添加到了组名中；二是解析得到了 not-exist-group 组，但是它却不属于 all 组。再来看 Ansible 1.2 中的测试结果：

```
(ansible1.2) [root@master ansible-1.1-test]# python
Python 2.7.18 (default, Oct 13 2020, 23:55:15)
[GCC 4.8.5 20150623 (Red Hat 4.8.5-39)] on linux2
Type "help", "copyright", "credits" or "license" for more information.
>>> from ansible.inventory.ini import InventoryParser
>>> parser = InventoryParser('/root/ansible-1.1-test/hosts')
>>> parser.groups
{'ungrouped': ungrouped, 'not-exist-group': not-exist-group, 'all': all,
'master': master, 'nodes': nodes}
>>> parser.groups['all'].child_groups
[ungrouped, master, nodes, not-exist-group]
```

以上结果说明 Ansible 1.2 正好解决了上面两个问题，修正的代码正是 ini.py 中改变的部分。但是在 Ansible 1.2 中还存在一个问题，只需要调整 hosts 中注释的位置，即可复现该 Bug，具体如下：

```
(ansible1.2) [root@master ansible-1.1-test]# cat /root/ansible-1.1-test/
test_hosts
[nodes]
ceph-[1:3]

[not-exist-group:vars]                        # 用于测试
test=123

[nodes:vars]
ansible_ssh_user=root
ansible_ssh_pass=@SHENcong19920522
ansible_ssh_connection=ssh
```

再进入 Ansible 1.2 的虚拟环境中执行前面的操作：

```
(ansible1.2) [root@master ansible-1.1-test]# python
Python 2.7.18 (default, Oct 13 2020, 23:55:15)
[GCC 4.8.5 20150623 (Red Hat 4.8.5-39)] on linux2
Type "help", "copyright", "credits" or "license" for more information.
>>> from ansible.inventory.ini import InventoryParser
>>> parser = InventoryParser('/root/ansible-1.1-test/test_hosts')
Traceback (most recent call last):
  File "<stdin>", line 1, in <module>
  File "/root/.pyenv/versions/ansible1.2/lib/python2.7/site-packages/
ansible/inventory/ini.py", line 40, in __init__
    self._parse()
  File "/root/.pyenv/versions/ansible1.2/lib/python2.7/site-packages/
ansible/inventory/ini.py", line 46, in _parse
    self._parse_group_variables()
```

```
    File "/root/.pyenv/versions/ansible1.2/lib/python2.7/site-packages/
ansible/inventory/ini.py", line 159, in _parse_group_variables
    raise errors.AnsibleError("can't add vars to undefined group: %s" % line)
ansible.errors.AnsibleError: can't add vars to undefined group: not-exist-
group                                                    # 用于测试
```

上面的报错信息非常明显，即无法给组"not-exist-group # 用于测试"添加变量，这里组后面的中文同样没有被去掉，导致任务执行失败。此时只需要查看下代码就能明白：

```
# 源码包：ansible 1.2
# 源码位置：lib/ansible/inventory/ini.py
# ...

class InventoryParser(object):

    # ...

    def _parse_group_variables(self):
        group = None
        for line in self.lines:
            line = line.strip()
            if line.startswith("[") and line.find(":vars]") != -1:
                # 出现问题的行，没有调用split()方法去掉注释内容
                line = line.replace("[","").replace(":vars]","")
                # 搜索 line 对应的组，line 表示组名
                group = self.groups.get(line, None)
                if group is None:
                    # 这里抛出的异常
                    raise errors.AnsibleError("can't add vars to undefined
group: %s" % line)
            elif line.startswith("#"):
                pass
            elif line.startswith("["):
                group = None
            elif line == '':
                pass
            elif group:
                # ...
```

根据上面的代码很快就能找到出错的语句，就是获取 line 时没有使用 split()方法去掉注释从而导致任务执行失败。只需简单修改下即可，具体如下：

```
line = line.split(' #')[0].replace("[","").replace(":vars]","")
```

再次执行前面的操作后，程序又恢复正常了。

继续对比 utils 目录，该目录下共有 4 个文件。其中，module_docs.py 只有一行微小变动，可忽略不计，plugins.py 文件的变动也不大，增加了一些缓存变量来缓存数据，对比结果如图 5-12 所示。

__init__.py 和 template.py 文件的改动是对原有代码的优化，以解决部分潜在的问题。图 5-13 中展示了两个版本中 parse_yaml()方法的结果对比。在 Ansible 1.2 中对该方法的执

行结果进行了进一步优化，使其可以处理返回结果为列表、字典或者模板变量（带双花括号）的情况。

图 5-12　两个版本中的 plugins.py 文件对比

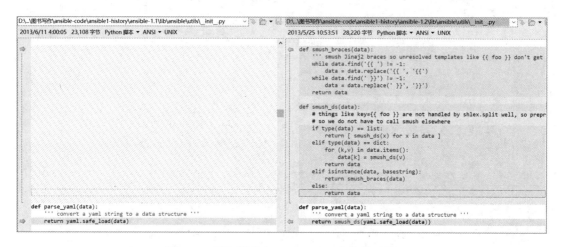

图 5-13　两个版本中 parse_yaml()方法的对比

在 Ansible 1.2 的 template.py 文件中重点优化了 template()和 template_from_file()方法，同时还新增了 template_from_string()方法，这些方法通过 Jinja2 模块实现了文本渲染功能。最后是__init__.py 文件，它的改动也不大，比较重要的有以下两处：

- 优化了_execute_module()方法。在最后执行模块的命令中考虑是否 sudo 提权及用户是否为 root 的情况，决定是否将运行模块和清除模块的命令放在一起执行。
- 优化了_executor_internal()方法中对返回结果的处理。

以上两处优化的代码对比结果如图 5-14 和图 5-15 所示。

接下来就是 Ansible 中的 playbook 目录，该目录下的代码文件主要负责解析 Playbook 文件。从文件变动中可以看出，Ansible 1.2 相比 Ansible 1.1 的一个重大升级：角色。我们先从 playbook 目录下定义的 3 个基本类，即 Task、Play 和 Playbook 讲起。Task 类描述了一个待执行的 task，该类的定义位于 task.py 文件中，它的改动较小，只是增加了一个属性 any_errors_fatal。task.py 文件的变动对比如图 5-16 所示。

图 5-14　Ansible 1.2 中优化的_execute_module()方法对比 1

图 5-15　Ansible 1.2 中优化的_executor_internal()方法对比 2

图 5-16　task.py 文件变动对比

　　然后是 Play 类，该类的定义位于 play.py 文件中，从两个版本的 play.py 文件对比中可以发现，Ansible 1.2 中的 Play 类中新增了 any_errors_fatal、roles、pre_tasks 和 post_tasks 4 个属性，也就是说 Play 类的编写支持以下 4 种语法：

- any_errors_fatal：遇到任何错误即停止执行该 Play。
- roles：支持导入角色中的任务。
- pre_tasks：前置 tasks。
- post_tasks：最后执行的 tasks。

　　有了这些新语法支持后，Ansible 1.2 在 Playbook 编写上更为灵活。除了在 Play 类中增加属性设置以外，最重要的是增加了_load_roles()方法。该方法根据数据的角色最终得到角色下的 ds 值，该值中包括解析的 tasks、variables 等信息。Ansible 1.2 中新增的_load_roles()方法如图 5-17 所示。

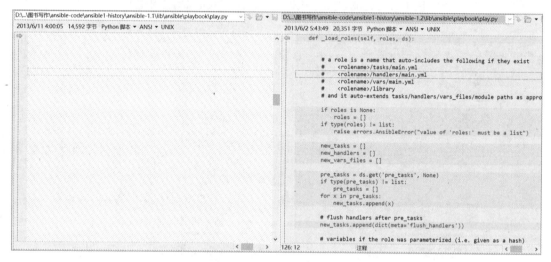

图 5-17　Ansible 1.2 中新增的_load_roles()方法

　　然后是__init__.py 文件中的 Playbook 类，该类中新增了 any_errors_fatal 属性并优化了部分 task 的执行顺序。此外，该类还新增了生成重试文件的 generate_retry_inventory()方法。

　　最后笔者对整个 Ansible 的核心目录做个对比。为了简单起见，这里不会对所有的文件进行对比，只选择相对典型的文件即可，其余文件的更新读者可以按照相同的方式自行分析。由于 runner 目录下的 connection.py、poller.py 和 return_data.py 文件改动较少（尤其是 poller.py 文件，只修正了一个拼写错误），这里不再说明。

　　首先来看 runner/action_plugins 目录，该目录下的文件与 Ansible 的所有模块的动作插件对应，这里以 fetch 模块为例介绍 Ansible 1.2 中 fetch 模块的升级功能。打开两个版本的 fetch.py 文件进行对比，如图 5-18 所示。

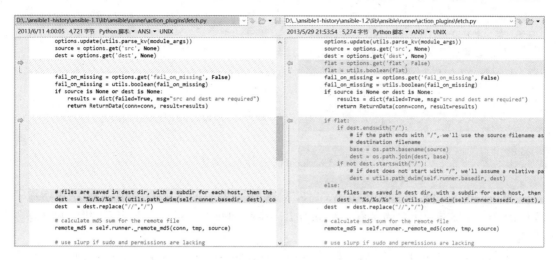

图 5-18 动作插件中的 fetch.py 文件对比

可以看到，Ansible 1.2 中的 fetch 模块多了一个 flat 参数，该参数用于控制下载远端文件到本地的路径，设置 flat=yes 时将不再分多层目录（默认路径为 hostname/path/to/file）存放下载文件，而是直接按照 dest 参数的值将文件保存到本地。下面来看 Ansible 1.2 中 flat 模块的示例，包含有和没有 flat 参数两种情况，观察对应命令执行成功后文件的保存地址：

```
(ansible1.2) [root@master ~]# ls /tmp/
(ansible1.2) [root@master ~]# ansible ceph-1 -i hosts -m fetch -a "src=
/root/test_copy.txt dest=/tmp"
ceph-1 | success >> {
    "changed": true,
    "dest": "/tmp/ceph-1/root/test_copy.txt",
    "md5sum": "00cd67a22f6c5083b24a0388607a57de"
}
(ansible1.2) [root@master ~]# ansible ceph-1 -i hosts -m fetch -a "src=
/root/test_copy.txt dest=/tmp/test_copy.txt flat=yes"
ceph-1 | success >> {
    "changed": true,
    "dest": "/tmp/test_copy.txt",
    "md5sum": "00cd67a22f6c5083b24a0388607a57de"
}
```

可以看到，在不加 flat 参数的情况下，文件会被保存到 "$dest/节点名/$src" 目录下，而添加了 flat=yes 后，则会直接保存到 dest 路径上，不会有任何中间嵌套。这正是 Ansible 1.2 中对 fetch 模块的功能升级，变动的代码则位于 fetch 模块的动作插件中。

下面对比 connection_plugins 目录。该目录中的几个文件是 Ansible 1.1 的底层通信插件，有本地通信（local.py）插件、基于 SSH 命令的通信（ssh.py）插件、基于 Paramiko 模块封装的远程通信插件等。该目录的改动较少，local.py 中只调整了 Connection 类的 __init__()方法的一个参数，如图 5-19 所示。

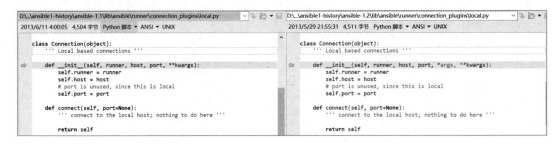

图 5-19　local.py 文件变动对比

最核心的 ssh.py 和 paramiko.py 文件的改动也不大，主要是在__init__()方法中增加了 private_key_file 参数。总的来看，Ansible 1.1 和 Ansible 1.2 的底层通信连接代码并没有多少变化。

再接着是过滤器插件代码，位于 lib/ansible/runner/filter_plugins 目录下。对外提供过滤器接口的是 core.py 文件，通过对比该文件可知，Ansible 1.2 相比 Ansible 1.1 提供了更多的过滤器供 Playbook 编写。两个版本中过滤器模块变动对比如图 5-20 所示。

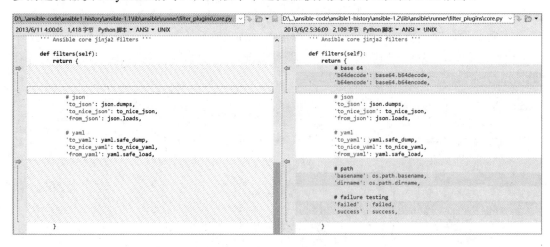

图 5-20　过滤器模块变动对比

最后一个目录是 lookup 插件，这里改动的代码较多，解读方式和动作插件类似，对比特定 lookup 插件的代码便能了解该插件的功能变化情况，这里不再过多讲述。

至此，Ansible 1.1 和 Ansible 1.2 的代码对比就结束了。从上面的对比结果来看，Ansible 1.2 在单个模块执行的代码中改变并不大，最重要的升级是在 Playbook 中引入了 roles 语法，使得 Playbook 编写的可扩展性更强。有兴趣的读者可以考虑以同样的方式对比 Ansible 1.2 与 Ansible 1.3、Ansible 1.3 与 Ansible 1.4 的变化，以此类推，直到 Ansible 1 的最后一个版本 Ansible 1.9.6。这样的方式可以让读者追踪 Ansible 1 源码的演进过程，体验一个经典的 Python 项目从小到大、由简入繁的完整过程，而这样的过程对于一个成长期的程序员而言至关重要。

5.2　Ansible 2.8 中的部分 Bug 溯源

本节将探讨 Ansible 2.8 中的部分 Bug 的溯源过程，这些过程均来自 GitHub 网站及 Ansible 2.8 本身自带的 ChangeLog。截至 2020 年 11 月份，Ansible 2.8 已经迭代到了 v2.8.16 版本。

5.2.1　Ansible 2.8 中的 ChangeLog

在每个 Ansible 2.8 的小版本包中都有详细的更改记录，其中记录了当前小版本中所有 的改动情况。例如，对于 Ansible 2.8.2 而言，该源码包中的 ChangeLog 就会记录 2.8.2、 2.8.1 及 2.8.0 三个版本关于 Minor Changes、Security Fixes 和 Bugfixes 方面的改动情况。 以下是 ChangeLog 中关于 2.8.16 的更新说明：

```
<!--文件位置: changelogs/CHANGELOG-v2.8.rst-->
==============================================
Ansible 2.8 "How Many More Times" Release Notes
==============================================


.. contents:: Topics


v2.8.16
=======


Release Summary
---------------


| Release Date: 2020-10-05
| `Porting Guide <https://docs.ansible.com/ansible/devel/porting_guides.
html>`__


Minor Changes
-------------


- ansible-test - Added CI provider support for Azure Pipelines.
- ansible-test - Added support for Ansible Core CI request signing for
Shippable.
- ansible-test - Allow custom ``--remote-stage`` options for development
and testing.
- ansible-test - Refactored CI related logic into a basic provider
```

```
abstraction.
- ansible-test - Remove the discontinued ``us-east-2`` choice from the
``--remote-aws-region`` option.
- ansible-test - Request remote resources by provider name for all provider
types.
- ansible-test - Show a warning when the obsolete ``--remote-aws-region``
option is used.
- ansible-test - Support custom remote endpoints with the ``--remote-
endpoint`` option.
- ansible-test - Update built-in service endpoints for the ``--remote``
option.
- ansible-test - Use new endpoint for Parallels based instances with the
``--remote`` option.

Security Fixes
--------------

- kubectl - connection plugin now redact kubectl_token and kubectl_password
in console log (https://github.com/ansible-collections/community.kubernetes/
issues/65) (CVE-2020-1753).

Bugfixes
--------

- ansible-test - Change classification using ``--changed`` now consistently
handles common configuration files for supported CI providers.
- ansible-test - Disabled the ``duplicate-code`` and ``cyclic-import``
checks for the ``pylint`` sanity test due to inconsistent results.
- ansible-test - The ``resource_prefix`` variable provided to tests running
on Azure Pipelines is now converted to lowercase to match other CI providers.
- ansible-test - for local change detection, allow to specify branch to
compare to with ``--base-branch`` for all types of tests (https://github.
com/ansible/ansible/pull/69508).

<!--忽略其余部分-->
```

  这里重点关注 Minor Changes 和 Bugfixes 部分的信息，因为 Security Fixes 部分难度较大，并不适合初学者研究。从 Minor Changes 部分的内容中可以看到，本次小版本之间（v2.8.15 到 v2.8.16）的升级主要是新增了 ansible-test 命令的部分功能，如增加了--remote-stage、--remote-endpoint 选项，更新了--remote 选项等。此外，从 Bugfixes 部分的内容中可以看到，Ansible 2.8.16 修复了 ansible-test 命令中的相关 Bug。其中，最后一个Bug 在 GitHub 上有相应记录，因此从 GitHub 上可以找到该 Bug 的提出、讨论及解决等相关过程，如图 5-21 所示。

  在图 5-21 中单击 Files changed 标签，可以看到涉及最后一个 Bug 的修复文件一共有4 个，如图 5-22 所示。

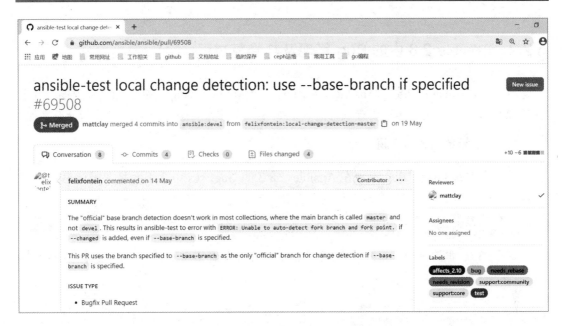

图 5-21　最后一个 Bug 的 pull 记录

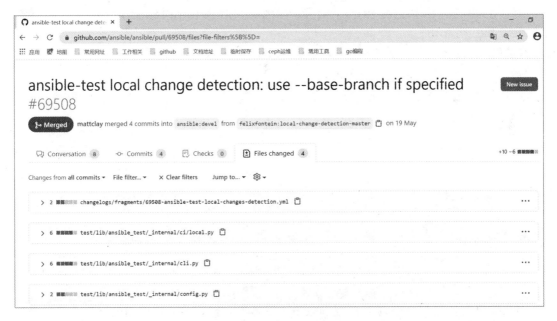

图 5-22　涉及最后一个 Bug 的修复文件

通过这种方式能方便地查看该 Bug 的发现过程及相关的修复代码，从而更好地掌握 Ansible 源码。

5.2.2　Ansible 2.8 中的 Bug 复现及其修复过程

本节将从 Ansible 2.8 到 Ansible 2.8.16 的整个 Minor Changes 和 Bugfixes 中抽取几项容易理解的 Bug 修复工作进行详细介绍，这将是一次非常有趣且有意义的源码分析之旅。

1．一个简单的选项问题

第一个追踪的历史 Bug 出现在 Ansible 2.8.2 的 Bugfixes 中，下面是关于该 Bug 的说明与记录：

```
Fix --diff to produce output when creating a new file (https://github.
com/ansible/ansible/issues/57618)
```

这个 Bug 非常典型，直接去对应的 issues 地址查看该 Bug 的完整信息，如图 5-23 所示。

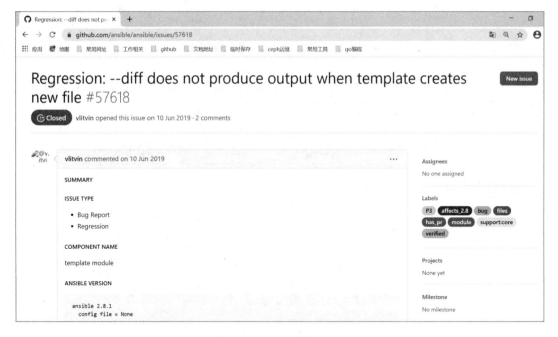

图 5-23　--diff 选项 issue

从标题上看是--diff 选项在应用到 template 模块时没有生效，问题的发现者也详细记录了这个问题并给出了复现测试。首先应了解 ansible 命令中--diff 选项的作用，其通常会配合--check 进行测试，用于显示文件在操作前与操作后的内容变化。

下面在 Ansible 2.8.16 的虚拟环境中进行以下操作：

```
(ansible2.8.16) [root@master playbook]# ansible ceph-1 -i hosts -m copy -a
"src=/etc/hosts dest=/tmp/hosts" --check --diff
```

```
--- before
+++ after: /etc/hosts
@@ -0,0 +1,9 @@
+127.0.0.1   localhost localhost.localdomain localhost4 localhost4.
localdomain4
+::1         localhost localhost.localdomain localhost6 localhost6.
localdomain6
+
+
+192.168.26.110 master
+192.168.26.120 ceph-1
+192.168.26.121 ceph-2
+192.168.26.122 ceph-3
+

ceph-1 | CHANGED => {
    "changed": true
}
```

上面使用--check 选项使得任务并没有真正执行，并且使用--diff 选项对比了复制前与复制成功后/tmp/hosts 文件的内容。然而有人发现这个功能在 Ansible 2.8.1 的 template 模块中却失效了，并以此内容提交了一个 issue。为了能复现该 Bug，先要创建 Ansible 2.8.1 和 Ansible 2.8.2 两个版本的虚拟环境，具体操作如下：

```
(ansible2.8.16) [root@master playbook]# pyenv virtualenv 3.8.6 ansible2.8.1
Looking in links: /tmp/tmp_46vo0hg
Requirement already satisfied: setuptools in /root/.pyenv/versions/3.8.6/
envs/ansible2.8.1/lib/python3.8/site-packages (49.2.1)
Requirement already satisfied: pip in /root/.pyenv/versions/3.8.6/envs/
ansible2.8.1/lib/python3.8/site-packages (20.2.1)
(ansible2.8.16) [root@master playbook]# pyenv virtualenv 3.8.6 ansible2.8.2
Looking in links: /tmp/tmpbop6uhe6
Requirement already satisfied: setuptools in /root/.pyenv/versions/3.8.6/
envs/ansible2.8.2/lib/python3.8/site-packages (49.2.1)
Requirement already satisfied: pip in /root/.pyenv/versions/3.8.6/envs/
ansible2.8.2/lib/python3.8/site-packages (20.2.1)
```

接着分别开两个 xshell 窗口并使用 pyenv activate 命令激活对应的虚拟环境，之后使用 pip 命令安装对应的 Ansible 版本即可。接下来准备一个模板文件，内容如下：

```
# ansible2.8.1 的虚拟环境中
(ansible2.8.1) [root@master playbook]# cat template.j2
{{ inventory_hostname }}
# ansible2.8.2 的虚拟环境中
(ansible2.8.2) [root@master playbook]# cat template.j2
{{ inventory_hostname }}
```

在 Ansible 2.8.1 的虚拟环境中执行 template 模块并结合--diff 选项，具体命令如下：

```
# Ansible 2.8.1 的虚拟环境
(ansible2.8.1) [root@master playbook]# ansible ceph-1 -i hosts -m template
-a "src=template.j2 dest=/tmp/template" --check --diff

ceph-1 | CHANGED => {
```

```
        "changed": true
}
```

可以看到，在 Ansible 2.8.1 中针对 template 模块而言--diff 确实没有起到相应的作用。再来看下该命令在修复版本中的执行结果，具体如下：

```
(ansible2.8.2) [root@master playbook]# ansible ceph-1 -i hosts -m template
-a "src=template.j2 dest=/tmp/template" --check --diff
--- before
+++ after: /root/.ansible/tmp/ansible-local-108876uuxxdwdl/tmp3xw3l990/
template.j2
@@ -0,0 +1 @@
+ceph-1

ceph-1 | CHANGED => {
        "changed": true
}
```

上面的结果表明，在 Ansible 2.8.2 中，--diff 选项的问题已经修复，出现了 diff 结果。现在请读者思考一下，如果是你，该如何修复这个 Bug 呢？很明显，为了解决上面的问题，需要完成以下几项工作：

（1）找出--diff 选项打印差异的代码。

（2）找出 Ansible 2.8.1 中使用--diff 选项无法实现打印差异的原因。

在第 4 章中笔者完整分析了 Ansible 模块的执行流程，下面会基于这个执行流程，不断在运行时添加 print()方法，帮助找出问题的根源并解决，至于寻找源码 Bug 原因的技术细节则不再详细解答。

第一步需要梳理 Ansible 的 template 模块的执行过程，对应的源码有两处：

（1）template 模块的源码。

（2）template 模块的动作插件源码。

其中，template 模块的源码（lib/ansible/modules/files/template.py）中只有说明和示例信息，并没有实质的代码内容。template 模块的动作插件源码如下：

```
# 源码位置: lib/ansible/plugins/action/template.py
# ...

class ActionModule(ActionBase):

    def run(self, tmp=None, task_vars=None):

        # 模块变量处理
        # ...

        # 模板源文件路径
        source = self._task.args.get('src', None)
        # ...


        try:
```

```
        try:
            tmp_source = self._loader.get_real_file(source)
        except AnsibleFileNotFound as e:
            # 抛出异常
            # ...
        b_tmp_source = to_bytes(tmp_source, errors='surrogate_or_strict')

        try:
            with open(b_tmp_source, 'rb') as f:
                try:
                    # 读取模板文件内容
                    template_data = to_text(f.read(), errors='surrogate_
or_strict')

                except UnicodeError:
                    # 编码错误，抛出异常
                    # ...

            # ...

            # 添加 Ansible 的模板变量
            temp_vars = task_vars.copy()
            temp_vars.update(generate_ansible_template_vars(source, dest))

            old_vars = self._templar._available_variables
            self._templar.set_available_variables(temp_vars)
            # 使用 Templar 对象的 do_template() 方法将模板数据进行渲染，得到最终
              的内容
            resultant = self._templar.do_template(template_data, preserve_
trailing_newlines=True,
                            escape_backslashes=False)
            self._templar.set_available_variables(old_vars)
        except AnsibleAction:
            raise
        except Exception as e:
            raise AnsibleActionFail("%s: %s" % (type(e).__name__, to_
text(e)))
        finally:
            self._loader.cleanup_tmp_file(b_tmp_source)

        # ...

        # 复制一个新的任务
        new_task = self._task.copy()

        # ...

        try:
            result_file = os.path.join(local_tempdir, os.path.basename
(source))
            # 将渲染后的文件写入本地的临时文件中
            with open(to_bytes(result_file, errors='...'), 'wb') as f:
                f.write(to_bytes(resultant, encoding=output_encoding,
errors='...'))
```

```
                    # 更新任务
                    new_task.args.update(
                        dict(
                            src=result_file,
                            dest=dest,
                            follow=follow,
                        ),
                    )
                    # 得到 copy 模块的动作插件
                    copy_action = self._shared_loader_obj.action_loader.get('copy',
                        task=new_task,connection=self._connection,
                        play_context=self._play_context,loader=self._loader,
                        templar=self._templar,shared_loader_obj=self._shared_
loader_obj)
                    # 最终调用 copy 模块的动作插件将本地渲染好的临时文件上传到远端主机上
                    result.update(copy_action.run(task_vars=task_vars))
                finally:
                    shutil.rmtree(to_bytes(local_tempdir, errors='surrogate_or
_strict'))

        except AnsibleAction as e:
            result.update(e.result)
        finally:
            self._remove_tmp_path(self._connection._shell.tmpdir)

        return result
```

上面的代码中给出了详细的注释，基本说明了 template 模块的动作插件的执行过程：渲染本地的模板文件保存到临时文件中，然后调用 copy 模块的动作插件将模板文件上传到远端的目标主机上。这里并没有涉及结果的处理部分，所以此时搜索的关注点应该放到 copy 模块的插件源码上，具体如下：

```
# 源码位置：lib/ansible/plugins/action/copy.py
# ...

class ActionModule(ActionBase):

    # ...

    def _copy_file(self, source_full, source_rel, content, content_tempfile,
                   dest, task_vars, follow):

        # ...

        result = {}
        # 保存上传文件内容与远端对应路径文件内容的差值
        result['diff'] = []

        # ...

        if local_checksum != dest_status['checksum']:
            # The checksums don't match and we will change or error out.
```

```python
        if self._play_context.diff and not raw:
            result['diff'].append(self._get_diff_data(dest_file, source_
full, task_vars))
            # 省略了上传文件、移动文件等一系列动作
            # ...

        else:
            # 不用上传文件动作
            # ...

        if not module_return.get('checksum'):
            module_return['checksum'] = local_checksum

        result.update(module_return)
        return result

    def run(self, tmp=None, task_vars=None):

        # ...

        # 最终得到需要执行上传的源文件
        if os.path.isdir(to_bytes(source, errors='surrogate_or_strict')):
            # ...
        else:
            source_files['files'] = [(source, os.path.basename(source))]

        # ...

        implicit_directories = set()
        for source_full, source_rel in source_files['files']:
            if source_files['directories']:
                follow = False
            else:
                follow = boolean(self._task.args.get('follow', False), strict=
False)

            # 在 _copy_file() 方法中会上传本地文件到远端节点, 同时对比本地文件与远端文
            # 件内容, 得到两者内容的差值, 即 diff。如果远端文件不存在, 则 diff 为本地
            # 文件的全部内容
            module_return = self._copy_file(source_full, source_rel, content,
content_tempfile, dest, task_vars, follow)
            # 添加 print() 语句, 打印结果
            # print('result diff:{}'.format(module_return['diff']))
            if module_return is None:
                continue

            # ...

        # ...
```

从上面的代码中终于找到想要的内容了：copy 模块执行结果中会包含 diff 字段，该字段的值正是想要的本地文件与远端文件的内容差值。笔者在调用_copy_file()方法之后添加了一条 print()语句，用于打印 diff 的结果（代码中已经添加打印注释，请参考）。在虚拟环境中修改相应的代码后再次运行之前的命令，结果如下：

```
(ansible2.8.1) [root@master playbook]# ansible ceph-1 -i hosts -m template
-a "src=template.j2 dest=/tmp/template" --check --diff
result diff:[{'after_header': '/root/.ansible/tmp/ansible-local-111856j
uj08f6h/tmpqtjb6l8j/template.j2', 'after': 'ceph-1\n'}]

ceph-1 | CHANGED => {
    "changed": true
}
```

上面的结果表明，在 Ansible 2.8.1 中其实已经包含 diff 结果，但是没有打印出来，这是什么原因呢？因此必须找到打印 diff 结果的地方。在 Ansible 中会有许多回调设置，例如在 Playbook 开始执行时、任务执行失败或成功时均会设置回调方法，而对于存在需要打印 diff 的选项也不例外，与之对应的也有一个打印 diff 结果的回调方法，对应的代码如下：

```
# 源码位置: lib/ansible/plugins/strategy/__init__.py
# ...

class StrategyBase:

    @debug_closure
    def _process_pending_results(self, iterator, one_pass=False, max_
passes=None):

        # ...

        while True:
            try:
                self._results_lock.acquire()
                # 从进程队列中获取 Ansible 执行 task 的结果
                task_result = self._results.popleft()
            except IndexError:
                break
            finally:
                self._results_lock.release()

            # ...

            if task_result.is_failed():
                # 失败处理
                # ...
            elif task_result.is_unreachable():
                # 主机不可达处理
                # ...
```

```
        elif task_result.is_skipped():
            # 任务跳过执行处理
            # ...
        else:
            # 正常执行处理
            # ...

            # 如果有 diff 结果
            if 'diff' in task_result._result:
                if self._diff or getattr(original_task, 'diff', False):
                    # 执行回调，回调使用的是插件形式，位于 plugins/callback 目录下
                    # print("调用 callback 插件:{}".format(self._tqm._stdout_
callback))

                    self._tqm.send_callback('v2_on_file_diff', task_result)

                # ...

        cur_pass += 1

    return ret_results
```

从上面的代码中可以看到，当正常解析了一个 task 的返回结果后，会调用 self._tqm
属性的 send_callback()方法进行回调，回调的方法为 v2_on_file_diff()。在 Ansible 中，回调
模块同样做成了插件形式，并且 Ansible 支持多种回调插件，笔者在 self_tqm.send_callback
语句之前加上了一个 print()语句（参考上述代码）用于打印具体调用的回调插件路径。在
虚拟环境下修改上述说明的 Ansible 源码并添加 print()语句，最后执行 ansible 命令，结果
如下：

```
# 修改文件路径:
# ~/.pyenv/versions/ansible2.8.1/lib/python3.8/site-packages/ansible/
plugins/strategy/__init__.py
(ansible2.8.1) [root@master playbook]# ansible ceph-1 -i hosts -m template
-a "src=template.j2 dest=/tmp/template" --check --diff
result diff:[{'after_header': '/root/.ansible/tmp/ansible-local-112618u
sm26byp/tmpd7vsry4o/template.j2', 'after': 'ceph-1\n'}]
调用 callback 插件:<ansible.plugins.callback.minimal.CallbackModule object
at 0x7feadd8c9f40>

ceph-1 | CHANGED => {
    "changed": true
}
```

从上面的 print()方法的执行结果中可以看到，ansible 命令行最终回调的是 minimal.py
插件。继续跟踪 minimal.py 文件中的 v2_on_file_diff()方法，具体代码如下：

```
# 源码位置: lib/ansible/plugins/callback/minimal.py
# ...

class CallbackModule(CallbackBase):
```

```
        # ...

    def v2_on_file_diff(self, result):
        if 'diff' in result._result and result._result['diff']:
            self._display.display(self._get_diff(result._result['diff']))

        # ...
```

上述代码非常简单，就是根据 result._result['diff'] 的结果调用 Display 对象的 display() 方法即可。之前通过打印 result['diff'] 的值能看到是有内容的，但是为什么控制台没有 diff 的输出呢？注意到此之前还要经过 _get_diff() 方法处理，该方法的实现位于父类中，具体代码如下：

```
# 源码位置：lib/ansible/plugins/callback/__init__.py
# ...

class CallbackBase(AnsiblePlugin):
    def _get_diff(self, difflist):

        if not isinstance(difflist, list):
            difflist = [difflist]

        ret = []
        for diff in difflist:
            # ...

            # diff 中没有 before 字段，因此不会进入该 if 块中处理 diff 的值，最后得到
            的 ret 为空
            if 'before' in diff and 'after' in diff:
                # format complex structures into 'files'
                for x in ['before', 'after']:
                    # ...

            if 'prepared' in diff:
                ret.append(diff['prepared'])
        return u''.join(ret)
```

对比之前打印的 result['diff'] 结果就能明白了，为什么在经过 _get_diff() 方法处理后输出的是空字符串。具体结果如下：

```
[{'after_header': '/root/.ansible/tmp/ansible-local-111856juj08f6h/tmpq
tjb6l8j/template.j2', 'after': 'ceph-1\n'}]
```

这里 diff 结果中没有 before 字段，因此输出的 ret 为空数组，最终控制台上什么都没有打印。

📖 注意：笔者开始是在 send_callback 关键字中搜索所有在 Ansible 源码中发送的回调方法，最终找到了与 diff 相关的回调方法（v2_on_file_diff()）。在 VSCode 中全局搜索 send_callback 关键字的结果如图 5-24 所示。

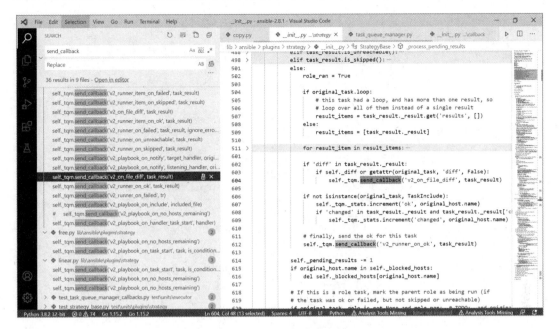

图 5-24　全局搜索 diff 回调

现在已经找到 diff 值不输出的原因了，那么该如何确定发生 Bug 的位置并解决它呢？之前分析过 Ansible 会在 copy 模块对应的动作插件（_copy_file()方法）中生成 result['diff'] 的结果，只需要追踪该值的生成即可。通过查看_copy_file()方法的源码可以看到，result['diff'] 为一个列表，其中的元素值主要是通过调用_get_diff_data()方法得到的，具体如下：

```
# 源码位置：lib/ansible/plugins/action/copy.py
result['diff'].append(self._get_diff_data(dest_file, source_full, task_vars))
```

_get_diff_data()方法在动作插件的父类中实现，其代码如下：

```
# 源码位置：lib/ansible/plugins/action/__init__.py
# ...

class ActionModule(ActionBase):

    # ...

    def _get_diff_data(self, destination, source, task_vars, source_file=
True):

        diff = {}
        display.debug("Going to peek to see if file has changed permissions")
        # 调用 file 模块获取远端文件信息
        peek_result = self._execute_module(module_name='file', module_
args=dict(path=destination, _diff_peek=True), task_vars=task_vars, persist_
files=True)
        # print('调用 file 获取结果：{}'.format(peek_result))
```

```
        if not peek_result.get('failed', False) or peek_result.get('rc', 0)
== 0:
            if peek_result.get('state') == 'absent':
                diff['before'] = u''
            elif peek_result.get('appears_binary'):
                diff['dst_binary'] = 1
            elif peek_result.get('size') and C.MAX_FILE_SIZE_FOR_DIFF > 0
and peek_result['size'] > C.MAX_FILE_SIZE_FOR_DIFF:
                diff['dst_larger'] = C.MAX_FILE_SIZE_FOR_DIFF
            else:
                # 对于一个不存在的文件，调用 slurp 模块会执行失败，结果中不会包含 content
                dest_result = self._execute_module(module_name='slurp', module_
args=dict(path=destination), task_vars=task_vars, persist_files=True)
                if 'content' in dest_result:
                    dest_contents = dest_result['content']
                    if dest_result['encoding'] == u'base64':
                        dest_contents = base64.b64decode(dest_contents)
                    else:
                        # 抛出异常
                        # ...
                    diff['before_header'] = destination
                    diff['before'] = to_text(dest_contents)

        if source_file:
            # 更新 diff 字典的 after_header 值和 after 值
            # ...
        else:
            diff['after_header'] = u'dynamically generated'
            diff['after'] = source
```

在_get_diff_data()方法中，第一步会调用 file 模块获取远端文件的信息，并以此生成diff 值。为了能更好地理解上面的判断过程，可以考虑打印调用 file 模块获取的结果。按上面的位置在虚拟环境中添加相应的 print()语句，然后执行 ansible 命令，结果如下：

```
(ansible2.8.1) [root@master playbook]# ansible ceph-1 -i hosts -m template
-a "src=template.j2 dest=/tmp/template" --check --diff
调用 file 获取结果:{'invocation': {'module_args': {'directory_mode': None,
'force': False, 'remote_src': None, '_original_basename': None, 'path':
'/tmp/template', 'owner': None, 'follow': True, 'group': None, 'unsafe_
writes': None, 'state': 'file', 'content': None, 'serole': None, 'selevel':
None, 'setype': None, 'access_time': None, 'access_time_format': '%Y%m%d%H%M.
%S', 'modification_time': None, 'regexp': None, 'src': None, 'seuser': None,
'recurse': False, '_diff_peek': 'True', 'delimiter': None, 'mode': None,
'modification_time_format': '%Y%m%d%H%M.%S', 'attributes': None, 'backup':
None}}, 'path': '/tmp/template', 'changed': False, 'appears_binary':
False, 'warnings': ["The value True (type bool) in a string field was
converted to u'True' (type string). If this does not look like what you expect,
quote the entire value to ensure it does not change."], '_ansible_parsed':
True, 'ansible_facts': {'discovered_interpreter_python': '/usr/bin/python'}}

ceph-1 | CHANGED => {
    "changed": true
}
```

从上面的结果中可以看到，peek_result 的值不会有 failed、rc 和 state 字段，所以代码最终会走到 if 的 else 分支（peek_result.get('state') = None）。由于是初次上传，相应的文件不存在，所以执行 slurp 模块会返回错误，即不会有 content 字段。因此，执行整段代码后，diff 中并不包含 before 字段，这是导致后面无法显示 diff 结果的根本原因。修复该 Bug 的方法也非常简单，对于文件不存在的情况，想办法让代码进入第一个 if 分支中，这样就可以执行 diff['before'] = u" 语句。因此，只需要简单修改即可，具体如下：

```
if peek_result.get('state') in (None, 'absent'):
```

修改虚拟环境下的代码，然后再次运行前面的命令，结果如下：

```
(ansible2.8.1) [root@master playbook]# ansible ceph-1 -i hosts -m template
-a "src=template.j2 dest=/tmp/template" --check --diff
--- before
+++ after: /root/.ansible/tmp/ansible-local-114084zb6t37xq/tmpoksdx47c/
template.j2
@@ -0,0 +1 @@
+ceph-1

ceph-1 | CHANGED => {
    "changed": true
}
```

可以看到，diff 结果已经打印出来了，这个 Bug 就算解决了，官方的 commit 结果也是这样简单的一行。通过上述 Bug 查找与修复的过程，读者是不是对 Ansible 的源码更加熟悉了呢？之前没有介绍 callback 的插件功能与使用，这里借助该 Bug 让我们了解了 Ansible 中的 callback 插件。至于该回调方法是如何工作的及多个回调插件之间有何区别等问题，则留给读者去研究了。

2．经常修复的模块-lineinfile

笔者快速浏览了 Ansible 2.8.16 的 changelogs/CHANGELOG-v2.8.rst 后发现：lineinfile 模块在多个版本的 Bugfixes 中出现过。此外，在第 3 章中已经对该模块的功能、源码等进行了详细解读，因此非常适合将其作为本次的 Bug 追踪样本。下面是 lineinfile 模块在 Ansible 2.8 中出现的所有 Bug 修复（v2.8.0 中修复的是上个版本问题，故忽略）情况：

```
v2.8.2
======

Bugfixes
--------

- lineinfile - fix a race / file descriptor leak when writing the file
(https://github.com/ansible/ansible/issues/57327)
```

在 v2.8.2 中修复的一个描述符占用问题，笔者在 CentOS 7.6 系统中并没有复现，故在此处也不详细介绍，这里重点看 v2.8.6 和 v2.8.7 中的 4 个 Bug 修复问题。下面是 lineinline 模块的 Bug 在 Ansible 2.8.6 中的修复信息：

```
v2.8.6
======

Bugfixes
--------

- lineinfile - fix bug that caused multiple line insertions (https://
github.com/ansible/ansible/issues/58923).
```

通过 Bugfixes 描述可知，本次版本修复了 lineinfile 模块中导致多行重复插入的 Bug，具体的细节请参考 issue 中的相关说明。笔者也在虚拟机中还原了这个异常现象，这里的示例来自该 issue 的创建者。

（1）在 ceph-1 节点上准备一个测试脚本/tmp/test_script.sh，具体内容如下：

```
#!/bin/sh

echo hello, world
```

（2）准备一个 Playbook 文件，用于复现 Bug。内容如下：

```
(ansible2.8.2) [root@master playbook]# cat test_lineinfile_bug1.yml
- hosts: ceph-1
  gather_facts: false
  tasks:
    - name: Set JVM opts
      lineinfile:
        path: /tmp/test_script.sh
        insertafter: '^#!/bin/sh'
        regexp: ^export FISHEYE_OPTS
        firstmatch: true
        line: export FISHEYE_OPTS="-Xmx4096m -Xms2048m"
```

（3）运行该 Playbook 并查看脚本内容。

```
(ansible2.8.2) [root@master playbook]# ansible-playbook -i hosts test_
lineinfile_bug1.yml

PLAY [ceph-1] ************************************************************

TASK [Set JVM opts] *****************************************************
changed: [ceph-1]

PLAY RECAP *************************************************************
# 忽略统计结果
# ...
```

在 ceph-1 上查看脚本，内容如下：

```
[root@ceph-1 ~]# cat /tmp/test_script.sh
#!/bin/sh
export FISHEYE_OPTS="-Xmx4096m -Xms2048m"

echo hello, world
```

（4）再次运行该 Playbook 后并查看脚本内容。

```
(ansible2.8.2) [root@master playbook]# ansible-playbook -i hosts test_
lineinfile_bug1.yml

PLAY [ceph-1] ****************************************************

TASK [Set JVM opts] *********************************************
changed: [ceph-1]

PLAY RECAP *****************************************************
# 忽略统计结果
# ...
```

可以看到，ceph-1 上的文件继续被改动。此时 ceph-1 上的/tmp/test_script.sh 文件内容
如下：

```
[root@ceph-1 ~]# cat /tmp/test_script.sh
#!/bin/sh
export FISHEYE_OPTS="-Xmx4096m -Xms2048m"
export FISHEYE_OPTS="-Xmx4096m -Xms2048m"

echo hello, world
```

问题就出在上面的再次运行的结果中，原 issue 的创建者认为，在 lineinfile 模块中同
时出现 insertafter 和 regexp 时，根据 Ansible 的文档描述：当 regexp 匹配时就会忽略
insertafter 选项。因此，这里预期的结果是，第一次运行 Playbook 后添加 export 语句，后
续多次运行时由于匹配到 regexp，则不会再执行插入动作，而真实的结果则是多次重复插
入相同的语句。

这种模块源码的问题比较好处理，因为知道问题所在的代码位置且非常容易分析。为
解决上面出现的 Bug，需要思考以下两个问题：

- 为什么出现这个重复插入的 Bug 呢？
- 如何解决这个 Bug？

回想前面分析的 lineinfile 模块源码，这里只需要关注 present()方法即可。该方法的实
现如下：

```python
# 源码位置：Ansible 2.8.2 中的 lineinfile.py 文件
# ...

def present(module, dest, regexp, line, insertafter, insertbefore, create,
            backup, backrefs, firstmatch):
    # 读取文件内容，组成行列表
    # ...

    if regexp is not None:
        bre_m = re.compile(to_bytes(regexp, errors='surrogate_or_strict'))
```

```
    # 针对 insertbefore 或者 insertafter 编译相应的正则表达式，最后得到统一的 bre_ins
    if insertafter not in (None, 'BOF', 'EOF'):
        bre_ins = re.compile(to_bytes(insertafter, errors='surrogate_or_
strict'))
    elif insertbefore not in (None, 'BOF'):
        bre_ins = re.compile(to_bytes(insertbefore, errors='surrogate_or_
strict'))
    else:
        bre_ins = None

    # index[0] 和 index[1] 的含义可参考 3.2.2 节的内容
    index = [-1, -1]
    m = None
    b_line = to_bytes(line, errors='surrogate_or_strict')
    for lineno, b_cur_line in enumerate(b_lines):
        if regexp is not None:
            match_found = bre_m.search(b_cur_line)
        else:
            match_found = b_line == b_cur_line.rstrip(b('\r\n'))
        if match_found:
            index[0] = lineno
            m = match_found
        elif bre_ins is not None and bre_ins.search(b_cur_line):
            if insertafter:
                # + 1 for the next line
                index[1] = lineno + 1
                if firstmatch:
                    break
            if insertbefore:
                # index[1] for the previous line
                index[1] = lineno
                if firstmatch:
                    break

    # ...
```

通过上面的这段代码便能想到导致该 Bug 的一个可能原因：regexp 的匹配和
insertbefore/insertafter 的匹配居然在同一个 for 循环中。前面测试 Bug 的文件内容如下：

```
[root@ceph-1 ~]# cat /tmp/test_script.sh
#!/bin/sh
export FISHEYE_OPTS="-Xmx4096m -Xms2048m"

echo hello, world
```

下面将以该文件为基础，分析为何会出现上述问题。下面继续查看 Lineinfile 模块源码中 for 循环之后的代码，厘清 Ansible 是如何优先使用 regexp 匹配的，具体代码如下：

```
# 源码位置：Ansible 2.8.2 中的 lineinfile.py 文件
# ...

def present(module, dest, regexp, line, insertafter, insertbefore, create,
            backup, backrefs, firstmatch):
    # 读取文件内容，组成行列表
    # ...

    # for 循环遍历文件每行，计算 index[0] 和 index[1]
    # ...

    if index[0] != -1:
        # regexp 匹配到结果，优先处理
        # ...
    elif backrefs:
        # ...
    elif insertbefore == 'BOF' or insertafter == 'BOF':
        # ...
    elif insertafter == 'EOF' or index[1] == -1:
        # ...
    else:
        b_lines.insert(index[1], b_line + b_linesep)
        msg = 'line added'
        changed = True

    # ...
```

可以看到，Ansible 是通过 index[0] 的值来判断是否优先处理 regexp 的匹配。那么前面的 for 循环是不是存在一种可能，即文本内容中既能匹配 regexp 又能匹配 insertbefore/insertafter，但是最终经过 for 循环后，index[0] 为 -1，而 index[1] 则能得到匹配行的行号呢？在这段 for 循环中，由于 firstmatch 参数的出现，使得这种情况成为可能。因为设置了 firstmatch 后，无论是 regexp 还是 insertbefore/insertafter 被匹配，会直接跳出 for 循环。因此，只需要设置文本内容，使 insertbefore/insertafter 表达式匹配成功，然后设置 firstmatch=yes 退出循环，这样即使后面行中有 regexp 匹配的行也无用了。上面的脚本正是这样的情况：第一行匹配 insertafter，第二行才匹配 regexp，并且在 task 中设置了 firstmatch=yes。因此，上述脚本内容在经过 lineinfile 模块中的 for 循环后，得到 index[0]=-1，index[1]=1，最终得到前面继续插入行的现象。

知道 Bug 产生的原因后，该如何修复代码呢？firstmatch 是必须保留的，因此只能将 regexp 的匹配过程直接拿出来先进行一次全文匹配，之后再根据 regexp 的匹配结果决定是否继续 insertbefore/insertafter 的匹配。在 Ansible 2.8.6 中的修复方法如图 5-25 所示。

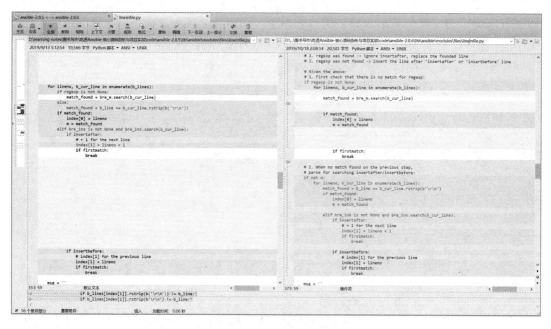

图 5-25　lineinfile 模块重复插入 Bug 修复代码

从图 5-25 中可以看到，Ansible 2.8.6 中的 lineinfile 模块相比 Ansible 2.8.5 会先对 regexp 进行处理，搜索文件行依次匹配 regexp 表达式，匹配后会更新 index[0] 的值。如果无法匹配 regexp，则继续匹配 insertbefore/insertafter 以计算 index[1] 的值。具体代码如下：

```python
# 源码位置: Ansible 2.8.6 中的 lineinfile.py 源码片段
# ...

def present(module, dest, regexp, line, insertafter, insertbefore, create,
        backup, backrefs, firstmatch):
    # ...

    # 先匹配 regexp，如果设置了 regexp 表达式
    if regexp is not None:
        for lineno, b_cur_line in enumerate(b_lines):
            match_found = bre_m.search(b_cur_line)
            if match_found:
                index[0] = lineno
                m = match_found
                if firstmatch:
                    break

    # 如果没有匹配到 regexp，则会继续遍历文件行，寻找 bre_ins 的匹配行更新 index[1]
    # 的值
    if not m:
        for lineno, b_cur_line in enumerate(b_lines):
            match_found = b_line == b_cur_line.rstrip(b'\r\n')
```

```
        if match_found:
            index[0] = lineno
            m = match_found

        elif bre_ins is not None and bre_ins.search(b_cur_line):
            if insertafter:
                # + 1 for the next line
                index[1] = lineno + 1
                if firstmatch:
                    break

            if insertbefore:
                # index[1] for the previous line
                index[1] = lineno
                if firstmatch:
                    break
    # ...
```

通过以上操作，就能得到符合原先功能设计的结果，而不会再出现前面重复插入的 Bug 了。通过这个 Bug 可以看到，与模块支持的参数增多之后，很容易遇到一些参数组合的 Bug，这些问题在开发或者学习代码时较难发现。

在 Ansible 2.8.7 的 Bugfixes 中描述修复了 3 个 lineinfile 模块的 Bug，具体内容如下：

```
v2.8.7
======


Bugfixes
--------

- lineinfile - don't attempt mkdirs when path doesn't contain directory path
- lineinfile - properly handle inserting a line when backrefs are enabled
  and the line already exists in the file (https://github.com/ansible/ansible/
  issues/63756)
- lineinfile - use correct index value when inserting a line at the end of
  a file (https://github.com/ansible/ansible/issues/63684)
```

下面笔者将通过对比代码文件的方式来查看 lineinfile 模块中源码的改动，从而映射到上面的 Bugfixes 中。首先从 Ansible 的历史版本库中下载 Ansible 2.8.6 和 Ansible 2.8.7 版本的源码包并解压，之后使用 Beyond Compare 4 工具对比两个版本的源码，主要看 lineinfile 模块的改动情况。

在 Ansible 2.8.7 的 lineinfile 模块的 present()方法中，相比 Ansible 2.8.6 版本多加了一个 b_destpath 的非空条件，如图 5-26 所示。这行改动对应 Bugfixes 中描述的第 1 个 Bug，这样做是为了使代码的健壮性更好，而且不会使模块出现异常。因为 path 会在最开始校验，只有确认该 path 路径的文件存在且是非目录，这样通过语句 os.path.basedirname()取出的文件的上一级目录才不为空。

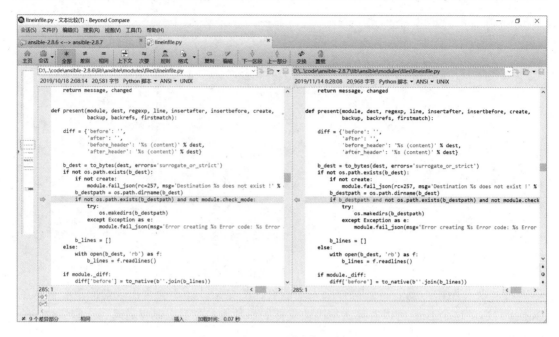

图 5-26　lineinfile 模块修复代码 1

第二部分对比结果如图 5-27 所示。

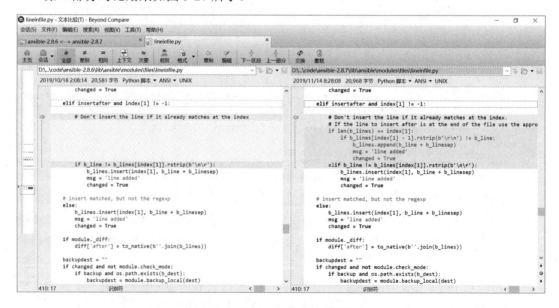

图 5-27　lineinfile 模块修复代码 2

在图 5-27 中，通过新增的注释可以看出对应新增的代码用于解决 lineinfile 模块中的第 3 个 Bug，即插入最后一行时抛出的索引异常问题。下面通过虚拟环境来复现该 Bug。

首先创建 Ansible 2.8.6 的虚拟环境并安装 Ansible 2.8.6，接着执行如下命令：

```
(ansible2.8.6) [root@master playbook]# ansible ceph-1 -i hosts -m lineinfile
-a "path=/tmp/test_script.sh line=xxx insertafter='^echo'"
ceph-1 | FAILED! => {
    "ansible_facts": {
        "discovered_interpreter_python": "/usr/bin/python"
    },
    "changed": false,
    "module_stderr": "Shared connection to ceph-1 closed.\r\n",
    "module_stdout": "Traceback (most recent call last):\r\n  File \"/root/
.ansible/tmp/ansible-tmp-1607722329.648007-105006101027633/AnsiballZ_
lineinfile.py\", line 114, in <module>\r\n    _ansiballz_main()\r\n  File \"
/root/.ansible/tmp/ansible-tmp-1607722329.648007-105006101027633/AnsiballZ_
lineinfile.py\", line 106, in _ansiballz_main\r\n    invoke_module(zipped_
mod, temp_path, ANSIBALLZ_PARAMS)\r\n  File \"/root/.ansible/tmp/ansible-
tmp-1607722329.648007-105006101027633/AnsiballZ_lineinfile.py\", line 49,
in invoke_module\r\n    imp.load_module('__main__', mod, module, MOD_DESC)
\r\n  File \"/tmp/ansible_lineinfile_payload_4_0Sd2/__main__.py\", line
567, in <module>\r\n  File \"/tmp/ansible_lineinfile_payload_4_0Sd2/__main__.
py\", line 558, in main\r\n  File \"/tmp/ansible_lineinfile_payload_4_
0Sd2/__main__.py\", line 413, in present\r\nIndexError: list index out of
range\r\n",
    "msg": "MODULE FAILURE\nSee stdout/stderr for the exact error",
    "rc": 1
}
```

可以看到，在 Ansible 2.8.6 中使用 lineinfile 模块向文件的最后一行插入数据时抛出了异常。由于显示的是完整的 lineinfile 模块代码抛出的异常，因此无法与源码中 lineinfile 模块的代码对应，但是从报错信息中可以清晰地看到这是一个数组越界的问题。下面分析图 5-27 中对应的代码，具体如下：

```
# Ansible 2.8.6 中的 lineinfile.py 模块源码
# ...


    # 这个分支是在 Ansible 2.8.6 之后添加的
    elif insertafter and index[1] != -1:

        # 如果待插入行内容与 index[1] 位置上的内容不同，则在 index[1] 位置后执行插入
        if b_line != b_lines[index[1]].rstrip(b'\n\r'):
            b_lines.insert(index[1], b_line + b_linesep)
            msg = 'line added'
            changed = True

# ...
```

上述代码中，index[1] 为计算出的待插入行的位置。如果要插入最后一行，而 b_lines 为原文件内容，则它的行索引范围是 0~index[1]-1，因此在上面的 if 判断中，b_lines[index[1]] 语句就会造成索引越界异常。解决方案也非常简单，具体代码如下：

```
# Ansible 2.8.7 中的 lineinfile.py 模块源码
# ...
```

```
elif insertafter and index[1] != -1:

    # 判断是否是最后一行
    if len(b_lines) == index[1]:
        # 确认待插入行和最后一行的内容是否相等
        if b_lines[index[1] - 1].rstrip(b'\r\n') != b_line:
            # 如果不相等，直接使用
            b_lines.append(b_line + b_linesep)
            msg = 'line added'
            changed = True
    elif b_line != b_lines[index[1]].rstrip(b'\n\r'):
        b_lines.insert(index[1], b_line + b_linesep)
        msg = 'line added'
        changed = True

# ...
```

上述修正代码只是在原来的 elif 内添加了一行代码 if len(b_lines) == index[1]，用于判断是否为最后一行。如果要插入最后一行，只需要确认插入行和文件的最后一行内容不相等后，直接使用 append()方法添加即可。这个 Bug 是在 Ansible 2.8.6 中修改 lineinfile 模块源码时引入的，在之前的版本中并不存在。

Bugfixes 中描述的第 2 个 lineinfile 模块的 Bug 是关于 backrefs 参数的，对应的代码对比如图 5-28 所示。

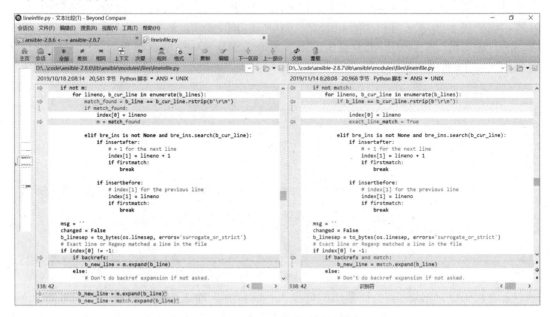

图 5-28　lineinfile 模块修复代码 3

分析图 5-28 中的代码对比结果，会发现一些细微的代码变化：

- 原来 m 作为匹配的结果被换成了 match 变量。

- 原来有一个 match_found 变量用于表示整行匹配的情况，而在后续的版本中被替换成了 exect_line_match 变量。

在代码的中间有这样一条语句 m = match_found，而这条语句正好引发了 lineinfile 模块的一个 Bug。该语句将 m 转换成了布尔类型，而后续正好有一个 m 变量的操作：

```
# 源码位置: Ansible 2.8.6 的 lineinfile 模块源码中
# ...

    if index[0] != -1:
        if backrefs:
            b_new_line = m.expand(b_line)
        else:
            # Don't do backref expansion if not asked.
            b_new_line = b_line

# ...
```

假设 m 前面被转换成了布尔类型，如果正好能执行 m.expand(b_line)，是不是会抛出异常？为了使得该 Bug 出现，需要满足以下两个条件：

- m 变量被转换成布尔类型变量，即需要 match_found 为 True，也就是 line 必须完全匹配文件中的一行。此外，还需要最初的 regexp 表达式不匹配文件中的任意一行。因为只有根据 regexp 匹配得到的 m 值最初为空时，代码才会进入设置 match_found 的语句中。
- 如果前面输入的 line 在文件中完整匹配，则 index[0] 为该匹配行的行号，此时只需要设置 backrefs 为 yes 即可进入 m.expand() 语句中。

上述条件似乎不难实现，我们在 ceph-1 节点上准备一个简单的测试文件，具体内容如下：

```
[root@ceph-1 tmp]# cat test_backrefs.txt
xxxxxx
chinatelecom123shenqicai
shenqicai19980214xxxtest
```

然后设置 lineinfile 模块的参数，即 line=xxxxxx（匹配整行）、regexp=^xyz$（没有匹配行）、backrefs=yes，执行结果如下：

```
(ansible2.8.6) [root@master ~]# ansible ceph-1 -i hosts -m lineinfile -a
"path=/tmp/test_backrefs.txt regexp=^xyz$ line=xxxxxx backrefs=yes"
ceph-1 | FAILED! => {
    "ansible_facts": {
        "discovered_interpreter_python": "/usr/bin/python"
    },
    "changed": false,
    "module_stderr": "Shared connection to ceph-1 closed.\r\n",
    "module_stdout": "Traceback (most recent call last):\r\n  File \"/root/
.ansible/tmp/ansible-tmp-1607735882.966169-176129582519029/AnsiballZ_
lineinfile.py\", line 114, in <module>\r\n    _ansiballz_main()\r\n  File
\"/root/.ansible/tmp/ansible-tmp-1607735882.966169-176129582519029/
AnsiballZ_lineinfile.py\", line 106, in _ansiballz_main\r\n    invoke_
module(zipped_mod, temp_path, ANSIBALLZ_PARAMS)\r\n  File \"/root/.ansible
/tmp/ansible-tmp-1607735882.966169-176129582519029/AnsiballZ_lineinfile.py\",
```

```
line 49, in invoke_module\r\n    imp.load_module('__main__', mod, module,
MOD_DESC)\r\n  File \"/tmp/ansible_lineinfile_payload_9dknuQ/__main__.py\",
 line 567, in <module>\r\n  File \"/tmp/ansible_lineinfile_payload_9dknuQ/
__main__.py\", line 558, in main\r\n  File \"/tmp/ansible_lineinfile_
payload_9dknuQ/__main__.py\", line 338, in present\r\nAttributeError:
'bool' object has no attribute 'expand'\r\n",
    "msg": "MODULE FAILURE\nSee stdout/stderr for the exact error",
    "rc": 1
}
```

上述命令正好复现了 'bool' object has no attribute 'expand' 异常，即布尔类型的变量没有 expand 属性，相应的修复代码如图 5-28 所示。

在 Ansible 2.8.7 中修复上述 Bug 的思路如下：

- 如果将完整行匹配标识 match_found 换成 exact_line_match，则该标识一定不要赋给相应的匹配结果。
- 在修复代码中将匹配结果 m 重新命名为了 match，此外在 backrefs 的判断中需要保证匹配成功（if backrefs 变成了 if backrefs and match）。

这个 Bug 警示 Python 程序员：不要随意变换变量的含义与类型，以免引发不必要的麻烦。

5.3　如何更深入地学习 Ansible 源码

在分析了 Ansible 的一些 Bug 后，我们可以总结一些学习开源项目的经验。

- 对于旧的、老的版本代码，由于缺乏相关文档和 GitHub 记录，只能通过代码对比来发现之前版本中的 Bug 及跟踪新版本的功能。刚开始这样操作时会比较困难，而且需要对版本的源码非常熟悉。好在旧版本的代码通常比较精简且容易理解，非常适合初级程序员阅读。
- 对于最新的、热门的开源项目而言，通常每个版本的发布都会有相应的 ChangeLog，一定要仔细研读这个 ChangeLog 文件，确保了解该版本的新增功能、修复的 Bug 等。
- 通过 GitHub 上的相关 issue 可以看到原问题创建者复现 Bug 并提交修正代码的记录。读者应先自行思考如何解决 Bug，然后再对比更新版本的解决代码，类似于自己的答案与标准答案的对比，也可能是自己的答案更优。

阅读 GitHub 上的相关记录会要求对源码有一定的了解。每次带着问题去阅读和调试源码，从中找到问题出现的原因并给出自己的解决方案，这是一种非常好的学习方式，也是笔者强烈推荐的进一步掌握 Ansible 源码的方法。当然，Ansible 源码模块众多，而为了兼容各种系统、各种云应用场景所编写的代码并不值得读者花费大量时间去阅读，读者只需要掌握 Ansible 的核心工作逻辑和核心模块的运行机制即可。

第 6 章　Playbook 项目实战

完成 Ansible 的核心源码剖析后，关于 Ansible 工具本身的源码剖析就告一段落了。本章将带领读者探索一个基于 Ansible 工具实现自动化部署 Ceph 集群的开源项目——ceph-ansible，读者从中能学到 Playbook 的很多高级语法，这对于初学者和渴望进阶的中级工程师而言是一份绝佳的学习资料。

6.1　ceph-ansible 项目介绍

Ceph 是美国加州大学圣克鲁兹分校的 Sage Weil 博士专门为博士论文设计的新一代自由软件分布式文件系统。从某种意义上来说，Ceph 是一种软件定义存储（SDS），可以运行在几乎所有主流的 Linux 发行版本上。由于 Ceph 具有分布式的特点，因此可以用于管理成百上千个节点，形成 PB 级以上规模的大存储集群。此外，Ceph 还是一个统一的存储系统，支持块存储、对象存储和文件存储。因此，Ceph 可以算得上是云计算时代的典型产物，也代表着云存储技术最新的发展方向。

早期在虚拟环境下部署 Ceph 存储集群时，笔者使用的是官方推荐的 ceph-deploy 工具。它是一个纯 Python 语言开发的工具，该工具的作者 Alfredo Deza 也是 Ceph 项目的代码贡献者之一。ceph-deploy 在部署 Ceph 集群时，会将搭建 Ceph 集群的动作分成若干个步骤，一步一步地搭建集群，这种方式比较适合新手搭建并测试集群。随着 Ansible 工具的流行，Ceph 社区创建了 ceph-ansible 项目并将搭建 Ceph 集群的工作进一步自动化，这使得开发者借助 ceph-ansible 工具，只需要指定部署的主机并简单修改配置文件后，即可实现一键部署 Ceph 集群。但是 ceph-ansible 将整个集群部署过程黑盒化，对于不熟悉 Ansible 的用户而言，偶尔的出错会让新手无所适从，不知从何查起。

此外，ceph-deploy 工具由于其性能和自动化程度不高等原因，并不适合在生产环境中部署大规模的 Ceph 集群，而且社区已经对 ceph-deploy 项目停止迭代，只对原有的 Bug 进行不定期更新。因此，ceph-ansible 工具成为了部署 Ceph 集群的主流工具，而该工具也正是本章的主角。本章将从以下几个方面完整剖析 ceph-ansible 工具：

（1）ceph-ansible 的目录结构。

（2）ceph-ansible 实战——搭建 Ceph 集群。

（3）基于实际需求改造 ceph-ansible。

（4）ceph-ansible 项目源码改造与简化。

完成以上 4 个部分内容的学习后，读者将会对 ceph-ansible 项目了如指掌，后续使用其部署 Ceph 集群时，对于出现的任何报错信息都可以很快定位异常位置并快速分析和解决。此外，ceph-ansible 项目可以说是 Playbook 编写的一个典范，它够复杂、够健壮且几乎囊括了大部分 Playbook 语法，是一份难得的学习 Playbook 语法的资料。笔者在编写天翼云科技有限公司内部的自动化部署脚本时，经常会打开 ceph-ansible 项目源码学习一些模块的优秀的编写方法，以提升自己编写 Playbook 的水平。

6.2　ceph-ansible 实战

本节主要演示 ceph-ansible 工具在部署 Ceph 集群方面的功能。至于 Ceph 相关的一些基础知识，则需要读者自行学习。

6.2.1　使用 ceph-ansible 部署 Ceph 集群

在部署 Ceph 集群之前，需要先明确部署工具的版本及部署 Ceph 集群的版本，这一点非常重要。笔者第一次使用 ceph-ansible 工具部署 Ceph 集群时，直接使用了 GitHub 上最新的 ceph-ansible 工具去部署 Ceph Mimic 版本，但总是运行失败且反复执行 Ceph 的命令，浪费了大量的时间，直到看到了官方的版本说明时才恍然大悟。官方的版本说明如图 6-1 所示。

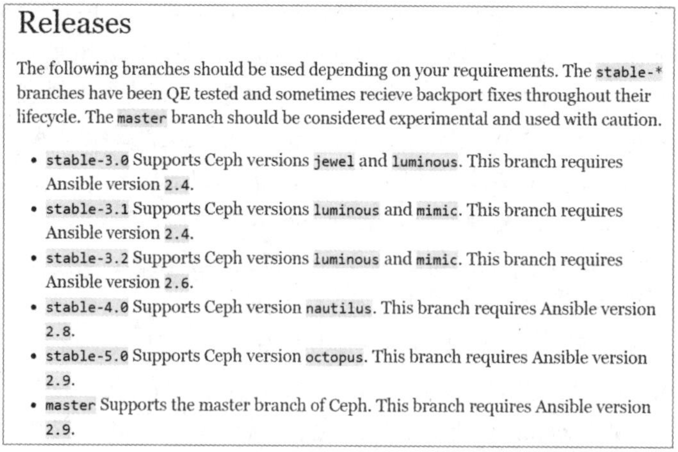

Releases

The following branches should be used depending on your requirements. The `stable-*` branches have been QE tested and sometimes recieve backport fixes throughout their lifecycle. The `master` branch should be considered experimental and used with caution.

- `stable-3.0` Supports Ceph versions `jewel` and `luminous`. This branch requires Ansible version `2.4`.
- `stable-3.1` Supports Ceph versions `luminous` and `mimic`. This branch requires Ansible version `2.4`.
- `stable-3.2` Supports Ceph versions `luminous` and `mimic`. This branch requires Ansible version `2.6`.
- `stable-4.0` Supports Ceph version `nautilus`. This branch requires Ansible version `2.8`.
- `stable-5.0` Supports Ceph version `octopus`. This branch requires Ansible version `2.9`.
- `master` Supports the master branch of Ceph. This branch requires Ansible version `2.9`.

图 6-1　ceph-ansible 的官方版本说明

从图 6-1 中可以知道，ceph-ansible 3.1 和 3.2 版本支持 Ceph 的 Luminous 和 Mimic 版本部署，而 ceph-ansible 4.0 版本只支持 Ceph 的 Nautilus 版本部署（笔者第一次使用 ceph-ansible 部署 Ceph 集群时，使用的是 ceph-ansible 4.0 版本部署 Ceph Mimic，所以一直不成功）。目前最新的 ceph-ansible 及 ceph-ansible 5.0 的稳定版只支持 Ceph 的最新版本 Octopus。

目前来说，Ceph 的发展速度很快，大约一年左右就会更新一个大版本。Luminous 和 Mimic 版本的 Ceph 目前还在各大互联网公司的生产环境中继续"服役"，Nautilus 版本则是近年来部署 Ceph 存储系统的优选版本。至于最新的 Octopus 版本，笔者从未在生产环境中使用过，也未见过有公司公开部署过该版本，故无法对该版本进行评论。综上所述，笔者决定使用 ceph-ansible 的 v4.0 版本搭建 Nautilus 版本的 Ceph 集群。

接下来按照笔者设定的步骤使用 ceph-ansible 来搭建 Ceph 集群。首先需要准备至少 3 台虚拟机且每台虚拟机上最好添加若干个虚拟硬盘用于安装 OSD 进程。在前期介绍 Ansible 的核心源码时，笔者准备了 4 台虚拟机，其虚拟硬件配置如下：

- master：1GB 内存，无额外虚拟硬盘，IP 地址为 192.168.26.110。
- ceph-1、ceph-2 和 ceph-3：1GB 内存，每台主机有 3 块 SATA 虚拟硬盘，每块硬盘的容量为 30GB，IP 地址为 192.168.26.120～192.168.26.122。

另外，需确保每台主机均可连通外网。

接着开始准备部署前的工作，具体操作如下：

```
(ansible2.8.16) [root@master ~]# mkdir ceph-cluster
(ansible2.8.16) [root@master ~]# cd ceph-cluster/
(ansible2.8.16) [root@master ceph-cluster]# wget
https://github.com/ceph/ceph-ansible/archive/v4.0.30.tar.gz
...
(ansible2.8.16) [root@master ceph-cluster]# tar -xzf v4.0.30.tar.gz
...
(ansible2.8.16) [root@master ceph-cluster]# cd ceph-ansible-4.0.30/
```

🔔注意：笔者计划使用 Ansible 2.8.16 部署 Ceph 集群，经过实际测试发现，ceph-ansible 4.0.34 之后的版本均要求运行的 Ansible 版本为 2.9，因此笔者选择了 ceph-ansible 4.0.30 版本来完成本次实践。经过测试，该版本的 Playbook 支持 Ansible 2.8 版本。

目前，搭建 Ceph 集群的核心节点角色可分为以下几类：

- MON 节点：部署 ceph-mon 进程的节点。该进程的主要任务是维护集群视图的一致性，在维护一致性的时候使用了 Paxos 协议，并将其实例化到数据库中，以方便后续访问。MON 节点是 Ceph 集群的核心节点，类似于领导地位。一般而言，MON 节点的数量必须为奇数个。在笔者公司部署的所有集群中，每个集群只有 3 个 MON 节点，但已经足以维持一个中等规模集群的数据。
- MGR 节点：部署 ceph-mgr 进程的节点。该进程主要用于分担和扩展 ceph-mon 进程的部分功能，减轻 ceph-mon 的负担，以便更好地管理存储集群。
- OSD 节点：部署 ceph-osd 进程的节点，这些节点可以称之为存储节点。一般而言，每个 ceph-osd 进程会对应一块裸盘。集群中的数据最终会落到这些 osd 进程对应的磁盘中进行保存。

- MDS 节点：部署 ceph-mds 进程的节点，它们是 Ceph 分布式文件系统的元数据服务器守护进程。一个或多个 ceph-mds 进程共同管理文件系统的命名空间，并且协调访问共享的 OSD 集群。
- RGW 节点：部署 ceph-radosgw 进程的节点。其实该进程就是 Rados 的一个网关，用于对外提供对象存储服务。

由于目前参与搭建集群的机器只有 3 台，笔者只能将这些 Ceph 进程全部部署在这 3 台机器上。计划部署的情况如下：

- ceph-[1:3]：3 台机器分别部署 ceph-mon、ceph-mgr 和 ceph-osd 进程。
- ceph-[1:2]：额外部署 ceph-radosgw 进程。

因此，ceph-ansible 中 hosts 文件的写法如下：

```
(ansible2.8.16) [root@master ceph-ansible-4.0.30]# cat hosts
[mons]
ceph-[1:3]

[mgrs]
ceph-[1:3]

[osds]
ceph-[1:3]

[rgws]
ceph-[1:2]
```

🔔注意：我们在第 2 章中已经设置了 master 节点对所有主机免密，因此这里不需要再输入相应的 SSH 密码。如果实验环境没有进行免密操作，则需要输入主机的 SSH 远程连接的账号（默认为当前用户）、密码及 SSH 开放端口（默认是 22）。

接着需要完成以下两个文件的复制：

```
(ansible2.8.16) [root@master ceph-ansible-4.0.30]# cp group_vars/all.yml.
sample group_vars/all.yml
(ansible2.8.16) [root@master ceph-ansible-4.0.30]# cp site.yml.sample
site.yml
```

🔔注意：group_vars/all.yml 是整个项目的全局变量文件，一般需要改动这里的配置以便 Playbook 能正常运行。site.yml 是 ceph-ansible 起始运行的 Playbook 文件，整个 Ceph 集群部署时执行的任务就是从这里开始的。

现在需要修改和设置 group_vars/all.yml 中的部分参数值。该文件中有非常多的变量定义，也设置了相应的默认值。在这个配置文件中将许多参数进行了注释并配上了相应的说明，开发者有需要时可以阅读这些说明并合理设置相应的值。但对于初学者而言，官方文档给出的说明并不详细，操作时经常会因为某些原因导致执行报错，这会打击使用者的信心。笔者先给出能完整运行的最小配置参数，具体如下：

```
# 配置位置: group_vars/all.yml
# ...
ceph_origin: repository
ceph_repository: community
# 指定安装 Ceph 包的源
ceph_mirror: https://mirrors.aliyun.com/ceph/
ceph_stable_key: "{{ ceph_mirror }}/keys/release.asc"
ceph_stable_release: nautilus
ceph_stable_repo: "{{ ceph_mirror }}/debian-{{ ceph_stable_release }}"
ceph_stable_distro_source: bionic

cephx: "true"

# 必须有，指定内部进程通信的网段
public_network: 192.168.26.0/24
cluster_network: 192.168.26.0/24

# ceph-ansible 会根据网卡参数获取相应的 IP 地址
monitor_interface: ens33
radosgw_interface: ens33
# 指定不配置防火墙，否则防火墙每次都会被打开
configure_firewall: False
osd_objectstore: bluestore
# 指定格式化的磁盘
devices:
  - /dev/sdb
  - /dev/sdc
  - /dev/sdd

# 不设置会报错
dashboard_enabled: False

# ...
```

上面的配置介绍如下：

- cept_mirror：使用国内的镜像源下载 Ceph 的相关包，这里使用的是阿里云的 Ceph 仓库。该配置会影响 Ceph 包的下载速度，十分重要。

- public_network 和 cluster_network：非常重要，它会生成 ceph.conf 中的核心配置参数，这些参数会直接影响 Ceph 内节点之间的通信。一般将其设置为集群的内网网段地址即可，表示这个网段内的节点可相互通信。

- monitor_interface 和 radosgw_interface：不设置也会导致执行失败。因为 ceph-ansible 需要根据设置的网卡接口获取相应的 IP 地址，从而完成后续的操作。

- configure_firewall：关闭防火墙的相关配置。如果不设置，每次运行时都会启动防火墙，这会导致一些不可预知的错误，主要在部署 ceph-mon 进程时出现。

- osd_objectstore：早期版本的默认值为 filestore，这会严重影响集群中 OSD 的性能。一般会使用 bluestore 方式创建 OSD 进程，在 v4.0.30 版本中默认使用 bluestore 方式，因此这里也可以不用设置。

- devices：用作保存数据节点的磁盘，必须是未格式化的空盘。
- dashboard_enabled：当没有要部署 grafana-server 节点的需求时，使用该参数的默认值（True）会在运行过程中抛出 "fail if [grafana-server] group doesn't exist" 的错误，如图 6-2 所示。此时必须要将该参数设置为 False 才能跳过这个 task。

实践是最好的老师，建议读者自行测试 configure_firewall 和 dashboard_enabled 参数的运行效果。

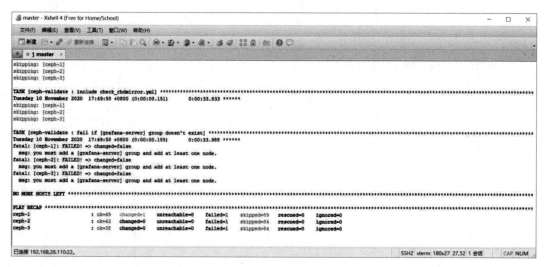

图 6-2　dashboard_enabled 参数默认为 True 时导致的报错信息

准备好 hosts、group_vars/all.yml 及 site.ym 后，就可以运行 ceph-ansible 项目了。进入 ceph-ansible 的源码目录，运行如下命令：

```
(ansible2.8.16) [root@master ceph-ansible-4.0.30]# ansible-playbook -i
hosts site.yml
PLAY [mons,osds,mdss,rgws,nfss,rbdmirrors,clients,mgrs,iscsigws,iscsi-
gws,grafana-server,rgwloadbalancers]
****************************************************************************

TASK [check for python]
****************************************************************************
Tuesday 10 November 2020  22:13:01 +0800 (0:00:00.117)   0:00:00.117 ****
fatal: [ceph-1]: FAILED! =>
  msg: The ips_in_ranges filter requires python's netaddr be installed on
the ansible controller.
fatal: [ceph-2]: FAILED! =>
  msg: The ips_in_ranges filter requires python's netaddr be installed on
the ansible controller.
fatal: [ceph-3]: FAILED! =>
  msg: The ips_in_ranges filter requires python's netaddr be installed on
the ansible controller.

NO MORE HOSTS LEFT
****************************************************************************
```

```
PLAY RECAP
******************************************************************
# 忽略统计结果
# ...
```

报错比较明显，需要安装 netaddr 模块。在 Python 2 中安装 netaddr 模块时总是会出现问题，而在 Python 3 中安装 netaddr 模块后就能正常运行 Playbook 了。因此，笔者推荐直接在 Python 3 中安装 Ansible 2.8，安装命令如下：

```
(ansible2.8.16) [root@master ceph-ansible-4.0.30]# pip install netaddr -i
https://pypi.tuna.tsinghua.edu.cn/simple
```

再次运行前面的 Playbook，等待几分钟之后，就可以看到 Ceph 集群部署成功的提示，如图 6-3 所示。

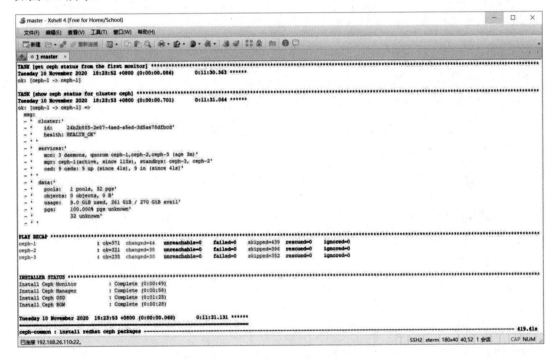

图 6-3　Ceph 集群部署成功

此时可以登录 ceph-1 节点并使用 ceph -s 命令查看集群状态，如果能够正常输出，则表示 Ceph 集群搭建成功。以下是笔者搭建环境的输出结果：

```
[root@ceph-1 ~]# ceph -s
  cluster:
    id:     24b2b885-2e87-4aed-a5ed-3d5ae78dfbcd
    health: HEALTH_OK

  services:
    mon: 3 daemons, quorum ceph-1,ceph-2,ceph-3 (age 52m)
```

```
    mgr: ceph-1(active, since 51m), standbys: ceph-3, ceph-2
    osd: 9 osds: 9 up (since 50m), 9 in (since 50m)
    rgw: 2 daemons active (ceph-1.rgw0, ceph-2.rgw0)

  task status:

  data:
    pools:   4 pools, 128 pgs
    objects: 187 objects, 1.2 KiB
    usage:   9.0 GiB used, 261 GiB / 270 GiB avail
    pgs:     128 active+clean
```

为了防止集群出现时钟不同步的情况，需要给每个节点添加一个定时任务，每分钟同步阿里云时钟，以确保集群内所有节点的时钟都一致。具体操作命令如下：

```
# 批量运行 cron 模块，添加同步时钟的定时任务
(ansible2.8.16) [root@master ceph-ansible-4.0.30]# ansible ceph-[1-9]* -i
hosts -m cron -a "name='sync time' job='/usr/sbin/ntpdate ntp1.aliyun.com'
minute='*/1'"
ceph-3 | CHANGED => {
    "changed": true,
    "envs": [],
    "jobs": [
        "sync time"
    ]
}
ceph-2 | CHANGED => {
    "changed": true,
    "envs": [],
    "jobs": [
        "sync time"
    ]
}
ceph-1 | CHANGED => {
    "changed": true,
    "envs": [],
    "jobs": [
        "sync time",
    ]
}
# 批量执行 crontab -l 命令，查看定时任务列表
(ansible2.8.16) [root@master ceph-ansible-4.0.30]# ansible ceph-[1-9]* -i
hosts -m shell -a "crontab -l"
ceph-3 | CHANGED | rc=0 >>
#Ansible: sync time
*/1 * * * * /usr/sbin/ntpdate ntp1.aliyun.com

ceph-2 | CHANGED | rc=0 >>
#Ansible: sync time
*/1 * * * * /usr/sbin/ntpdate ntp1.aliyun.com

ceph-1 | CHANGED | rc=0 >>
#Ansible: sync time
*/1 * * * * /usr/sbin/ntpdate ntp1.aliyun.com
```

6.2.2　ceph-ansible 的其他功能

除了完成 Ceph 集群的创建以外，ceph-ansible 还给使用者提供了各种"贴心"的功能，具体如下：

- Ceph 集群的容器化部署与移除。
- 一键清除 Ceph 集群的 Playbook：infrastructure-playbooks/purge-cluster.yml。
- 添加 MON、OSD 节点或者相应的进程。
- 缩减 MON、MDS、MGR、RGW 节点等。
- 一键移除 ISCSI 网关等。

本节将以 osd 进程的增、删需求为例，演示 ceph-ansible 中的其他功能。首先检查前面搭建好的 Ceph 集群，每个磁盘会对应一个 osd 进程，并且 Ceph 集群会给每个 osd 进程分配一个 osd-id。默认创建 osd 进程时没有指定相应的 osd-id 值，此时 Ceph 集群会为其分配一个当前系统未分配的最小的 ID 值。通过 ceph osd tree 命令可以看到所有 osd 的信息，具体如下：

```
[root@ceph-1 ~]# ceph osd tree
ID CLASS WEIGHT  TYPE NAME      STATUS REWEIGHT PRI-AFF
-1       0.26367 root default
-3       0.08789     host ceph-1
 1   hdd 0.02930         osd.1      up  1.00000 1.00000
 4   hdd 0.02930         osd.4      up  1.00000 1.00000
 8   hdd 0.02930         osd.8      up  1.00000 1.00000
-5       0.08789     host ceph-2
 2   hdd 0.02930         osd.2      up  1.00000 1.00000
 5   hdd 0.02930         osd.5      up  1.00000 1.00000
 7   hdd 0.02930         osd.7      up  1.00000 1.00000
-7       0.08789     host ceph-3
 0   hdd 0.02930         osd.0      up  1.00000 1.00000
 3   hdd 0.02930         osd.3      up  1.00000 1.00000
 6   hdd 0.02930         osd.6      up  1.00000 1.00000
[root@ceph-1 ~]# lsblk
NAME
MAJ:MIN RM  SIZE RO TYPE MOUNTPOINT
sda
8:0     0   30G 0 disk
├─sda1
8:1     0    1G 0 part /boot
└─sda2
8:2     0   29G 0 part
  ├─centos-root
253:0   0   27G 0 lvm  /
  └─centos-swap
253:1   0    2G 0 lvm  [SWAP]
sdb
```

```
8:16     0    30G   0 disk
└─ceph--c280f9c7--fdf4--40d8--aba7--b25f435e7ee1-osd--block--9264ad18-
32ea--417d--9ef6--5e176d0c33fc 253:2     0    30G   0 lvm
sdc
8:32     0    30G   0 disk
└─ceph--267375c5--677b--4528--8044--7c8f1af94076-osd--block--f06c761b-
2a64--42c5--bb1f--7944e7559c9f 253:3     0    30G   0 lvm
sdd
8:48     0    30G   0 disk
└─ceph-25eb7c52--e73b--4ff9--a7e5--d273e429488b-osd--block--99aac5ba-
87e3--462f--a104--9b23320e14d8 253:4     0    30G   0 lvm
sr0
11:0     1   942M   0 rom
```

以上是从集群获得的 osd 信息。ceph-1 节点下有 3 个 osd 进程，对应的 osd-id 分别为 1、4、8，且处于 UP 状态。

对于 Ceph 这样的分布式存储系统而言，扩容和缩容是一个非常重要的特点。ceph-ansible 也支持对 Ceph 集群进行扩缩容，并且实现方式非常简单，参考如下指令：

```
(ansible2.8.16) [root@master ceph-ansible-4.0.30]# ansible-playbook -i
hosts infrastructure-playbooks/shrink-osd.yml -e osd_to_kill=0,2,3,5,6,7
```

上面执行的是 infrastructure-playbooks/shrink-osd.yml 这个 Playbook，它通过 osd-id 指定需要移除的 osd 进程。可以使用-e 选项指定相应的参数及其值，也可以将参数及其值写入 group_vars/all.yml 中。此外，在执行该 Playbook 时会询问是否要执行，只有输入 yes 后才会继续执行。上面的命令是移除 ceph-2 和 ceph-3 上的所有 osd 进程，同时恢复磁盘原样。图 6-4 所示为该 Playbook 执行的最后结果。

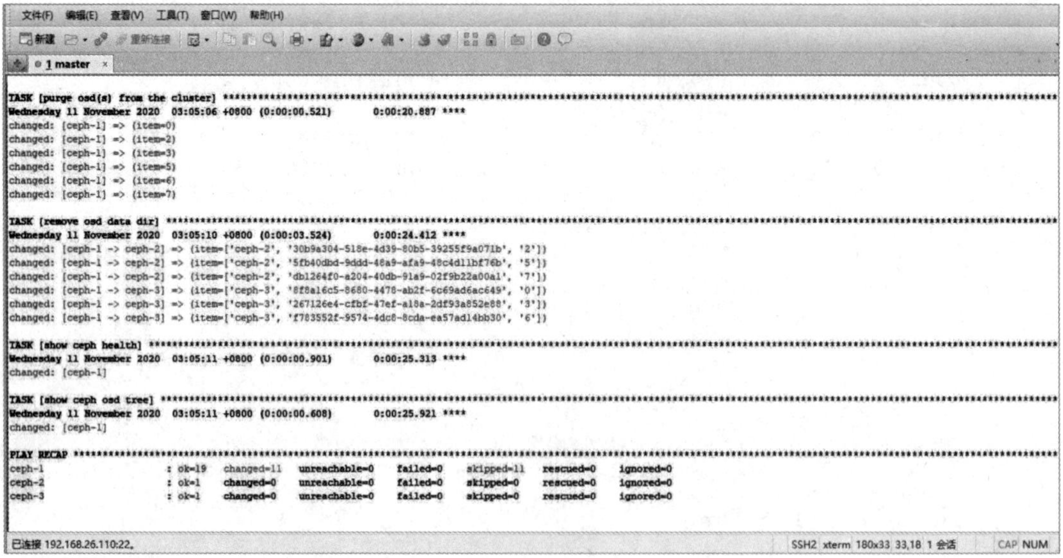

图 6-4　移除 osd 进程的 Playbook 运行结果

　　缩容的 Playbook 执行成功后，可以在 ceph-1 上执行 ceph osd tree 命令查看，可以看到 ceph-2 和 ceph3 上的 osd 信息已经全部被抹掉。此外，还可以登录到 ceph-2 和 ceph-3 上查看是否还有 osd 进程，或者使用 lsblk 命令查看磁盘情况，具体如下：

```
[root@ceph-2 ~]# ceph osd tree
ID CLASS WEIGHT  TYPE NAME       STATUS REWEIGHT PRI-AFF
-1       0.08789 root default
-3       0.08789     host ceph-1
 1  hdd 0.02930         osd.1       up  1.00000 1.00000
 4  hdd 0.02930         osd.4       up  1.00000 1.00000
 8  hdd 0.02930         osd.8       up  1.00000 1.00000
-5       0           host ceph-2
-7       0           host ceph-3
[root@ceph-2 ~]# lsblk
NAME            MAJ:MIN RM  SIZE RO TYPE MOUNTPOINT
sda             8:0      0   30G  0 disk
├─sda1          8:1      0    1G  0 part /boot
└─sda2          8:2      0   29G  0 part
  ├─centos-root 253:0    0   27G  0 lvm  /
  └─centos-swap 253:1    0    2G  0 lvm  [SWAP]
sdb             8:16     0   30G  0 disk
sdc             8:32     0   30G  0 disk
sdd             8:48     0   30G  0 disk
sr0             11:0     1  942M  0 rom
```

　　有缩容需求，自然也有扩容需求，这也是分布式集群中最常见的场景。ceph-ansible 中如何实现 Ceph 集群的存储节点的扩容呢？所谓存储扩容，就是在具有多块磁盘的主机上安装相应的 osd 进程并加入存储集群以扩大集群整体的存储容量。首先在 hosts 文件中添加新增节点的信息，具体如下：

```
(ansible2.8.16) [root@master ceph-ansible-4.0.30]# cat add_osds
[mons]
ceph-[1:3]

[osds]
ceph-[2:3]
```

　　📖注意：点的磁盘信息使用的是 group_vars/all.yml 中的 devices 参数，用于指定需要 osd 进程管理的磁盘。

　　接着执行 ceph-ansible 中相应的添加 osd 的 Playbook，具体如下：

```
(ansible2.8.16) [root@master ceph-ansible-4.0.30]# ansible-playbook -i
add_osds infrastructure-playbooks/add-osd.yml
```

　　输入确认执行命令（yes）后，该 Playbook 执行成功的结果如图 6-5 所示。

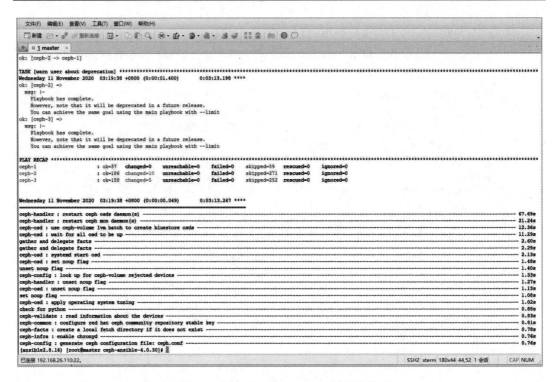

图 6-5　添加 osd 的 Playbook 运行结果

可以看到，此时集群的 OSD Map 中又有了 ceph-2 和 ceph-3 节点上的 osd 信息，即在 ceph-2 和 ceph-3 上各启动了 3 个 osd 进程，分别管理相应的磁盘，具体如下：

```
[root@ceph-3 ~]# ceph osd tree
ID CLASS WEIGHT  TYPE NAME      STATUS REWEIGHT PRI-AFF
-1     0.26367 root default
-3     0.08789     host ceph-1
 1 hdd 0.02930         osd.1      up  1.00000 1.00000
 4 hdd 0.02930         osd.4      up  1.00000 1.00000
 8 hdd 0.02930         osd.8      up  1.00000 1.00000
-5     0.08789     host ceph-2
 0 hdd 0.02930         osd.0      up  1.00000 1.00000
 3 hdd 0.02930         osd.3      up  1.00000 1.00000
 6 hdd 0.02930         osd.6      up  1.00000 1.00000
-7     0.08789     host ceph-3
 2 hdd 0.02930         osd.2      up  1.00000 1.00000
 5 hdd 0.02930         osd.5      up  1.00000 1.00000
 7 hdd 0.02930         osd.7      up  1.00000 1.00000

[root@ceph-3 ~]# ps -ef | grep ceph-osd | grep -v grep
ceph      23226     1  0 00:06 ?        00:00:02 /usr/bin/ceph-osd -f
--cluster ceph --id 2 --setuser ceph --setgroup ceph
ceph      23607     1  0 00:06 ?        00:00:02 /usr/bin/ceph-osd -f
--cluster ceph --id 5 --setuser ceph --setgroup ceph
ceph      23989     1  0 00:06 ?        00:00:02 /usr/bin/ceph-osd -f
--cluster ceph --id 7 --setuser ceph --setgroup ceph
```

这里只是简单实践了 ceph-ansible 中新增和删除 osd 进程的功能。ceph-ansible 的功能非常强大，读者如果有兴趣，可以尝试实践其他的功能，如容器化部署 Ceph、一键搭建 Ceph 集群监控页面等。

6.3　一个真实场景的改造

本节将介绍笔者在工作中部署 Ceph 集群时对 ceph-ansible 做的一个简单改造。通过这个案例来学习根据实际的业务需求改造开源项目的方法。本次改造除了要求读者熟悉 Playbook 项目的基本组成之外，还需要对 Ansible 中的自定义模块有一定的了解，这些内容均在前面介绍过。下面先来看一下本案例的业务需求。

6.3.1　业务需求

在 ceph-1 节点上执行 ceph osd tree 命令查看 osd-id 的分配情况，结果如下：

```
[root@ceph-1 ~]# ceph osd tree
ID CLASS WEIGHT  TYPE NAME      STATUS REWEIGHT PRI-AFF
-1       0.26367 root default
-3       0.08789     host ceph-1
 1  hdd 0.02930         osd.1      up  1.00000 1.00000
 4  hdd 0.02930         osd.4      up  1.00000 1.00000
 8  hdd 0.02930         osd.8      up  1.00000 1.00000
-5       0.08789     host ceph-2
 0  hdd 0.02930         osd.0      up  1.00000 1.00000
 3  hdd 0.02930         osd.3      up  1.00000 1.00000
 6  hdd 0.02930         osd.6      up  1.00000 1.00000
-7       0.08789     host ceph-3
 2  hdd 0.02930         osd.2      up  1.00000 1.00000
 5  hdd 0.02930         osd.5      up  1.00000 1.00000
 7  hdd 0.02930         osd.7      up  1.00000 1.00000
```

前面介绍过，在 ceph-ansible 中是批量格式化磁盘且在创建 osd 进程时会向集群申请 osd-id 值，集群会按照先后顺序分配当前集群中未分配的最小 osd-id 值（从 0 开始）。由于 ceph-ansible 是多进程执行且批量操作磁盘，因此会使 osd-id 的分配毫无规律。

笔者所在的公司希望能在部署 Ceph 集群时保证每台主机的存储节点都是连续的 osd-id 编号，并且编号随着节点的编号而连续。例如当前环境下的 3 台主机的存储节点 ceph-1～ceph-3，每个节点上有 3 块 30GB 的裸盘，因此分配的 osd 进程的 ID 值为 0～8。另外，本次部署希望 ceph-1 上的 osd 进程编号为 0～2，ceph-2 上的 osd 进程编号为 3～5，ceph-3 上的 osd 进程编号为 6～8。除此之外，还希望对应 osd 进程使用的数据盘也是有序的，即对于 ceph-1 节点而言，osd-id 为 0 的进程数据盘为/dev/sdb，osd-id 为 1 的进程数据盘为/dev/sdc，最后一个进程数据盘为/dev/sdd。无论有多少块数据盘，都将按照这种盘符顺序依次排列。

按照这样的模式部署后，后续该集群的运维和管理工作将会变得十分便捷。假设 osd-id 为 10 的进程处于 DOWN 状态，这样运维人员能迅速根据 osd-id 编号找到对应的问题节点及使用的数据盘，以便及时对该磁盘进行重启或者进行异常处理。这是集群部署方面的规范，虽无硬性要求，但是为了减少后续运维工作的复杂度，部署人员也应该按照这种规范开展相关工作。

6.3.2　手工实现

本节先介绍如何手工实现上述需求，然后才能追踪 ceph-ansible 的相关代码并进行改造，这需要读者对于搭建 Ceph 集群的 osd 服务有一定的了解，但在这里并不重要。笔者会指出需要调整哪里，并在 ceph-ansible 中找到创建 osd 进程时需要指定 osd-id 的地方，然后进行修改和测试，最后实现前面的需求。

对于创建 Ceph 的 osd 进程而言，无论是使用 ceph-deploy 工具还是 ceph-ansible 工具，都绕不开它们共同调用的一个工具——ceph-volume。它们最后创建 osd 进程的命令如下：

```
sudo ceph-volume --cluster 集群名称 lvm prepare --bluestore --data 数据盘
[--osd-id 指定 osd-id 值] --osd-fsid 随机数
```

💭说明：在 ceph-ansible 中可以指定创建 Ceph 集群的名称，但通常情况下使用默认值 ceph 即可；--data 参数指定创建 osd 进程关联的数据盘（如上面的/dev/sdb 等）；--osd-id 可以指定创建 osd 进程的 osd-id 值，直接指定时会报错，需要对 ceph-volume 代码进行一些调整；--osd-fsid 参数可以直接由 Linux 下的 uuidgen 命令生成，也可以任意指定一个唯一的字符串。

接下来将演示如何手工搭建集群的 osd 进程，实现前面的需求。首先将原先搭建的集群中的 osd 进程全部移除，具体命令如下：

```
(ansible2.8.16) [root@master ceph-ansible-4.0.30]# ansible-playbook -i
hosts infrastructure-playbooks/shrink-osd.yml -e osd_to_kill=0,1,2,3,4,
5,6,7,8
```

然后按照如下步骤依次创建 Ceph 集群的 osd 进程。

（1）安装 ceph-common 和 ceph-osd 包，直接用 yum 命令安装。

（2）复制 MON 节点上的两个密钥文件（/etc/ceph/ceph.conf 和/etc/ceph/ceph.client.admin.keyring）并放到需要安装 ceph-osd 进程的节点上，确保能连接创建的 Ceph 集群。

（3）复制 MON 节点上的/var/lib/ceph/bootstrap-osd/ceph.keyring 到安装 osd 节点的相同位置处。

（4）执行创建 osd 进程的脚本。

注意，本次实验中，ceph-1~ceph-3 三个节点共同部署了 ceph-mon 和 ceph-osd 进程，因此上面的步骤中前三步可以不用做，只需要执行创建 osd 进程的脚本即可。下面给出一个创建 osd 进程的脚本（该脚本由笔者同事谭伟杰提供，笔者在原脚本基础上进行了简单修改），内容如下：

```bash
#!/bin/bash
#
# 创建 osd 进程，脚本名称为 make_osds.sh
#
device_array[0]=/dev/sdb
device_array[1]=/dev/sdc
device_array[2]=/dev/sdd

if [ $# -ge 1 ]; then
   osd_id_begin=$1
else
   hdd_osd_count=${#device_array[@]}
   host_id=$(hostname -s|sed "s/ceph-\([0-9]*\)/\1/g")
   host_id_begin=$(expr $host_id - 1)
   osd_id_begin=$(expr $host_id_begin \* $hdd_osd_count)
fi

# 修正 ceph-volume 代码中的一个问题
file_path=$(sudo find /usr/lib -name prepare.py -type f | grep ceph_volume
| grep util)
sudo sed -i 's/if osd and/if not osd or/g' $file_path

create_hdd_osd(){
   sudo fdisk -l $1 > /dev/null 2> /dev/null
   if [ $? -ne 0 ]; then
     continue
   fi
   # 创建 osd 进程核心语句
   sudo ceph-volume --cluster ceph lvm prepare --bluestore --data $1 --osd-id
$2 --osd-fsid $3
   # 激活相应的 osd 进程
   sudo ceph-volume lvm activate $2 $3
   sleep 2
   # 主要是为了得到磁盘大小和单位，T 单位不用处理，G 单位需要转换成 T 单位
   disk_name=$(basename $1)
   disk_size=$(lsblk | grep "^$disk_name[^a-z]" | awk '{print $4}')
   len=$((${#disk_size}-1))
   size=${disk_size:0:-1}
   unit=${disk_size:$len:1}
   if [ $unit == "G" ]; then
     size=$(echo "scale=4; $size/1024" | bc)
   fi
   # 创建 crush map
   sudo ceph osd crush add osd.$2 $size host=`hostname -s`
}

for i in ${!device_array[@]}
do
   create_hdd_osd ${device_array[$i]} $osd_id_begin `uuidgen`
```

```
    osd_id_begin=$((osd_id_begin+1))
done
```

注意，在使用--osd-id 参数指定创建 osd 进程的 ID 编号时，需要调整 ceph-volume 工具中的一行代码，否则会抛出"The osd ID xxx is already in use or does not exist."这样的错误信息。需要修正的是 ceph-volume 工具中判断 osd_id 是否有效的语句，具体如下：

```python
# 源码位置：ceph_volume/util/prepare.py
# ...

def osd_id_available(osd_id):
    # ...

    if osd and osd[0].get('status') == "destroyed":
        return True
    return False
```

这里只需要将 if osd and 替换成 if not osd or 就可以让该函数返回 True，保证创建 osd 的过程可以继续下去。总的来说，上述脚本做了两件事情：

（1）计算当前 osd-id 的起始值，可以从外部输入（第一个参数），也可以根据当前节点名称进行计算。对于 ceph-2 而言，取出它的节点数字编号为 2，第一个节点有 3 个 osd 进程，占据 osd-id 的值为 0、1、2，于是对于编号为 2 的节点而言，其 osd-id 的值为 3、4、5，依此类推。

（2）基于 ceph-volume 工具和 Ceph 客户端命令，循环创建 osd 进程。

将 osd 进程的创建脚本复制到所有节点上，然后执行脚本（如果是带参数，那么第一个参数为创建 osd 进程的起始 ID 值）。以 ceph-2 节点为例，执行以下任意一行命令均可：

```
[root@ceph-2 ~]# sh make_osds.sh
# 或者带上参数
[root@ceph-2 ~]# sh make_osds.sh 3
```

3 个节点都执行完脚本命令后，可继续使用 ceph osd tree 命令查看 osd 进程信息，结果如下：

```
[root@ceph-1 ~]# ceph osd tree
ID CLASS WEIGHT  TYPE NAME     STATUS REWEIGHT PRI-AFF
-1       0.37999 root default
-3       0.08800     host ceph-1
 0  hdd 0.02899         osd.0      up  1.00000 1.00000
 1  hdd 0.02899         osd.1      up  1.00000 1.00000
 2  hdd 0.02899         osd.2      up  1.00000 1.00000
-5       0.14600     host ceph-2
 3  hdd 0.02899         osd.3      up  1.00000 1.00000
 4  hdd 0.02899         osd.4      up  1.00000 1.00000
 5  hdd 0.02899         osd.5      up  1.00000 1.00000
-7       0.14600     host ceph-3
 6  hdd 0.02899         osd.6      up  1.00000 1.00000
 7  hdd 0.02899         osd.7      up  1.00000 1.00000
 8  hdd 0.02899         osd.8      up  1.00000 1.00000
```

通过上述手动步骤已经基本实现了想要的部署结果。在正式的生产环境部署过程中，由于 ceph-mon 和 ceph-osd 服务分开部署且存储节点上的磁盘较多且容量较大,往往会针对创建 osd 的脚本和 ceph-ansible 进行很多优化。例如,去掉 ceph-ansible 代码中各种不必要的判断和无谓的检测(setup 模块),启动多进程加快 osd 的创建,以加快集群的创建等。当然这些需要依据具体的业务需求进行相关改造,不具有普适性。

6.3.3　改造 ceph-ansible

为了能改造 ceph-ansible 项目实现上述需求,需要对 ceph-ansible 的运行流程有一个大致的认识。前面创建集群运行的 Playbook 文件为 site.yml,该 Playbook 根据不同的组来执行相应角色的剧本。其中部署 osd 服务部分的代码如下:

```
# ceph-ansible/site.yml
# ...

- hosts: osds
  gather_facts: false
  become: True
  any_errors_fatal: true
  pre_tasks:
    - name: set ceph osd install 'In Progress'
      run_once: true
      set_stats:
        data:
          installer_phase_ceph_osd:
            status: "In Progress"
            start: "{{ lookup('pipe', 'date +%Y%m%d%H%M%SZ') }}"

  # 主要是执行角色任务
  tasks:
    - import_role:
        name: ceph-defaults
      tags: ['ceph_update_config']
    - import_role:
        name: ceph-facts
      tags: ['ceph_update_config']
    - import_role:
        name: ceph-handler
      tags: ['ceph_update_config']
    - import_role:
        name: ceph-config
      tags: ['ceph_update_config']
    - import_role:
        name: ceph-osd

  # ...
```

对于 ceph-ansible 项目来说,通过查看 ceph-ansible/roles 目录下的文件可知,它定义了很多部署角色。例如,ceph-defaults 用于加载全局 host 变量,ceph-config 用于处理 Ceph 配置文件的生成和下发等,这些都属于公共角色。而 ceph-mon、ceph-osd 及 ceph-rgw 等角色则专门用于创建对应的 Ceph 服务进程。例如，在上面的部署 osd 服务的 Play 中，最核心的角色就是 ceph-osd，可以直接深入该角色的任务集中寻找到相应的创建 osd 的 task。

接下来重点剖析 ceph-ansible 中的 ceph-osd 角色。该角色目录下的 tasks 目录内容如下：

```
[root@master ceph-ansible-4.0.30]# tree roles/ceph-osd/tasks/
roles/ceph-osd/tasks/
├── common.yml
├── container_options_facts.yml
├── crush_rules.yml
├── main.yml
├── openstack_config.yml
├── scenarios
│   ├── lvm-batch.yml
│   └── lvm.yml
├── start_osds.yml
├── systemd.yml
└── system_tuning.yml

1 directory, 10 files
```

下面先看 main.yml 文件（该文件位于角色任务的起点位置），内容如下：

```
# 源码位置: ceph-ansible/roles/ceph-osd/tasks/main.yml
# ...

- name: include_tasks scenarios/lvm.yml
  include_tasks: scenarios/lvm.yml
  when:
    - lvm_volumes|length > 0
    - not rolling_update|default(False) | bool

- name: include_tasks scenarios/lvm-batch.yml
  include_tasks: scenarios/lvm-batch.yml
  when:
    - devices|length > 0
    - not rolling_update|default(False) | bool

- name: include_tasks start_osds.yml
  include_tasks: start_osds.yml

# ...
```

main.yml 文件中定义的任务并不多，因此能容易地定位到上面代码中的 3 个任务。根据名称可知，第 3 个任务是启动 osd 服务，打开 start_osds.yml 文件即能看到该任务的执行过程，于是创建 osd 进程的任务必定在该任务前面。此外，devices 参数正是前面定义了存储节点上相应的数据盘列表，因此原先创建的 osd 服务的过程很有可能就在该任务中。至于为什么要保留该任务的上一个任务，在后续的分析中会解答。

下面是 lvm-batch.yml 文件的内容。

```
# 源码位置: ceph-ansible/roles/ceph-osd/tasks/lvm-batch.yml

---
```

```
- name: "use ceph-volume lvm batch to create {{ osd_objectstore }} osds"
  ceph_volume:
    cluster: "{{ cluster }}"
    objectstore: "{{ osd_objectstore }}"
    batch_devices: "{{ _devices }}"
    dmcrypt: "{{ dmcrypt|default(omit) }}"
    crush_device_class: "{{ crush_device_class|default(omit) }}"
    osds_per_device: "{{ osds_per_device }}"
    journal_size: "{{ journal_size }}"
    block_db_size: "{{ block_db_size }}"
    block_db_devices: "{{ dedicated_devices | unique if dedicated_devices
| length > 0 else omit }}"
    wal_devices: "{{ bluestore_wal_devices | unique if bluestore_wal_
devices | length > 0 else omit }}"
    action: "batch"
  environment:
    CEPH_VOLUME_DEBUG: "{{ ceph_volume_debug }}"
    CEPH_CONTAINER_IMAGE: "{{ ceph_docker_registry + '/' + ceph_docker_
image + ':' + ceph_docker_image_tag if containerized_deployment else
None }}"
    CEPH_CONTAINER_BINARY: "{{ container_binary }}"
    PYTHONIOENCODING: utf-8
  when: _devices | default([]) | length > 0
```

在上面的代码中可以看到一个自定义的 Ansible 模块——ceph_volume。通过该模块传入的参数基本可以断定正是在该模块中创建的 osd 服务。ceph-ansible 中自定义的模块库位置可以从 ceph-ansible/ansible.cfg 文件中找到，具体如下：

```
[defaults]
ansible_managed = Please do not change this file directly since it is managed
by Ansible and will be overwritten
library = ./library
# ...
```

以下是自定义模块 ceph-volume 的具体内容：

```
# 源码位置：ceph-ansible/library/ceph-volume.py
# ...

def run_module():
    module_args = dict(
        cluster=dict(type='str', required=False, default='ceph'),
        objectstore=dict(type='str', required=False, choices=[
                    'bluestore', 'filestore'], default='bluestore'),
        action=dict(type='str', required=False, choices=[
                'create', 'zap', 'batch', 'prepare', 'activate', 'list',
                'inventory'], default='create'),  # noqa 4502
        data=dict(type='str', required=False),
        data_vg=dict(type='str', required=False),
        journal=dict(type='str', required=False),
        journal_vg=dict(type='str', required=False),
        db=dict(type='str', required=False),
        db_vg=dict(type='str', required=False),
        wal=dict(type='str', required=False),
```

```
            wal_vg=dict(type='str', required=False),
            crush_device_class=dict(type='str', required=False),
            dmcrypt=dict(type='bool', required=False, default=False),
            batch_devices=dict(type='list', required=False, default=[]),
            osds_per_device=dict(type='int', required=False, default=1),
            journal_size=dict(type='str', required=False, default='5120'),
            block_db_size=dict(type='str', required=False, default='-1'),
            block_db_devices=dict(type='list', required=False, default=[]),
            wal_devices=dict(type='list', required=False, default=[]),
            report=dict(type='bool', required=False, default=False),
            containerized=dict(type='str', required=False, default=False),
            osd_fsid=dict(type='str', required=False),
            destroy=dict(type='bool', required=False, default=True),
    )

    module = AnsibleModule(
        argument_spec=module_args,
        supports_check_mode=True
    )

    # ...

    # get the desired action
    action = module.params['action']

    # ...

    if action == 'create' or action == 'prepare':
        # 创建 osd
        # ...

        # Prepare or create the OSD
        rc, cmd, out, err = exec_command(
            module, prepare_or_create_osd(module, action, container_image))

    elif action == 'activate':
        # 激活 osd
        # ...

    elif action == 'zap':
        # 擦除 osd
        # ...

    elif action == 'list':
        # ...

    elif action == 'inventory':
        # ...

    elif action == 'batch':
        # Batch prepare AND activate OSDs
        report = module.params.get('report', None)

        # Add --report flag for the idempotency test
        report_flags = [
```

```
            '--report',
            '--format=json',
        ]

        cmd = batch(module, container_image)
        batch_report_cmd = copy.copy(cmd)
        batch_report_cmd.extend(report_flags)

        # 先使用批量的--report 参数，看看会发生什么
        rc, cmd, out, err = exec_command(
            module, batch_report_cmd)
        try:
            report_result = json.loads(out)
        except ValueError:
            # 返回结果不是 JSON 格式的数据
            # ...

        if not report:
            # 如果是真的需要执行创建 osd 服务
            if 'changed' in report_result:
                changed = report_result['changed']
                if changed:
                    # 批量创建 osd 服务，不再带--report 参数
                    rc, cmd, out, err = exec_command(
                        module, batch(module, container_image))
            else:
                # 批量创建 osd 服务，不再带--report 参数
                rc, cmd, out, err = exec_command(
                    module, batch(module, container_image))
        else:
            cmd = batch_report_cmd

    else:
        module.fail_json(
            msg='State must either be "create" or "prepare" or "activate" or
"list" or "zap" or "batch" or "inventory".', changed=False, rc=1)  # noqa
E501

    # 准备返回数据
    # ...


def main():
    run_module()


if __name__ == '__main__':
    main()
```

前面我们已经分析过不少 Ansible 的模块源码了，因此对自定义模块 ceph_volume 源码的分析应该轻车熟路了。先看一下 main()方法，然后进入 run_module()方法，会看到十分熟悉的 create、prepare、activate 等选项，这些正是 ceph-volume 工具的子命令。这里以用于创建 osd 服务的 create 和 prepare 选项为例分析，执行代码如下：

```python
# 源码位置: ceph-ansible/library/ceph-volume.py
# ...

def build_cmd(action, container_image, cluster='ceph', binary='ceph-volume'):
    '''
    构建 ceph-volume 命令
    '''

    _binary = binary

    if container_image:
        # 容器化
        cmd = container_exec(
            binary, container_image)
    else:
        binary = [binary]
        cmd = binary

    if _binary == 'ceph-volume':
        cmd.extend(['--cluster', cluster])

    cmd.extend(action)

    return cmd

# ...

def prepare_or_create_osd(module, action, container_image):
    '''
    生成创建 osd 进程的命令并返回
    '''

    # 获取模块参数
    cluster = module.params['cluster']
    objectstore = module.params['objectstore']
    data = module.params['data']
    data_vg = module.params.get('data_vg', None)
    data = get_data(data, data_vg)
    journal = module.params.get('journal', None)
    journal_vg = module.params.get('journal_vg', None)
    db = module.params.get('db', None)
    db_vg = module.params.get('db_vg', None)
    wal = module.params.get('wal', None)
    wal_vg = module.params.get('wal_vg', None)
    crush_device_class = module.params.get('crush_device_class', None)
    dmcrypt = module.params.get('dmcrypt', None)

    # 构建执行命令
    action = ['lvm', action]
    cmd = build_cmd(action, container_image, cluster)
    cmd.extend(['--%s' % objectstore])
    cmd.append('--data')
    cmd.append(data)
```

```
    if journal and objectstore == 'filestore':
        journal = get_journal(journal, journal_vg)
        cmd.extend(['--journal', journal])

    if db and objectstore == 'bluestore':
        db = get_db(db, db_vg)
        cmd.extend(['--block.db', db])

    if wal and objectstore == 'bluestore':
        wal = get_wal(wal, wal_vg)
        cmd.extend(['--block.wal', wal])

    if crush_device_class:
        cmd.extend(['--crush-device-class', crush_device_class])

    if dmcrypt:
        cmd.append('--dmcrypt')

    return cmd
# ...
```

看完上述代码，读者应该对 ceph-volume 模块的功能有了一定的了解了。build_cmd()
方法首先组成了 ceph-volume 的前半部分命令，而 prepare_or_create_osd()方法则根据相关
参数结合前面的命令一起组合成最终执行创建 osd 服务的命令。

对上面自定义的 ceph-volume 模块有了初步了解后，还需要厘清在 ceph-ansible 中创
建 osd 进程的过程。在 ceph-ansible 中，存储磁盘的列表定义在 devices 变量（group_vars/
all.yml）中，在 ceph-osd 角色的任务集中会调用 ceph-volume 模块的任务去创建 osd 进程。
注意，ceph-osd/tasks/scenarios/lvm-batch.yml 文件中只定义了唯一的任务，该任务使用自
定义的 ceph-volume 模块且其 action 参数值为"batch"，因此该任务是调用 ceph-volume 模
块源码中的 batch()函数得到批量创建 osd 进程的命令，最后执行这个命令去创建 osd 进程。
具体代码如下：

```
# 源码位置：ceph-ansible/library/ceph-volume.py
# ...

def batch(module, container_image):
    '''
    批量创建 osd 进程
    '''

    # 获取模块变量
    cluster = module.params['cluster']
    objectstore = module.params['objectstore']
    batch_devices = module.params.get('batch_devices', None)
    crush_device_class = module.params.get('crush_device_class', None)
    journal_size = module.params.get('journal_size', None)
    block_db_size = module.params.get('block_db_size', None)
    block_db_devices = module.params.get('block_db_devices', None)
    wal_devices = module.params.get('wal_devices', None)
```

```
    dmcrypt = module.params.get('dmcrypt', None)
    osds_per_device = module.params.get('osds_per_device', 1)

    # 处理一些异常情况
    if not osds_per_device:
        fatal('osds_per_device must be provided if action is "batch"',
module)

    if osds_per_device < 1:
        fatal('osds_per_device must be greater than 0 if action is "batch"',
module)  # noqa E501

    if not batch_devices:
        fatal('batch_devices must be provided if action is "batch"', module)

    # 合成相应的命令
    action = ['lvm', 'batch']
    cmd = build_cmd(action, container_image, cluster)
    cmd.extend(['--%s' % objectstore])
    cmd.append('--yes')

    if container_image:
        cmd.append('--prepare')

    if crush_device_class:
        cmd.extend(['--crush-device-class', crush_device_class])

    if dmcrypt:
        cmd.append('--dmcrypt')

    if osds_per_device > 1:
        cmd.extend(['--osds-per-device', str(osds_per_device)])

    if objectstore == 'filestore':
        cmd.extend(['--journal-size', journal_size])

    if objectstore == 'bluestore' and block_db_size != '-1':
        cmd.extend(['--block-db-size', block_db_size])

    cmd.extend(batch_devices)

    if block_db_devices and objectstore == 'bluestore':
        cmd.append('--db-devices')
        cmd.extend(block_db_devices)

    if wal_devices and objectstore == 'bluestore':
        cmd.append('--wal-devices')
        cmd.extend(wal_devices)

    return cmd

# ...
```

参数 batch_devices 为 osd 的数据盘列表。通过查看上面的代码，可以发现这样一条语句：cmd.extend(batch_devices)。该语句使笔者无法像前面那样手动为每个创建 osd 进程的 ceph-volume 命令添加--osd-id 参数。那么该如何将批量创建 osd 进程改造成一个个地创建呢？这里就用到了批处理任务的上一个任务，具体代码如下：

```
# 源码位置：ceph-ansible/roles/ceph-osd/tasks/main.yml
# ...

- name: include_tasks scenarios/lvm.yml
  include_tasks: scenarios/lvm.yml
  when:
    - lvm_volumes|length > 0
    - not rolling_update|default(False) | bool

# ...
```

继续看 lvm.yml 文件的内容，具体如下：

```
# 源码位置：ceph-ansible/roles/ceph-osd/tasks/scenarios/lvm.yml
---
- name: "use ceph-volume to create {{ osd_objectstore }} osds"
  ceph_volume:
    cluster: "{{ cluster }}"
    objectstore: "{{ osd_objectstore }}"
    data: "{{ item.data }}"
    data_vg: "{{ item.data_vg|default(omit) }}"
    journal: "{{ item.journal|default(omit) }}"
    journal_vg: "{{ item.journal_vg|default(omit) }}"
    db: "{{ item.db|default(omit) }}"
    db_vg: "{{ item.db_vg|default(omit) }}"
    wal: "{{ item.wal|default(omit) }}"
    wal_vg: "{{ item.wal_vg|default(omit) }}"
    crush_device_class: "{{ item.crush_device_class|default(omit) }}"
    dmcrypt: "{{ dmcrypt|default(omit) }}"
    action: "{{ 'prepare' if containerized_deployment else 'create' }}"
  environment:
    CEPH_VOLUME_DEBUG: "{{ ceph_volume_debug }}"
    CEPH_CONTAINER_IMAGE: "{{ ceph_docker_registry + '/' + ceph_docker_
image + ':' + ceph_docker_image_tag if containerized_deployment else
None }}"
    CEPH_CONTAINER_BINARY: "{{ container_binary }}"
    PYTHONIOENCODING: utf-8
  with_items: "{{ lvm_volumes }}"
  tags: prepare_osd
```

这里使用的依旧是自定义的 ceph_volume 模块。注意，该任务使用了 with_items 语法，类似于编程中的 for 循环，lvm_volumes 参数用于指定 osd 的数据盘，这不正是前面想要单独创建一个个 osd 进程的地方吗？ lvm_volumes 参数的示例写法如下：

```
# 源码位置：ceph-ansible/group_vars/osds.yml.sample
# ...


# Use ceph-volume to create OSDs from logical volumes.
# lvm_volumes is a list of dictionaries.
```

```
# ...
# Bluestore: Each dictionary must contain at least data. When defining wal or
# db, it must have both the lv name and vg group (db and wal are not required).
# This allows for four combinations: just data, data and wal, data and wal and
# db, data and db.
# For example:
# lvm_volumes:
#   - data: data-lv1
#     data_vg: vg1
#     wal: wal-lv1
#     wal_vg: vg1
#     crush_device_class: foo
#   - data: data-lv2
#     db: db-lv2
#     db_vg: vg2
#   - data: data-lv3
#     wal: wal-lv1
#     wal_vg: vg3
#     db: db-lv3
#     db_vg: vg3
#   - data: data-lv4
#     data_vg: vg4
#   - data: /dev/sda
#   - data: /dev/sdb1

#lvm_volumes: []
#crush_device_class: ""
#osds_per_device: 1
```

很明显，在当前的硬件环境下，lvm_volumes 参数的写法如下：

```
lvm_volumes:
  - data: /dev/sdb
  - data: /dev/sdc
  - data: /dev/sdd
```

读者可以尝试使用该参数代替前面的 devices 参数去创建 osd 进程。该参数对应的任务正是根据设置的磁盘数一个个创建 osd 进程，且该任务在 ceph-volume 模块源码中正是调用 prepare_or_create_osd()方法来生成创建 osd 的命令。因此，想要实现创建的 osd 进程的 id 值连续就需要从该方法入手。

那么如何在 prepare_or_create_osd()方法中添加--osd-id 参数以及如何确定该参数的值呢？在前面手工创建 osd 的步骤中，笔者使用了 sed 命令根据主机名获取相应的节点编号，然后计算当前节点 osd-id 编号的起始值。对于节点上的磁盘而言，就非常好划分，假设每个节点上有 10 块数据盘，一个 osd 进程对应一个数据盘，则第 6 个存储节点的 osd-id 的起始值为 10 * (6-1) = 50。此外，该节点上第 5 块盘对应的 osd 进程的 id 值应为 50 + (5-1) = 54。因此，从上面的分析可知，如果想要给执行的 ceph-volume 命令添加--osd-id 参数并得到该参数的值，必须先得到以下三个值：

- 获取 OSD 节点编号，通常根据主机名获取。例如 ceph-1，表示安装 osd 进程的第一个节点，即 host_id=1。因此，这里会涉及业务相关的信息。

- 获取每个节点上数据盘数目（osds_per_node），所有的存储节点需保持一致，否则会出现 osd-id 不连续甚至冲突的情况。
- 获取当前节点创建 osd 进程所使用数据盘的位置（current_num）。

下面做一个简单说明：假设本次创建的 osd 进程将使用 ceph-2 节点上的 sdc 盘作为数据盘，那么对应的 osd-id 值应该是多少呢？节点的编号为 2（host_id=2），每个节点上有 3 块盘（osds_per_node=3），ceph-2 上的磁盘 sdc 是第 2 块盘（current_num=1，从 0 开始），因此对应 osd 进程的 id 值应为(host_id−1) * osds_per_node + current_num=4。

因此，为了能将上面这些信息传入 ceph_volume 模块中，笔者对原来的 task 进行了调整并添加了 3 个参数：

```yaml
# 源码位置：ceph-ansible/roles/ceph-osd/tasks/scenarios/lvm.yml
---
- name: "use ceph-volume to create {{ osd_objectstore }} osds"
  ceph_volume:
    cluster: "{{ cluster }}"
    objectstore: "{{ osd_objectstore }}"
    data: "{{ item.data }}"
    data_vg: "{{ item.data_vg|default(omit) }}"
    journal: "{{ item.journal|default(omit) }}"
    journal_vg: "{{ item.journal_vg|default(omit) }}"
    db: "{{ item.db|default(omit) }}"
    db_vg: "{{ item.db_vg|default(omit) }}"
    wal: "{{ item.wal|default(omit) }}"
    wal_vg: "{{ item.wal_vg|default(omit) }}"
    crush_device_class: "{{ item.crush_device_class|default(omit) }}"
    dmcrypt: "{{ dmcrypt|default(omit) }}"
    action: "{{ 'prepare' if containerized_deployment else 'create' }}"
    # 添加 3 个参数，用于创建连续的 osd-id
    host_regex: "{{ host_regex }}"
    osds_per_node: "{{ lvm_volumes|length }}"
    current_num: "{{ index }}"
  environment:
    CEPH_VOLUME_DEBUG: "{{ ceph_volume_debug }}"
CEPH_CONTAINER_IMAGE: "{{ ceph_docker_registry + '/' + ceph_docker_image
+ ':' + ceph_docker_image_tag if containerized_deployment else None }}"
    CEPH_CONTAINER_BINARY: "{{ container_binary }}"
    PYTHONIOENCODING: utf-8
  # 修改为 loop 语法以便能获取循环的索引值
  loop: "{{ lvm_volumes }}"
  loop_control:
    index_var: index
  tags: prepare_osd
```

在模块中添加了 3 个参数，意味着在模块的源码中就需要接收这 3 个参数。下面来看调整后的 ceph_volume 模块源码：

```python
# 代码位置：library/ceph_volume.py
# ...

def run_module():
```

```
    module_args = dict(
        # ...
        # 添加 3 个新的参数
        host_regex=dict(type='str', required=False),
        osds_per_node=dict(type='int', required=False, default=1),
        current_num=dict(type='int', required=False, default=0),
    )

# ...
```

接着给 prepare_or_create_osd()方法生成的 ceph-volume 命令添加--osd-id 参数及其值：

```
# 代码位置：library/ceph_volume.py
#   -*- coding: utf-8  -*-
import re
# ...

def prepare_or_create_osd(module, action, container_image):
    # ...

    # 添加的--osd_id 选项及参数，如果需要自行添加 osd-id 选项，需要修改 ceph-volume
      中的代码
    osd_id = get_osd_id(module)
    if osd_id >= 0:
        # 注意，这里不能输入整数，必须是字符串形式，否则执行命令时会报错
        cmd.extend(['--osd-id', str(osd_id)])

    return cmd

def get_osd_id(module):
    pattern = module.params.get('host_regex', None)
    osds_per_node = module.params.get('osds_per_node', 1)
    current_num = module.params.get('current_num', 0)
    # 通过主机名获取相应顺序
    rc, out, _ = module.run_command('hostname')
    if rc == 0 and pattern:
        hostname = out.rstrip("\r\n")
        m = re.search(pattern, hostname, flags=0)
        return (int(m.group(1)) - 1) * osds_per_node + current_num if m else -1
    # 注意不能用 None 返回并用于 if 判断，因为 0 和 None 的判断可能会混淆
    return -1
```

注意：记得要导入 re 模块，并且在导入语句的前一行一定要指定文件编码，避免中文注释报错。

此外，还需要调整 ceph-volume 模块的代码，避免添加的--osd-id 参数在执行时报错。可以将前面修正 ceph-volume 代码的 shell 命令写成 tasks 并放在 ceph_volume 模块执行之前，具体代码如下：

```
# 源码位置：ceph-ansible/roles/ceph-osd/tasks/scenarios/lvm.yml
---
# 添加 osd-id 参数，为使执行生效，需要修改 ceph-volume 模块中的一段代码
- name: find python file to modify
```

```
  shell: "find /usr/lib -name prepare.py -type f | grep ceph_volume | grep
util"
  register: path
  become: true

- name: set file_path fact
  set_fact:
    file_path: "{{ path.stdout }}"

- name: fix the bug prepare.py
  shell: "sed -i 's/if osd and/if not osd or/g' {{ file_path }}"
  become: true
```

接着要在 group_vars/all.yml 中填写相应的参数值，主要是 lvm_volumes 及相应的获取 DN 节点编号的正则表达式 host_regex，具体代码如下：

```
# 源码位置: ceph-ansible/group_vars/all.yml
# ...

#devices:
#  - /dev/sdb
#  - /dev/sdc
#  - /dev/sdd
host_regex: ceph-([0-9]+)
lvm_volumes:
  - data: /dev/sdb
  - data: /dev/sdc
  - data: /dev/sdd

# ...
```

⚠注意：一定要去掉原先的 devices 参数值，因为 devices 和 lvm_volumes 分别表示两种不同的格式化磁盘的方式。

最后使用 ansible-playbook 命令运行创建集群的 Playbook：

```
(ansible2.8.16) [root@master ceph-ansible-4.0.30]# ansible-playbook -i
hosts site.yml
```

大约几分钟后，就可以看到 Playbook 运行成功的输出提示了。接着登录到 ceph-1 节点执行 ceph osd tree 命令查看 osd 进程的 ID 分布，结果如下：

```
[root@ceph-1 ~]# ceph osd tree
ID CLASS WEIGHT  TYPE NAME      STATUS REWEIGHT PRI-AFF
-1       0.26367 root default
-3       0.08789     host ceph-1
 0   hdd 0.02930         osd.0      up  1.00000 1.00000
 1   hdd 0.02930         osd.1      up  1.00000 1.00000
 2   hdd 0.02930         osd.2      up  1.00000 1.00000
-5       0.08789     host ceph-2
 3   hdd 0.02930         osd.3      up  1.00000 1.00000
 4   hdd 0.02930         osd.4      up  1.00000 1.00000
 5   hdd 0.02930         osd.5      up  1.00000 1.00000
-7       0.08789     host ceph-3
 6   hdd 0.02930         osd.6      up  1.00000 1.00000
```

```
    7    hdd 0.02930        osd.7      up  1.00000 1.00000
    8    hdd 0.02930        osd.8      up  1.00000 1.00000
```

可以看到，结果正好符合预期，表明这次改造的目的达到了。

以上是对部署工具 ceph-ansible 的探索和改造过程。通过一个简单的改造需求，往往可以让使用者进一步理解 ceph-ansible 的运行过程，但这还远远不够。接下来将深度解析和改造 ceph-ansible 项目，去掉臃肿的判断，实现以最小的 Playbook 部署 Ceph 集群。通过这个过程，读者可以完全掌握 ceph-ansible。

6.4　ceph-ansible 项目源码探索

本节将分析 ceph-ansible 工具的源码，同时结合项目场景对 ceph-ansible 项目进行进一步的学习和改造。一方面，通过学习 ceph-ansible 的代码，可以掌握不少 Playbook 的优秀写法；另一方面，能够从 ceph-ansible 中梳理出 Ceph 集群部署 MON、OSD 和 RGW 等服务的步骤，为后续实现自己的自动化部署工具打好基础。

6.4.1　追踪 ceph-mon 服务的搭建过程

6.3 节中的内容涉及 ceph-ansible 中的 Playbook 文件及自定义的模块源码。ceph-ansible 项目中最核心的部分就是 roles 目录下定义的角色了。roles 目录下共有 24 个角色目录，每种角色代表某一类的功能。例如 ceph-common 角色，主要是给全部的节点装上搭建集群所依赖的包，并且对于搭建不同服务的节点安装相应的 Ceph 包，ceph-mon 则是专门用于搭建 ceph-mon 服务的角色。在 6.3 节中，笔者简单变动了搭建 osd 服务的过程，实现了一个特定的需求。而在本节中，笔者将追踪 ceph-mon 服务的部署过程，并以 CentOS 系统和非容器化部署方式为例，整理部署 ceph-mon 服务的全过程，最终完成一个简易的 Playbook。

ceph-ansible 的运行入口依旧是 site.yml 文件（从 site.yml.sample 复制过来），部署剧本集的内容如下：

```
---

- hosts:
  - mons
  - osds
  - mdss
  - rgws
  - nfss
  - rbdmirrors
  - clients
  - mgrs
  - iscsigws
  - iscsi-gws # for backward compatibility only!
  - grafana-server
```

```yaml
    - rgwloadbalancers

  gather_facts: false
  any_errors_fatal: true
  become: true

  tags: always

  vars:
    delegate_facts_host: True

  pre_tasks:
    # 如果远端系统没有 Python，那么后面的执行会保存。这里的检查将直接失败
    - import_tasks: raw_install_python.yml

    - name: gather facts
      setup:
        gather_subset:
          - 'all'
          - '!facter'
          - '!ohai'
      when:
        - not delegate_facts_host | bool or inventory_hostname in groups.get
(client_group_name, [])

    - name: gather and delegate facts
      setup:
        gather_subset:
          - 'all'
          - '!facter'
          - '!ohai'
      delegate_to: "{{ item }}"
      delegate_facts: True
      with_items: "{{ groups['all'] | difference(groups.get('clients',
[])) }}"
      run_once: true
      when: delegate_facts_host | bool

  # 核心角色
  tasks:
    - import_role:
        name: ceph-defaults
    - import_role:
        name: ceph-facts
    - import_role:
        name: ceph-validate
    - import_role:
        name: ceph-infra
    - import_role:
        name: ceph-common

# 部署 MON 节点
- hosts: mons
  gather_facts: false
  become: True
```

```yaml
      any_errors_fatal: true
      pre_tasks:
        - name: set ceph monitor install 'In Progress'
          run_once: true
          set_stats:
            data:
              installer_phase_ceph_mon:
                status: "In Progress"
                start: "{{ lookup('pipe', 'date +%Y%m%d%H%M%SZ') }}"

      tasks:
        - import_role:
            name: ceph-defaults
          tags: ['ceph_update_config']
        - import_role:
            name: ceph-facts
          tags: ['ceph_update_config']
        - import_role:
            name: ceph-handler
          tags: ['ceph_update_config']
        - import_role:
            name: ceph-config
          tags: ['ceph_update_config']
        - import_role:
            name: ceph-mon
      # 如果没有指定部署 MGR 节点，则在 MON 节点上同时部署 ceph-mgr 服务
        - import_role:
            name: ceph-mgr
          when: groups.get(mgr_group_name, []) | length == 0

      post_tasks:
        - name: set ceph monitor install 'Complete'
          run_once: true
          set_stats:
            data:
              installer_phase_ceph_mon:
                status: "Complete"
                end: "{{ lookup('pipe', 'date +%Y%m%d%H%M%SZ') }}"

# 部署其他角色
# ...

# 最后检查集群搭建情况
- hosts: mons
  gather_facts: false
  become: True
  any_errors_fatal: true
  tasks:
    - import_role:
        name: ceph-defaults
    - name: get ceph status from the first monitor
      command: ceph --cluster {{ cluster }} -s
      register: ceph_status
      changed_when: false
      delegate_to: "{{ groups[mon_group_name][0] }}"
```

```
    run_once: true

  - name: "show ceph status for cluster {{ cluster }}"
    debug:
      msg: "{{ ceph_status.stdout_lines }}"
    delegate_to: "{{ groups[mon_group_name][0] }}"
    run_once: true
```

上面的执行流程非常清晰，ceph-ansible 先是在所有节点上执行公共的角色剧本，完成加载统一的主机变量、安装各自公共依赖包以及相应的 Ceph 包、配置好相应的防火墙及信息校验等工作，确保初始工作顺利、有序地执行。然后按节点组依次执行一个或者多个角色的 Playbook，这里会有一些先后顺序的要求。最后在集群搭建完毕后，又回到 MON 节点上执行 ceph -s 命令，输出集群的状态。需要注意 mons 组的节点导入 ceph-mgr 角色的前提是没有定义 mgrs 组。也就是说，当搭建集群的 ceph-mon 服务时，一定要在集群中同时部署 ceph-mgr 服务。如果没有指定相应的 MGR 节点，则会在 MON 节点上同时部署 ceph-mgr 服务。

为了能整理出 ceph-mon 搭建的所有必要步骤，下面逐个介绍这些公共角色的 Playbook。为了简单起见，直接去掉非 CentOS 系统及容器化部署要执行的 tasks。

1. ceph-default角色

ceph-default 角色的 tasks 目录下并没有任何 YAML 文件，该角色的目的只是用来加载 defaults 目录下 main.yml 文件中的全部变量。

2. ceph-facts角色

ceph-facts 角色主要是设置节点的各种变量，因此这里会大量使用 set_fact 语法。这里不会介绍设置的全部变量，只是挑出后续会用到的变量进行说明。先看下面一组 tasks：

```
# 源码位置: ceph-ansible/roles/ceph-facts/tasks/facts.yml
# ...

- name: fsid related tasks
  when:
    - generate_fsid | bool
    - ceph_current_status.fsid is undefined
    - not rolling_update | bool
  block:
  - name: generate cluster fsid
    command: "{{ hostvars[groups[mon_group_name][0]]['discovered_interpreter_
python'] }} -c 'import uuid; print(str(uuid.uuid4()))'"
    register: cluster_uuid
    delegate_to: "{{ groups[mon_group_name][0] }}"
    run_once: true

  - name: set_fact fsid
    set_fact:
      fsid: "{{ cluster_uuid.stdout }}"
```

```
# ...
```

上面的 tasks 主要是生成一个随机的 ID 值作为后续创建集群的 ID 并将该值赋给 fsid 变量。从这里可以看到，集群的 fsid 生成的方式竟然如此简单，直接使用 Python 中的 uuid 模块即可。

```
>>> import uuid
>>> print(str(uuid.uuid4()))
65c960f8-9266-4eb3-8cb6-134f2815d6da
```

再来看后面的几个 tasks，用于设置 NTP 服务名称。这里为了兼容各类 Linux，进行了系统类型判断，具体代码如下：

```
# 源码位置：ceph-ansible/roles/ceph-facts/tasks/facts.yml
# ...

- name: set ntp service name depending on OS family
  block:
  - name: set ntp service name for Debian family
    set_fact:
      ntp_service_name: ntp
    when: ansible_os_family == 'Debian'

  - name: set ntp service name for Red Hat family
    set_fact:
      ntp_service_name: ntpd
    when: ansible_os_family in ['RedHat', 'Suse']

# ...
```

最后再来看一组较为复杂的 tasks，它位于 ceph-facts/tasks/set_monitor_address.yml 文件中，内容如下：

```
# 源码位置：ceph-ansible/roles/ceph-facts/tasks/set_monitor_address.yml
# ...

- name: set_fact _monitor_address to monitor_interface - ipv4
  set_fact:
    _monitor_addresses: "{{ _monitor_addresses | default([]) + [{ 'name':
item, 'addr': hostvars[item]['ansible_' + (hostvars[item]['monitor_
interface']|replace('-', '_'))][ip_version]['address'] | ipwrap }] }}"
  with_items: "{{ groups.get(mon_group_name, []) }}"
  when:
    - "item not in _monitor_addresses | default([]) | selectattr('name',
'defined') | map(attribute='name') | list"
    - ip_version == 'ipv4'
    - hostvars[item]['monitor_address_block'] | default('subnet') ==
'subnet'
    - hostvars[item]['monitor_address'] | default('x.x.x.x') == 'x.x.x.x'
    - hostvars[item]['monitor_interface'] | default('interface') !=
'interface'

# ...
```

在 6.2 节中，笔者只设置了 monitor_interface 值为相应的网卡名，而 monitor_address_block 和 monitor_address 没有设置，它们的默认值如下：

```
# 源码位置：ceph-ansible/roles/ceph-defaults/defaults/main.yml
# ...

# 会被 group_vars/all.yml 中的 monitor_interface 值覆盖
monitor_interface: interface
monitor_address: x.x.x.x
monitor_address_block: subnet

# ...
```

为了帮助读者理解上面的语法，需要知道 setup 模块的输出结果，首先执行如下命令：

```
(ansible2.8.16) [root@master ~]# ansible ceph-1 -i hosts -m setup > setup.txt
```

先看 hostvars[item]['monitor_interface']|replace('-', '_')语句的含义，其中，item 表示 mons 组下的节点，以 ceph-1 为例，这个值的最终结果为 ens33，也就是设置的网卡接口变量值。于是 hostvars[item]['ansible_' + (hostvars[item]['monitor_interface']|replace('-', '_'))][ip_version]['address'] | ipwrap }]表示的是 hostvars[item]['ansible_ens33']['ipv4']的值。根据前面得到的 ceph-1 节点的探测结果（保存在 setup.exe 中），找到 ansible_ens33 对应的数据：

```
#setup.exe 中的内容
"ansible_facts": {
    ...

    "ansible_ens33": {
        "active": true,
        "device": "ens33",
        "features": {
            ...
        },
        "ipv4": {
            "address": "192.168.26.120",
            "broadcast": "192.168.26.255",
            "netmask": "255.255.255.0",
            "network": "192.168.26.0"
        },

    ...

    }

}
```

由以上代码可知，第一次循环后 _monitor_addresses 的值为[{ 'name': 'ceph-1', 'addr':'192.168.26.120'}]，第二次循环后，该值又加上了 ceph-2 节点的地址信息为[{ 'name': 'ceph-1', 'addr':'192.168.26.120'},{ 'name': 'ceph-2', 'addr':'192.168.26.121'}]。最终，_monitor_addresses 的值为 mons 组下所有主机的地址信息集合。在本次的部署环境下，得到的 _monitor_addresses 的值如下：

```
[{ 'name': 'ceph-1', 'addr':'192.168.26.120'},{ 'name': 'ceph-2', 'addr':
'192.168.26.121'},{ 'name': 'ceph-3', 'addr':'192.168.26.122'}]
```

为了验证结果，可以简单地在该任务后加上两个 task，一个用于打印_monitor_addresses 的结果，另一个用于停止运行 ceph-ansible，具体代码如下：

```yaml
# 源码位置：ceph-ansible/roles/ceph-facts/tasks/set_monitor_address.yml
# ...

- name: set_fact _monitor_address to monitor_interface - ipv4
  set_fact:
    _monitor_addresses: "{{ _monitor_addresses | default([]) + [{ 'name':
item, 'addr': hostvars[item]['ansible_' + (hostvars[item]['monitor_
interface']|replace('-', '_'))][ip_version]['address'] | ipwrap }] }}"
  with_items: "{{ groups.get(mon_group_name, []) }}"
  when:
    - "item not in _monitor_addresses | default([]) | selectattr('name',
'defined') | map(attribute='name') | list"
    - ip_version == 'ipv4'
    - hostvars[item]['monitor_address_block'] | default('subnet') ==
'subnet'
    - hostvars[item]['monitor_address'] | default('x.x.x.x') == 'x.x.x.x'
    - hostvars[item]['monitor_interface'] | default('interface') !=
'interface'

# 打印_monitor_addresses 变量结果
- name: debug _monitor_addresses
  debug:
    msg: _monitor_addresses = "{{ _monitor_addresses }}"

# 停止运行
- name: stop here
  fail:
    msg: "now we stop here and see _monitor_addresses"

# ...
```

可以直接执行 ceph-ansible，最后会停在上面的 stop here 任务处。相应的结果如下：

```
(ansible2.8.16) [root@master ceph-ansible-4.0.30]# ansible-playbook -i
hosts site.yml
...

TASK [ceph-facts : debug _monitor_addresses]
**********************************************************************
Thursday 12 November 2020  09:06:03 +0800 (0:00:00.474)   0:00:30.022 *****
ok: [ceph-1] =>
  msg: '_monitor_addresses = "[{''name'': ''ceph-1'', ''addr'': ''192.168.
26.120''}, {''name'': ''ceph-2'', ''addr'': ''192.168.26.121''}, {''name'':
''ceph-3'', ''addr'': ''192.168.26.122''}]"'
ok: [ceph-2] =>
  msg: '_monitor_addresses = "[{''name'': ''ceph-1'', ''addr'': ''192.168.
26.120''}, {''name'': ''ceph-2'', ''addr'': ''192.168.26.121''}, {''name'':
''ceph-3'', ''addr'': ''192.168.26.122''}]"'
ok: [ceph-3] =>
```

```
  msg: '_monitor_addresses = "[{''name'': ''ceph-1'', ''addr'': ''192.168.
26.120''}, {''name'': ''ceph-2'', ''addr'': ''192.168.26.121''}, {''name'':
''ceph-3'', ''addr'': ''192.168.26.122''}]"'

TASK [ceph-facts : stop here]
*********************************************************************
Thursday 12 November 2020  09:06:03 +0800 (0:00:00.265)   0:00:30.287 *****
fatal: [ceph-1]: FAILED! => changed=false
  msg: now we stop here and see _monitor_addresses
fatal: [ceph-2]: FAILED! => changed=false
  msg: now we stop here and see _monitor_addresses
fatal: [ceph-3]: FAILED! => changed=false
  msg: now we stop here and see _monitor_addresses
```

这样就验证了前面的分析结果，这种调试方式在 Playbook 中十分有用。

3．ceph-validate角色

根据 ceph-validate 的名称就能知道该角色的作用，即用于校验各种数据的有效性。例如，校验该 ceph-ansible 版本支持的 Ansible 版本、系统版本及 CPU 架构等，这部分校验位于 ceph-validate 目录下的 check_system.yml 文件中。具体代码如下：

```
# 源码位置：ceph-ansible/roles/ceph-validate/tasks/check_system.yml

---
# 校验 Ansible 的大版本，必须是 2 和 2 以上的版本
- name: fail on unsupported ansible version (1.X)
  fail:
    msg: "Ansible version must be >= 2.x, please update!"
  when: ansible_version.major|int < 2

# 校验 Ansible 的小版本，必须是 2.8 或者 2.9 版本
- name: fail on unsupported ansible version
  fail:
    msg: "Ansible version must be 2.8 or 2.9!"
  when:
    - ansible_version.major|int == 2
    - ansible_version.minor|int not in [8, 9]

# 校验操作系统，非 Linux 系统不支持
- name: fail on unsupported system
  fail:
    msg: "System not supported {{ ansible_system }}"
  when: ansible_system not in ['Linux']

# 校验 Ceph 节点的架构
- name: fail on unsupported architecture
  fail:
    msg: "Architecture not supported {{ ansible_architecture }}"
  when: ansible_architecture not in ['x86_64', 'ppc64le', 'armv7l', 'aarch64']

- name: fail on unsupported distribution
  fail:
```

```
      msg: "Distribution not supported {{ ansible_os_family }}"
    when: ansible_os_family not in ['Debian', 'RedHat', 'ClearLinux', 'Suse']

# 校验 RedHat Linux 系统版本
- name: red hat based systems tasks
  when:
    - ceph_repository == 'rhcs'
    - ansible_distribution == 'RedHat'
  block:
    - name: fail on unsupported distribution for red hat ceph storage
      fail:
        msg: "Distribution not supported {{ ansible_distribution_version }}
by Red Hat Ceph Storage, only RHEL 8 (>= 8.1) or RHEL 7 (>= 7.7)"
      when: (ansible_distribution_major_version | int == 8 and ansible_
distribution_version is version('8.1', '<')) or
            (ansible_distribution_major_version | int == 7 and ansible_
distribution_version is version('7.7', '<'))

    - name: subscription manager related tasks
      when: ceph_repository_type == 'cdn'
      block:
        - name: determine if node is registered with subscription-manager
          command: subscription-manager identity
          register: subscription
          changed_when: false
          failed_when: false
          check_mode: no

        - name: fail on unregistered red hat rhcs linux
          fail:
            msg: "You must register your machine with subscription-manager"
          when: subscription.rc != 0

- name: fail on unsupported distribution for ubuntu cloud archive
  fail:
    msg: "Distribution not supported by Ubuntu Cloud Archive: {{ ansible_
distribution }}"
  when:
    - ceph_repository == 'uca'
    - ansible_distribution != 'Ubuntu'

- name: "fail on unsupported openSUSE distribution (only 15.x supported)"
  fail:
    msg: "Distribution not supported: {{ ansible_distribution }}"
  when:
    - ansible_distribution == 'openSUSE Leap'
    - ansible_distribution_major_version != '15'

- name: fail if systemd is not present
  fail:
    msg: "Systemd must be present"
  when: ansible_service_mgr != 'systemd'
```

上面的代码中大量使用了 fail 模块和 when 语法，任务逻辑非常清晰。下面依据介绍 ceph-facts 角色时使用 setup 模块探测的主机结果来进行校验，例如校验系统大类

（ansible_system）和系统版本（ansible_os_family）等。其他的校验包括参数是否设置校验、osds 组中的磁盘参数及磁盘校验、NTP 服务类型校验等，但这些校验在部署过程中一般会去掉，即对于一些已经确定的情况，避免再次校验浪费时间。由于部署时大部分组下并没有设置相应的节点，所以在后续的校验任务中会被跳过，没有执行。

4．ceph-infra角色

ceph-infra 角色非常简单，里面的 tasks 也较少。下面是该角色的几个核心任务，具体代码如下：

```
# 源码位置: ceph-ansible/roles/ceph-infra/tasks/main.yml
# ...

- name: include_tasks configure_firewall.yml
  include_tasks: configure_firewall.yml
  when:
    - configure_firewall | bool
    - ansible_os_family in ['RedHat', 'Suse']
  tags: configure_firewall

- name: include_tasks setup_ntp.yml
  include_tasks: setup_ntp.yml
  when: ntp_service_enabled | bool
  tags: configure_ntp

# 容器化部署任务
# ...
```

第一个任务是转向配置防火墙的 Playbook 文件，第二个任务是转向设置 NTP 服务的 Playbook 文件。因此，ceph-infra 角色的主要功能是搭建防火墙和 NTP 服务这两个基础设施。从 when 语法中可以看到，这两个任务是否执行，分别由 configure_firewall 和 ntp_service_enabled 参数来控制。在笔者的公司实施的集群部署中，这两个参数都设置为 False，即不会执行相应的任务，这是因为生产环境的主机都不会联网，无须配置防火墙，额外的防火墙配置可能会引发各种问题。此外，对于集群内节点的时钟同步并不会使用这里的 NTP 服务来完成，因为对于服务而言，经常会出现"挂死"现象，所以部署时一般会选择使用系统的 crontab 服务来定时同步跳板机时钟，这样的方式简单且有效。

5．ceph-common角色

ceph-common 角色中主要是安装 Ceph 的相关依赖包。这里继续从 tasks 目录下的 main.yml 文件开始，找出 CentOS 系统需要执行的相应任务，具体代码如下：

```
# 源码位置: ceph-ansible/roles/ceph-common/tasks/main.yml
---
- name: include_tasks installs/install_on_redhat.yml
  include_tasks: installs/install_on_redhat.yml
  when: ansible_os_family == 'RedHat'
  tags: package-install
```

```
# 其他系统
# ...

- name: get ceph version
  command: ceph --version
  changed_when: false
  check_mode: no
  register: ceph_version

- name: set_fact ceph_version
  set_fact:
    ceph_version: "{{ ceph_version.stdout.split(' ')[2] }}"

# override ceph_stable_release for ceph_dev and rhcs installations since
ceph_stable_release is not mandatory
- name: include release-rhcs.yml
  include_tasks: release-rhcs.yml
  when: ceph_repository in ['rhcs', 'dev']
      or
      ceph_origin == 'distro'
  tags: always

- name: set_fact ceph_release - override ceph_release with ceph_stable_
release
  set_fact:
    ceph_release: "{{ ceph_stable_release }}"
  when:
    - ceph_origin == 'repository'
    - ceph_repository not in ['dev', 'rhcs', 'custom']
  tags: always

# 创建一些与 Ceph 相关的目录
- name: include create_rbd_client_dir.yml
  include_tasks: create_rbd_client_dir.yml

# 设置集群名称，将 Ceph 集群名写入配置文件中
- name: include configure_cluster_name.yml
  include_tasks: configure_cluster_name.yml

# 设置 bluestore 时，该任务不会执行
- name: include configure_memory_allocator.yml
  include_tasks: configure_memory_allocator.yml
  when:
    - (ceph_tcmalloc_max_total_thread_cache | int) > 0
    - osd_objectstore == 'filestore'
    - (ceph_origin == 'repository' or ceph_origin == 'distro')
```

主要安装的 Ceph 包在第一个 include 语句中，继续追踪 installs/install_on_redhat.yml 文件，代码如下：

```
# 源码位置：ceph-ansible/roles/ceph-common/tasks/installs/install_on_redhat.yml

---
```

```
- name: include configure_redhat_repository_installation.yml
  include_tasks: configure_redhat_repository_installation.yml
  when: ceph_origin == 'repository'

- name: include configure_redhat_local_installation.yml
  include_tasks: configure_redhat_local_installation.yml
  when: ceph_origin == 'local'

- name: include install_redhat_packages.yml
  include_tasks: install_redhat_packages.yml
  when: (ceph_origin == 'repository' or ceph_origin == 'distro')
```

来看最后一个 include 的 YAML 文件，该文件中包含的任务正是在对应组的节点上安装相应的 Ceph 包。例如，对于 mons 组的节点，就安装 ceph-mon 包，对于 osds 组的节点，则安装 ceph-osd 包等。具体代码如下：

```
# 源码位置：ceph-ansible/roles/ceph-common/tasks/installs/install_redhat_
packages.yml
---
- name: install redhat dependencies
  package:
    name: "{{ redhat_package_dependencies }}"
    state: present
  register: result
  until: result is succeeded
  when: ansible_distribution == 'RedHat'

- name: install centos dependencies
  yum:
    name: "{{ centos_package_dependencies }}"
    state: present
  register: result
  until: result is succeeded
  when: ansible_distribution == 'CentOS'

- name: install redhat ceph packages
  package:
    name: "{{ redhat_ceph_pkgs | unique }}"
    state: "{{ (upgrade_ceph_packages|bool) | ternary('latest','present') }}"
  register: result
  until: result is succeeded
```

在 ceph-common 角色中查看 centos_package_dependencies、redhat_package_dependencies 和 redhat_ceph_pkgs 这 3 个变量的值，它们分别位于 defaults 和 vars 目录下的 main.yml 文件中，具体如下：

```
# 源码位置：ceph-ansible/roles/ceph-common/vars/main.yml
# ...

centos_package_dependencies:
  - epel-release
  - libselinux-python

redhat_package_dependencies: []
```

```
# ...

redhat_ceph_pkgs:
  - "{{ (ceph_test) | ternary('ceph-test', 'ceph-common') }}"
  - "ceph-common"
  - "{{ (mon_group_name in group_names) | ternary('ceph-mon', 'ceph-
common') }}"
  - "{{ (osd_group_name in group_names) | ternary('ceph-osd', 'ceph-
common') }}"
  - "{{ (client_group_name in group_names) | ternary('ceph-fuse', 'ceph-
common') }}"
  - "{{ (client_group_name in group_names) | ternary('ceph-base', 'ceph-
common') }}"
  - "{{ (rgw_group_name in group_names) | ternary('ceph-radosgw', 'ceph-
common') }}"

# ...
```

上面这种变量的写法非常值得借鉴，这也是一种高级写法，个人的 Playbook 很难写出这样的语法。ternary()方法将根据过滤器（1）前一个表达式的结果的真假返回相应的值。例如对于 MON 节点而言，redhat_ceph_pkgs 变量的结果为：

```
redhat_ceph_pkgs:
  - "ceph-common"
  - "ceph-common"
  - "ceph-mon"
  - "ceph-common"
  - "ceph-common"
  - "ceph-common"
  - "ceph-common"
```

🔔注意：默认设置 ceph_test 参数值为 False。

综上所述，ceph-common 角色的工作有（针对 CentOS 系统和非容器化部署方式）：

- 安装必要的依赖包，公共的包有 epel-release、libselinux-python 和 ceph-common。不同组的节点，需要安装不同的 Ceph 包。
- 创建与 Ceph 相关的目录，修改与 Ceph 相关的配置文件，设置集群名称。

6. ceph-config角色

ceph-config 角色负责生成 Ceph 集群的配置文件 ceph.conf，并且创建各 Ceph 节点需要的目录。下面是该角色的核心任务：

```
# 源码位置: ceph-ansible/roles/ceph-config/tasks/main.yml
# ...

# 对 RGW 节点创建相关的实例目录
- name: include_tasks rgw_systemd_environment_file.yml
  include_tasks: rgw_systemd_environment_file.yml
  when: inventory_hostname in groups.get(rgw_group_name, [])
```

```
# 对数据盘进行校验
# ...

# ceph-common
- name: config file operation for non-containerized scenarios
  when: not containerized_deployment | bool
  block:
  # 创建 Ceph 配置目录
  - name: create ceph conf directory
    file:
      path: "/etc/ceph"
      state: directory
      owner: "ceph"
      group: "ceph"
      mode: "{{ ceph_directories_mode | default('0755') }}"

  # 生成配置文件 ceph.conf
  - name: "generate ceph configuration file: {{ cluster }}.conf"
    # 自定义 action
    action: config_template
    args:
      src: ceph.conf.j2
      dest: /etc/ceph/{{ cluster }}.conf
      owner: "ceph"
      group: "ceph"
      mode: "0644"
      # 注意这个参数
      config_overrides: "{{ ceph_conf_overrides }}"
      config_type: ini
    # 需要重新启动所有与 Ceph 相关的服务
    notify:
      - restart ceph mons
      - restart ceph osds
      - restart ceph mdss
      - restart ceph rgws
      - restart ceph mgrs
      - restart ceph rbdmirrors
      - restart ceph rbd-target-api-gw

  - name: "ensure fetch directory exists"
    run_once: true
    become: false
    file:
      path: "{{ fetch_directory }}/{{ fsid }}/etc/ceph"
      state: directory
      mode: "{{ ceph_directories_mode | default('0755') }}"
    delegate_to: localhost
    when: ceph_conf_local | bool

  # 重新生成一份 ceph.conf 并放到本地 fetch 目录下
  - name: "generate {{ cluster }}.conf configuration file locally"
    config_template:
    become: false
    run_once: true
```

```
    delegate_to: localhost
    args:
      src: "ceph.conf.j2"
      dest: "{{ fetch_directory }}/{{ fsid }}/etc/ceph/{{ cluster }}.conf"
      config_overrides: "{{ ceph_conf_overrides }}"
      config_type: ini
    when:
      - inventory_hostname in groups.get(mon_group_name, [])
      - ceph_conf_local | bool

# 容器化任务
# ...
```

可以看到，这里的 Playbook 主要是创建了 Ceph 的配置目录/etc/ceph，并依据设置的
ceph_conf_overrides 参数生成集群的 ceph.conf 文件，然后通知所有 Ceph 服务后续将要执
行重启操作。

7. ceph-mon角色

继续看 ceph-mon/tasks 目录下的 main.yml 文件内容，具体如下：

```
# 源码位置：ceph-ansible/roles/ceph-mon/tasks/main.yml

---
- name: set_fact container_exec_cmd
  set_fact:
    container_exec_cmd: "{{ container_binary }} exec ceph-mon-{{ ansible_hostname }}"
  when: containerized_deployment | bool

- name: include deploy_monitors.yml
  include_tasks: deploy_monitors.yml
  when:
    # we test for both container and non-container
    - (mon_socket_stat is defined and mon_socket_stat.get('rc') != 0) or
(ceph_mon_container_stat is defined and ceph_mon_container_stat.get('stdout_lines', []))|length == 0)
    - not switch_to_containers | default(False) | bool

- name: include start_monitor.yml
  include_tasks: start_monitor.yml

- name: include_tasks ceph_keys.yml
  include_tasks: ceph_keys.yml
  when: not switch_to_containers | default(False) | bool

- name: include secure_cluster.yml
  include_tasks: secure_cluster.yml
  when:
    - secure_cluster | bool
    - inventory_hostname == groups[mon_group_name] | first
```

注意，第一个任务和最后一个任务在默认的配置中不会执行，因此 ceph-mon 角色的

主要工作是部署和启动 monitor 服务，然后设置相关变量。下面来看 deploy_monitors.yml
文件中的任务，具体代码如下：

```
# 源码位置: ceph-ansible/roles/ceph-mon/tasks/deploy_monitors.yml
---
- name: check if monitor initial keyring already exists
  command: >
    {{ _container_exec_cmd | default('') }} ceph --cluster {{ cluster }}
--name mon. -k
    /var/lib/ceph/mon/{{ cluster }}-{{ hostvars[groups[mon_group_name][0]
if running_mon is undefined else running_mon]['ansible_hostname'] }}/
keyring
    auth get-key mon.
  register: initial_mon_key
  run_once: True
  delegate_to: "{{ groups[mon_group_name][0] if running_mon is undefined
else running_mon }}"
  when: ceph_current_status.fsid is defined

- name: generate monitor initial keyring
  command: >
    {{ hostvars[groups[mon_group_name][0] if running_mon is undefined else
running_mon]['discovered_interpreter_python'] }} -c "import os ; import
struct ;
    import time; import base64 ; key = os.urandom(16) ;
    header = struct.pack('<hiih',1,int(time.time()),0,len(key)) ;
    print(base64.b64encode(header + key).decode())"
  register: monitor_keyring
  run_once: True
  delegate_to: "{{ groups[mon_group_name][0] if running_mon is undefined
else running_mon }}"
  when:
    - initial_mon_key.skipped is defined
    - ceph_current_status.fsid is undefined

- name: get initial keyring when it already exists
  set_fact:
    monitor_keyring: "{{ initial_mon_key.stdout if monitor_keyring.skipped
is defined else monitor_keyring.stdout if initial_mon_key.skipped is
defined }}"

# ...
```

上面 3 个任务的主要目的是生成 monitor 的 keyring 值。通过第二个任务可知，ceph-
ansible 是通过在 mons 组的第一个节点上执行一段 Python 代码得到的。现在可以在 Python
交互模式下手工执行这段 Python 代码，具体如下：

```
>>> import os
>>> import struct
>>> import time
>>> import base64
>>> key = os.urandom(16)
>>> header = struct.pack('<hiih',1,int(time.time()),0,len(key))
```

```
>>> print(base64.b64encode(header + key).decode())
AQAuarNfAAAAABAADpbkUVlrNb/b5ON4VvOxtw==
```

继续看接下来的任务：

```
# 源码位置: ceph-ansible/roles/ceph-mon/tasks/deploy_monitors.yml
# ...

- name: create monitor initial keyring
  ceph_key:
    name: mon.
    state: present
    dest: "/var/lib/ceph/tmp/"
    secret: "{{ monitor_keyring }}"
    cluster: "{{ cluster }}"
    caps:
      mon: allow *
    import_key: False
    owner: "{{ ceph_uid if containerized_deployment else 'ceph' }}"
    group: "{{ ceph_uid if containerized_deployment else 'ceph' }}"
    mode: "0400"
  environment:
    CEPH_CONTAINER_IMAGE: "{{ ceph_docker_registry + '/' + ceph_docker_
image + ':' + ceph_docker_image_tag if containerized_deployment else
None }}"
    CEPH_CONTAINER_BINARY: "{{ container_binary }}"

# 容器化任务
# ...

- name: create (and fix ownership of) monitor directory
  file:
    path: /var/lib/ceph/mon/{{ cluster }}-{{ monitor_name }}
    state: directory
    owner: "{{ ceph_uid if containerized_deployment else 'ceph' }}"
    group: "{{ ceph_uid if containerized_deployment else 'ceph' }}"
    mode: "{{ ceph_directories_mode | default('0755') }}"
    recurse: true

# ...
```

上面同样涉及一个自定义模块——ceph_key，该模块用于生成 Ceph 集群的密钥文件，并且创建相应的 Ceph 目录。为了方便理解 ceph_key 模块，下面做一个简单的测试，具体操作如下：

```
(ansible2.8.16) [root@master ceph-ansible-4.0.30]# cat hosts
[master]
master

[all:vars]
ansible_connection = local
(ansible2.8.16) [root@master ceph-ansible-4.0.30]# cat test_ceph_key.yml
---
- hosts: master
  remote_user: root
```

```
    gather_facts: false
    tasks:
      - name: create ceph admin key
        ceph_key:
          name: client.admin
          state: present
          secret: AQAin8tU2DsKFBAAFIAzVTzkL3+gtAjjpQiomw==
          caps:
            mon: allow *
            osd: allow *
            mgr: allow *
            mds: allow
          dest: /tmp
          mode: 0600
          import_key: False
(ansible2.8.16) [root@master ceph-ansible-4.0.30]# ls /tmp/
(ansible2.8.16) [root@master ceph-ansible-4.0.30]# ansible-playbook -i
hosts test_ceph_key.yml
# ...
(ansible2.8.16) [root@master ceph-ansible-4.0.30]# cat /tmp/ceph.client.
admin.keyring
[client.admin]
    key = AQAin8tU2DsKFBAAFIAzVTzkL3+gtAjjpQiomw==
    caps mds = "allow"
    caps mgr = "allow *"
    caps mon = "allow *"
    caps osd = "allow *"
```

上述操作演示了通过 ceph_key 模块生成 Ceph 集群密钥文件 ceph.client.admin.keyring 的过程。有兴趣的读者可以阅读 ceph-ansible/library 目录下的 ceph_key.py 文件，学习该密钥文件的生成代码，这里就不再深入解析模块代码了。

继续看后面的任务：

```
# 源码位置: ceph-ansible/roles/ceph-mon/tasks/deploy_monitorsy.yml
# ...

- name: set_fact ceph-mon container command
  set_fact:
    ceph_mon_cmd: "{{ container_binary + ' run --rm --net=host -v /var/lib/
ceph/:/var/lib/ceph:z -v /etc/ceph/:/etc/ceph/:z --entrypoint=ceph-mon '
+ ceph_client_docker_registry + '/' + ceph_client_docker_image + ':'
+ceph_client_docker_image_tag if containerized_deployment else 'ceph-
mon' }}"

- name: ceph monitor mkfs with keyring
  command: >
    {{ ceph_mon_cmd }}
    --cluster {{ cluster }}
    --setuser "{{ ceph_uid if containerized_deployment else 'ceph' }}"
    --setgroup "{{ ceph_uid if containerized_deployment else 'ceph' }}"
    --mkfs
    -i {{ monitor_name }}
    --fsid {{ fsid }}
    --keyring /var/lib/ceph/tmp/{{ cluster }}.mon..keyring
```

```
  args:
    creates: /var/lib/ceph/mon/{{ cluster }}-{{ monitor_name }}/keyring
  when: cephx | bool

  # ...
```

🔔**注意**：部分未执行的任务没有在此列出，笔者是根据默认配置进行判断的。例如 name: create custom admin keyring 任务，由于没有设置 admin_secret 值，默认值为 'admin_secret'，这会使得条件判断语句 admin_secret != 'admin_secret'为 False，该任务会被跳过。同样被忽略的任务还有 name: import admin keyring into mon keyring 等，这些都要根据实际的配置值进行判断。

上面的两个任务看似复杂，实际上使用非容器化方式部署 Ceph 集群时，ceph_mon_cmd 的值为 ceph-mon。因此，这两个任务主要是在 mons 组内的所有主机上执行如下命令：

```
ceph-mon --cluster ceph --setuser ceph --setgroup ceph --mkfs -i 对应的 mon
主机名 --fsid 集群的 fsid --keyring /var/lib/ceph/tmp/ceph.mon..keyring
```

🔔**注意**：/var/lib/ceph/tmp/ceph.mon..keyring 这个密钥文件就是在 name: create monitor initial keyring 任务中通过 ceph_key 模块生成的。

再来看启动 monitor 进程的 Playbook 文件，其中的任务如下：

```
# 源码位置: ceph-ansible/roles/ceph-mon/tasks/start_monitor.yml
---
- name: ensure systemd service override directory exists
  file:
    state: directory
    path: "/etc/systemd/system/ceph-mon@.service.d/"
  when:
    - not containerized_deployment | bool
    # 默认 ceph_mon_systemd_overrides 没有设置，所以该任务不会执行
    - ceph_mon_systemd_overrides is defined
    - ansible_service_mgr == 'systemd'

- name: add ceph-mon systemd service overrides
  config_template:
    src: "ceph-mon.service.d-overrides.j2"
    dest: "/etc/systemd/system/ceph-mon@.service.d/ceph-mon-systemd-
overrides.conf"
    config_overrides: "{{ ceph_mon_systemd_overrides | default({}) }}"
    config_type: "ini"
  when:
    - not containerized_deployment | bool
    - ceph_mon_systemd_overrides is defined
    - ansible_service_mgr == 'systemd'

- name: include_tasks systemd.yml
  include_tasks: systemd.yml
  when: containerized_deployment | bool
```

```
- name: start the monitor service
  systemd:
    name: ceph-mon@{{ monitor_name if not containerized_deployment else
ansible_hostname }}
    state: started
    enabled: yes
    masked: no
    daemon_reload: yes
```

注意上面的参数 ceph_mon_systemd_overrides，默认没有设置，所以前两个任务不会执行。此外，由于是非容器化部署，所以第三个 include_tasks 也不会执行。最后执行的只有一个 name: start the monitor service 任务，用于启动 ceph-mon 服务并加入系统的自启动模式中。

再来看 ceph_keys.yml 的 Playbook，其中只有两个任务，主要是等待 ceph-mon 服务全部正常启动，这里用到了一个 until 语法，具体代码如下：

```
# 源码位置：ceph-ansible/roles/ceph-mon/tasks/ceph_keys.yml
---
- name: waiting for the monitor(s) to form the quorum...
  command: >
    {{ container_exec_cmd }}
    ceph
    --cluster {{ cluster }}
    -n mon.
    -k /var/lib/ceph/mon/{{ cluster }}-{{ ansible_hostname }}/keyring
    mon_status
    --format json
  register: ceph_health_raw
  run_once: true
  until: >
    (ceph_health_raw.stdout | length > 0) and (ceph_health_raw.stdout |
default('{}') | from_json)['state'] in ['leader', 'peon']
  retries: "{{ handler_health_mon_check_retries }}"
  delay: "{{ handler_health_mon_check_delay }}"
  changed_when: false

- name: fetch ceph initial keys
  ceph_key:
    state: fetch_initial_keys
    cluster: "{{ cluster }}"
    owner: "{{ ceph_uid if containerized_deployment else 'ceph' }}"
    group: "{{ ceph_uid if containerized_deployment else 'ceph' }}"
    mode: "0400"
  environment:
    CEPH_CONTAINER_IMAGE: "{{ ceph_docker_registry + '/' + ceph_docker_
image + ':' + ceph_docker_image_tag if containerized_deployment else
None }}"
    CEPH_CONTAINER_BINARY: "{{ container_binary }}"
    CEPH_ROLLING_UPDATE: "{{ rolling_update }}"
  when:
    - cephx | bool
```

为了帮助读者理解第一个任务，笔者先找了一台已经部署了 Ceph 集群的主机（6.2

节中搭建好的集群），来看看第一个任务执行的结果，具体如下：

```
# 这里笔者随便找了一个内部测试集群环境，读者可以在前面搭建好的集群上测试该命令
[root@server220 ~]# ceph --cluster ceph -n mon. -k /var/lib/ceph/mon/ceph-
server220/keyring mon_status --format json

{"name":"server220","rank":0,"state":"leader","election_epoch":18,"quorum":
[0,1,2],"quorum_age":16896,"features":{"required_con":"2449958747315912708",
"required_mon":["kraken","luminous","mimic","osdmap-prune","nautilus"],
"quorum_con":"4611087854035861503","quorum_mon":["kraken","luminous",
"mimic","osdmap-prune","nautilus"]},"outside_quorum":[],"extra_probe_
peers":[],"sync_provider":[],"monmap":{"epoch":1,"fsid":"18be3a79-ef39-
4b74-aba2-fba61a3d49c2","modified":"2020-11-16 20:14:26.485891","created":
"2020-11-16 20:14:26.485891","min_mon_release":14,"min_mon_release_name":
"nautilus","features":{"persistent":["kraken","luminous","mimic","osdmap-
prune","nautilus"],"optional":[]},"mons":[{"rank":0,"name":"server220",
"public_addrs":{"addrvec":[{"type":"v2","addr":"192.168.115.220:3300",
"nonce":0},{"type":"v1","addr":"192.168.115.220:6789","nonce":0}]},"addr":
"192.168.115.220:6789/0","public_addr":"192.168.115.220:6789/0"},{"rank":1,
"name":"server223","public_addrs":{"addrvec":[{"type":"v2","addr":"192.
168.115.223:3300","nonce":0},{"type":"v1","addr":"192.168.115.223:6789",
"nonce":0}]},"addr":"192.168.115.223:6789/0","public_addr":"192.168.115
.223:6789/0"},{"rank":2,"name":"server224","public_addrs":{"addrvec":
[{"type":"v2","addr":"192.168.115.224:3300","nonce":0},{"type":"v1","addr":
"192.168.115.224:6789","nonce":0}]},"addr":"192.168.115.224:6789/0",
"public_addr":"192.168.115.224:6789/0"}]},"feature_map":{"mon":[{"features":
"0x3ffddff8ffecffff","release":"luminous","num":1}],"osd":[{"features":
"0x3ffddff8ffecffff","release":"luminous","num":20}],"client":[{"features":
"0x27018fb86aa42ada","release":"jewel","num":1},{"features":"0x3ffddff8
ffecffff","release":"luminous","num":3}],"mgr":[{"features":"0x3ffddff8
ffecffff","release":"luminous","num":2}]}}}
```

可以看到，当 monitor 服务都正常时，上述命令输出的字典中的 state 字段值为 leader。ceph-ansible 会判断该值是否在['leader', 'peon']中，当任意一个值都表明服务正常时，才会继续下一个任务。第二个任务非常关键，它会生成访问 Ceph 集群的密钥文件 ceph.client.admin.keyring。该密钥文件的生成细节就在 library/ceph_key.py 文件中，有兴趣的读者可以研究相关代码，找出生成密钥文件的命令。

8. ceph-mgr角色

接下来看 ceph-mgr/tasks 目录下的 main.yml 文件，其核心的任务如下：

```
# 源码位置：ceph-ansible/roles/ceph-mgr/tasks/main.yml
# ...

- name: include common.yml
  include_tasks: common.yml

- name: include pre_requisite.yml
  include_tasks: pre_requisite.yml
  when: not containerized_deployment | bool

- name: include start_mgr.yml
```

```
  include_tasks: start_mgr.yml

# 跳过的其他任务
# ...
```

首先来看 common.yml 文件中的任务，主要包括创建相应的 mgr 目录、复制 MON 节点上的密钥到 MGR 节点上等任务，具体代码如下：

```
# 源码位置: ceph-ansible/roles/ceph-mgr/tasks/common.yml
---

# 创建 MGR 节点的相关目录
- name: create mgr directory
  file:
    path: /var/lib/ceph/mgr/{{ cluster }}-{{ ansible_hostname }}
    state: directory
    owner: "{{ ceph_uid if containerized_deployment else 'ceph' }}"
    group: "{{ ceph_uid if containerized_deployment else 'ceph' }}"
    mode: "{{ ceph_directories_mode | default('0755') }}"

# 得到 MGR 节点的 keyring，通过 ceph_key 模块实现
- name: fetch ceph mgr keyring
  ceph_key:
    name: "mgr.{{ ansible_hostname }}"
    state: present
    caps:
      mon: allow profile mgr
      osd: allow *
      mds: allow *
    cluster: "{{ cluster }}"
    secret: "{{ (mgr_secret != 'mgr_secret') | ternary(mgr_secret, omit) }}"
    owner: "{{ ceph_uid if containerized_deployment else 'ceph' }}"
    group: "{{ ceph_uid if containerized_deployment else 'ceph' }}"
    mode: "0400"
    dest: "/var/lib/ceph/mgr/{{ cluster }}-{{ ansible_hostname }}/keyring"
  environment:
    CEPH_CONTAINER_IMAGE: "{{ ceph_docker_registry + '/' + ceph_docker
_image + ':' + ceph_docker_image_tag if containerized_deployment else
None }}"
    CEPH_CONTAINER_BINARY: "{{ container_binary }}"
  when: groups.get(mgr_group_name, []) | length == 0 # the key is present
already since one of the mons created it in "create ceph mgr keyring(s)"

# 创建和复制相应的密钥文件
- name: create and copy keyrings
  when: groups.get(mgr_group_name, []) | length > 0
  block:
    - name: create ceph mgr keyring(s) on a mon node
      ceph_key:
        name: "mgr.{{ hostvars[item]['ansible_hostname'] }}"
        state: present
        caps:
          mon: allow profile mgr
          osd: allow *
```

```
      mds: allow *
    cluster: "{{ cluster }}"
    secret: "{{ (mgr_secret != 'mgr_secret') | ternary(mgr_secret, omit) }}"
    owner: "{{ ceph_uid if containerized_deployment else 'ceph' }}"
    group: "{{ ceph_uid if containerized_deployment else 'ceph' }}"
    mode: "0400"
  environment:
    CEPH_CONTAINER_IMAGE: "{{ ceph_docker_registry + '/' + ceph_docker_
image + ':' + ceph_docker_image_tag if containerized_deployment else None }}"
    CEPH_CONTAINER_BINARY: "{{ container_binary }}"
  with_items: "{{ groups.get(mgr_group_name, []) }}"
  run_once: True
  # 在第一个 MON 节点上生成 mgr 密钥文件
  delegate_to: "{{ groups[mon_group_name][0] }}"

- name: set_fact _mgr_keys
  set_fact:
    _mgr_keys:
      - { 'name': 'client.admin', 'path': '/etc/ceph/{{ cluster }}.
client.admin.keyring", 'copy_key': copy_admin_key }
      - { 'name': "mgr.{{ ansible_hostname }}", 'path': "/var/lib/ceph/
mgr/{{ cluster }}-{{ ansible_hostname }}/keyring", 'copy_key': true }

  # 从第一个 MON 节点上获取相应的密钥文件
- name: get keys from monitors
  command: "{{ _container_exec_cmd | default('') }} ceph --cluster
{{ cluster }} auth get {{ item.name }}"
  register: _mgr_keys
  with_items: "{{ _mgr_keys }}"
  delegate_to: "{{ groups[mon_group_name][0] if running_mon is undefined
else running_mon }}"
  when:
    - cephx | bool
    - item.copy_key | bool

  # 将密钥匙复制到所有的 MGR 节点上
- name: copy ceph key(s) if needed
  copy:
    dest: "{{ item.item.path }}"
    content: "{{ item.stdout + '\n' }}"
    owner: "{{ ceph_uid if containerized_deployment else 'ceph' }}"
    group: "{{ ceph_uid if containerized_deployment else 'ceph' }}"
    mode: "{{ ceph_keyring_permissions }}"
  with_items: "{{ _mgr_keys.results }}"
  when:
    - cephx | bool
    - item.item.copy_key | bool

# 设置 mgr 密钥文件权限
- name: set mgr key permissions
  file:
    path: /var/lib/ceph/mgr/{{ cluster }}-{{ ansible_hostname }}/keyring
    owner: "{{ ceph_uid if containerized_deployment else 'ceph' }}"
    group: "{{ ceph_uid if containerized_deployment else 'ceph' }}"
```

```
      mode: "{{ ceph_keyring_permissions }}"
  when: cephx | bool
```

上面的任务中均添加了相应注释，内容并不复杂，后续会直接复制相关任务。

再来看 pre_requisite.yml 文件中的任务，这里将只有一个任务被运行，具体代码如下：

```
# 源码位置: ceph-ansible/roles/ceph-mgr/tasks/pre_requisite.yml
# ...

- name: install ceph-mgr packages on RedHat or SUSE
  package:
    name: '{{ ceph_mgr_packages }}'
    state: "{{ (upgrade_ceph_packages|bool) | ternary('latest','present') }}"
  register: result
  until: result is succeeded
  when: ansible_os_family in ['RedHat', 'Suse']

# ...
```

以上任务的主要目的就是安装 ceph-mgr 相关的依赖包，其中，**ceph_mgr_packages** 参数的默认值如下：

```
# 源码位置: ceph-ansible/roles/ceph-mgr/defaults/main.yml
# ...

ceph_mgr_packages:
  - ceph-mgr
  - ceph-mgr-dashboard
  - ceph-mgr-diskprediction-local

# ...
```

start_mgr.yml 文件中的任务也非常简单且同样只有一个任务被运行，具体代码如下：

```
# 源码位置: ceph-ansible/roles/ceph-mgr/tasks/start_mgr.yml
# ...

# 启动 ceph-mgr 服务
- name: systemd start mgr
  systemd:
    name: ceph-mgr@{{ ansible_hostname }}
    state: started
    enabled: yes
    masked: no
    daemon_reload: yes
```

至此，部署 monitor 节点依赖的所有角色都已经介绍完了，接下来就是在 CentOS 7 环境下对 MON 节点的部署过程进行重写，最终简化 Ceph 集群的部署代码。

6.4.2　在 CentOS 7 下部署 Ceph 集群的简化版 Playbook

在 6.4.1 节中我们基本梳理了 MON 节点的部署过程，下面对整个过程做一个总结，并在此基础上简化和生成新 Ceph 集群部署的 Playbook 文件。根据前面分析的结果，比较重要的基础角色有 ceph-common、ceph-config 和 ceph-fact，其作用分别如下：

- ceph-common：为不同角色节点安装相应的依赖包及相应的 Ceph 包。
- ceph-fact：得到一些主机变量，比如生成集群的 fsid 值，用于后面的 monitor 服务部署，再比如 _monitor_addresses 值用于生成 ceph.conf 等。
- ceph-config：创建相应的 Ceph 目录，此外还会生成 ceph.conf 文件并下发到所有的节点中。

对于 ceph-mon 和 ceph-mgr 两个角色中的任务集，去掉一些特殊情况下执行的任务及容器化部署时执行的任务，保留其核心的任务即可。下面手动实现一个简化部署 monitor 节点的工作。首先创建一个新的集群部署目录，具体操作如下：

```
(ansible2.8.16) [root@master ceph-cluster]# mkdir ceph_mon_deploy
(ansible2.8.16) [root@master ceph-cluster]# cd ceph_mon_deploy/
```

由于这里生成密钥文件的核心部分依赖于 ceph-ansible 中自定义的 ceph_key 模块，因此可以直接复制 ceph-ansible-4.0.30/library 目录，具体操作如下：

```
(ansible2.8.16) [root@master my_ceph_deploy]# cp -r ../ceph-ansible-4.0.30/
library .
```

另外，在 ansible.cfg 文件中指定了自定义模块文件所在的目录及优化的 SSH 连接参数，因此可以将其复制过来直接使用，具体操作如下：

```
(ansible2.8.16) [root@master my_ceph_deploy]# cp -r ../ceph-ansible-
4.0.30/ansible.cfg .
(ansible2.8.16) [root@master ceph_mon_deploy]# cat ansible.cfg | grep
library
library = ./library
```

接着需要思考如何实现 YUM 安装 Ceph 的包。这里可以直接使用阿里源，为此需要准备两个 repo 文件：CentOS-Base.repo 和 ceph.repo。具体操作如下：

```
(ansible2.8.16) [root@master ceph_mon_deploy]# wget -O /etc/yum.repos.d/
CentOS-Base.repo http://mirrors.aliyun.com/repo/Centos-7.repo
(ansible2.8.16) [root@master ceph_mon_deploy]# cat ceph.repo
[ceph_stable]
baseurl = https://mirrors.aliyun.com/ceph//rpm-nautilus/el7/$basearch
gpgcheck = 1
gpgkey = https://mirrors.aliyun.com/ceph//keys/release.asc
name = Ceph Stable $basearch repo
priority = 2

[ceph_stable_noarch]
baseurl = https://mirrors.aliyun.com/ceph//rpm-nautilus/el7/noarch
gpgcheck = 1
gpgkey = https://mirrors.aliyun.com/ceph//keys/release.asc
name = Ceph Stable noarch repo
priority = 2
```

说明：CentOS-Base.repo 文件可以从阿里云镜像网站上获取，而 ceph.repo 文件可以由前面使用 ceph-ansible 搭建 Cceph 集群时生成的 ceph_stable.repo 得到。

首先新建一个 common.yml 文件，简化 ceph-common 角色，只实现下发 repo 文件及安装相应的 Ceph 包的任务，具体内容如下：

```
---
# 安装服务
- name: copy repo files to all hosts
  copy:
    src: "{{ item }}"
    dest: "/etc/yum.repos.d/{{ item }}"
  with_items:
    - "CentOS-Base.repo"
    - "ceph.repo"
  become: true

- name: yum install require packages
  yum:
    name: "{{ require_packages }}"
  vars:
    require_packages:
      - libselinux-python
      - ceph-common
      - ceph-mon
      - ceph-mgr
  become: true
```

接着简化 ceph-facts 角色中的任务，为此准备一个 facts.yml 文件，其内容如下：

```
---
- name: generate cluster fsid
  command: "/bin/python -c 'import uuid; print(str(uuid.uuid4()))'"
  register: cluster_uuid
  delegate_to: "{{ groups[mon_group_name][0] }}"
  run_once: true

- name: set_fact fsid
  set_fact:
    fsid: "{{ cluster_uuid.stdout }}"

- name: get monitor ipv4 address
  shell: "cat /etc/hosts | grep {{ inventory_hostname }} | awk '{print $1}'"
  register: shell_out

- name: set_fact monitor_ipv4
  set_fact:
    monitor_ipv4: "{{ shell_out.stdout }}"

- name: set_fact _monitor_address to monitor_interface - ipv4
  set_fact:
    _monitor_addresses: "{{ _monitor_addresses | default([]) + [{ 'name':
item, 'addr': hostvars[item]['monitor_ipv4'] }] }}"
  with_items: "{{ groups.get(mon_group_name, []) }}"
```

下面来看看 ceph-ansible 中 ceph.conf.j2 文件的内容：

```
#jinja2: trim_blocks: "true", lstrip_blocks: "true"
# {{ ansible_managed }}
```

```
[global]
{# 忽略这里的配置 #}
{# ... #}

{% if nb_mon > 0 and inventory_hostname in groups.get(mon_group_name, []) %}
mon initial members = {% for host in groups[mon_group_name] %}
    {% if hostvars[host]['ansible_fqdn'] is defined and mon_use_fqdn -%}
      {{ hostvars[host]['ansible_fqdn'] }}
    {%- elif hostvars[host]['ansible_hostname'] is defined -%}
      {{ hostvars[host]['ansible_hostname'] }}
    {%- endif %}
    {%- if not loop.last %},{% endif %}
  {% endfor %}

osd pool default crush rule = {{ osd_pool_default_crush_rule }}
{% endif %}

fsid = {{ fsid }}
mon host = {% if nb_mon > 0 %}
{% for host in _monitor_addresses -%}
{% if msgr2_migration | default(False) %}
[{{ "v2:" + host.addr + mon_host_v2.suffix }},{{ "v1:" + host.addr + mon_
host_v1.suffix }}]
{%- elif not msgr2_migration | default(False) and rolling_update -%}
{{ host.addr }}
{%- else -%}
{% if mon_host_v1.enabled %}
{% set _v1 = ',v1:' + host.addr + mon_host_v1.suffix %}
{% endif %}
[{{ "v2:" + host.addr + mon_host_v2.suffix }}{{ _v1 | default('') }}]
{%- endif %}
{%- if not loop.last -%},{%- endif %}
{%- endfor %}
{% elif nb_mon == 0 %}
{{ external_cluster_mon_ips }}
{% endif %}

{% if public_network is defined %}
public network = {{ public_network | regex_replace(' ', '') }}
{% endif %}
{% if cluster_network is defined %}
cluster network = {{ cluster_network | regex_replace(' ', '') }}
{% endif %}

{# 忽略后面的配置 #}
{# ... #}
```

一般而言，ceph.conf 中需要动态调整的参数有 mon_initial_members、fsid、mon_host 及 public_network/cluster_network 等，其余涉及集群性能的参数通常会有专门的运维人员写好并加到该模板文件后面。下面是笔者准备的 ceph.conf.j2 文件，其中省略了 Ceph 相关的调优参数。

```
[global]
fsid = {{ fsid }}
mon_host = {% for host in _monitor_addresses -%}
{% if mon_host_v1.enabled %}
{% set _v1 = ',v1:' + host.addr + mon_host_v1.suffix %}
{% endif %}
[{{ "v2:" + host.addr + mon_host_v2.suffix }}{{ _v1 | default('') }}]
{%- if not loop.last -%},{%- endif %}
{%- endfor %}

mon_initial_members = {% for host in groups[mon_group_name] %}{{ host }}{%-
if not loop.last %},{% endif %}{% endfor %}

public_network = {{ public_network | regex_replace(' ', '') }}
cluster_network = {{ public_network | regex_replace(' ', '') }}

max_open_files = 1310720
mon_max_pg_per_osd = 512
osd_crush_update_on_start = false
ms_bind_port_min = 10000
ms_bind_port_max = 15000
mon_osd_down_out_subtree_limit = host

{# 忽略集群调优参数 #}
```

上面的 fsid 和 _monitor_addresses 变量从 facts.yml 文件中得到，而其余变量将在变量文件中给出。为此，新建 group_vars/all.yml 文件并从 ceph-ansible 中找到原 mon_host_v1 的参数值，同时在 group_vars/all.yml 文件中设置 public_network 的值，具体操作如下：

```
(ansible2.8.16) [root@master ceph_mon_deploy]# cat group_vars/all.yml

cluster: ceph
mon_group_name: mons
public_network: "192.168.26.0/24"

mon_host_v1:
  enabled: True
  suffix: ':6789'
mon_host_v2:
  suffix: ':3300'

# Monitor handler checks
handler_health_mon_check_retries: 10
handler_health_mon_check_delay: 20
```

由此就可以得到简化 ceph-config 角色的任务了，将其写入文件 config.yml 中。具体内容如下：

```
---
- name: create ceph initial directories
  file:
    path: "{{ item }}"
    state: directory
    owner: "ceph"
    group: "ceph"
```

```
      mode: 0755
    loop:
      - /etc/ceph
      - /var/lib/ceph/
      - /var/lib/ceph/mon
      - /var/lib/ceph/osd
      - /var/lib/ceph/mds
      - /var/lib/ceph/tmp
      - /var/lib/ceph/radosgw
      - /var/lib/ceph/bootstrap-rgw
      - /var/lib/ceph/bootstrap-mgr
      - /var/lib/ceph/bootstrap-mds
      - /var/lib/ceph/bootstrap-osd
      - /var/lib/ceph/bootstrap-rbd
      - /var/lib/ceph/bootstrap-rbd-mirror
      - /var/run/ceph
      - /var/log/ceph

  - name: generate ceph configuration file - ceph.conf
    template:
      src: ceph.conf.j2
      dest: /etc/ceph/ceph.conf
      owner: ceph
      group: ceph
    become: true
```

最后就是复制 ceph-mon 和 ceph-mgr 角色中的核心任务了，这里不再继续分析，直接给出两个简化后的 Playbook 文件——mon.yml 和 mgr.yml。首先来看 mon.yml 文件的内容，具体如下：

```
  ---
  - name: generate monitor initial keyring
    command: >
      /bin/python -c "import os ; import struct ;
      import time; import base64 ; key = os.urandom(16) ;
      header = struct.pack('<hiih',1,int(time.time()),0,len(key)) ;
      print(base64.b64encode(header + key).decode())"
    register: monitor_keyring
    run_once: True
    delegate_to: "{{ groups[mon_group_name][0] }}"

  - name: get initial keyring
    set_fact:
      monitor_keyring: "{{ monitor_keyring.stdout }}"

  - name: create monitor initial keyring
    ceph_key:
      name: mon.
      state: present
      dest: "/var/lib/ceph/tmp/"
      secret: "{{ monitor_keyring }}"
      cluster: "{{ cluster }}"
      caps:
        mon: allow *
      import_key: False
```

```yaml
        owner: "ceph"
        group: "ceph"
        mode: "0400"

  - name: create (and fix ownership of) monitor directory
    file:
      path: /var/lib/ceph/mon/ceph-{{ inventory_hostname }}
      state: directory
      owner: "ceph"
      group: "ceph"
      mode: "{{ ceph_directories_mode | default('0755') }}"
      recurse: true

- name: set_fact ceph-mon container command
  set_fact:
    ceph_mon_cmd: "ceph-mon"

- name: ceph monitor mkfs with keyring
  command: >
    {{ ceph_mon_cmd }}
    --cluster {{ cluster }}
    --setuser ceph
    --setgroup "ceph"
    --mkfs
    -i {{ inventory_hostname }}
    --fsid {{ fsid }}
    --keyring /var/lib/ceph/tmp/ceph.mon..keyring
  args:
    creates: /var/lib/ceph/mon/ceph-{{ inventory_hostname }}/keyring

# 启动 ceph-mon 服务
- name: start the monitor service
  systemd:
    name: ceph-mon@{{ inventory_hostname }}
    state: started
    enabled: yes
    masked: no
    daemon_reload: yes

- name: waiting for the monitor(s) to form the quorum...
  command: >
    ceph
    --cluster {{ cluster }}
    -n mon.
    -k /var/lib/ceph/mon/{{ cluster }}-{{ inventory_hostname }}/keyring
    mon_status
    --format json
  register: ceph_health_raw
  run_once: true
  until: >
    (ceph_health_raw.stdout | length > 0) and (ceph_health_raw.stdout |
  default('{}') | from_json)['state'] in ['leader', 'peon']
  retries: "{{ handler_health_mon_check_retries }}"
  delay: "{{ handler_health_mon_check_delay }}"
  changed_when: false
```

```
  - name: fetch ceph initial keys
    ceph_key:
      state: fetch_initial_keys
      cluster: "{{ cluster }}"
      owner: "ceph"
      group: "ceph"
      mode: "0400"
```

再来看简化的 mgr.yml 文件内容，具体如下：

```
---
- name: create mgr directory
  file:
    path: /var/lib/ceph/mgr/{{ cluster }}-{{ inventory_hostname }}
    state: directory
    owner: "ceph"
    group: "ceph"
    mode: "{{ ceph_directories_mode | default('0755') }}"

- name: fetch ceph mgr keyring
  ceph_key:
    name: "mgr.{{ inventory_hostname }}"
    state: present
    caps:
      mon: allow profile mgr
      osd: allow *
      mds: allow *
    cluster: "{{ cluster }}"
    secret: "{{ omit }}"
    owner: "ceph"
    group: "ceph"
    mode: "0400"
    dest: "/var/lib/ceph/mgr/{{ cluster }}-{{ inventory_hostname }}/keyring"

- name: set mgr key permissions
  file:
    path: /var/lib/ceph/mgr/{{ cluster }}-{{ inventory_hostname }}/keyring
    owner: "ceph"
    group: "ceph"
    mode: "0600"

- name: install ceph-mgr packages on RedHat or SUSE
  package:
    name: '{{ ceph_mgr_packages }}'
    state: "present"
  register: result
  until: result is succeeded

- name: systemd start mgr
  systemd:
    name: ceph-mgr@{{ inventory_hostname }}
    state: started
    enabled: yes
    masked: no
    daemon_reload: yes
```

有了这些 Playbook 文件后，可以用 include 语法将它们组合成一个最终的 Playbook 并命名为 deploy_mon.yml，该文件的内容如下：

```
---
- hosts: mons
  gather_facts: false

  tasks:
    - name: include common.yml
      include_tasks: common.yml

    - name: include facts.yml
      include_tasks: facts.yml

    - name: include config.yml
      include_tasks: config.yml

    - name: include mon.yml
      include_tasks: mon.yml

    - name: include mgr.yml
      include_tasks: mgr.yml
```

最后，使用 ansible-playbook 命令运行这个最终的 Playbook，操作如下：

```
(ansible2.8.16) [root@master ceph_mon_deploy]# ansible-playbook -i hosts
deploy_mon.yml
```

在网络通畅的情况下，对于小规模集群，大概几分钟后就可以看到部署成功的输出信息了，如图 6-6 所示。

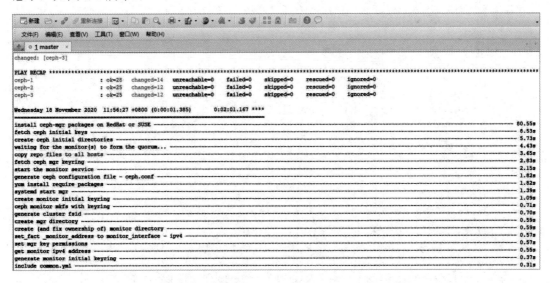

图 6-6　执行自定义搭建 Ceph 集群的 Playbook 的输出信息

接着登录到 ceph-1 节点，使用 ceph -s 命令查看集群情况，具体输出结果如下：

```
[root@ceph-1 ~]# sudo ceph -s
  cluster:
    id:     21e62f57-fc7d-4eb0-bd1c-a77256b4fb46
    health: HEALTH_OK

  services:
    mon: 3 daemons, quorum ceph-1,ceph-2,ceph-3 (age 100s)
    mgr: ceph-3(active, since 6s), standbys: ceph-1, ceph-2
    osd: 0 osds: 0 up, 0 in

  data:
    pools:   0 pools, 0 pgs
    objects: 0 objects, 0 B
    usage:   0 B used, 0 B / 0 B avail
    pgs:
```

以上输出结果说明 Ceph 集群部署成功，也证明了前面改造工程的有效性。终于可以不用大费周章使用那么复杂的剧本来部署 Ceph 集群了。

🔔注意：如果主机名包含有 .后缀，如主机名为 x01.gz.cn~x03.gz.cn，直接执行上面的 Playbook 时会报错。有兴趣的读者可以尝试复现这个问题。报错的原因在于密钥文件没有生成，从而导致后续执行的 ceph 命令报无法连接集群。那么密钥为什么没有生成呢？由前面的分析可知，/etc/ceph 目录下的 ceph.client.admin.keyring 文件是由 task="fetch ceph initial keys"生成的，它是 mon.yml 文件中的最后一个任务。该任务执行的是自定义模块，同时传入参数 state="fetch_initial_keys"。
ceph_key 模块的源码如下：

```
# 代码位置：library/ceph_key.py
# ...

def run_module():

# ...

    if state in ["present", "update"]:
        # ...

    elif state == "absent":
        # ...
    elif state == "info":
        # ...
    elif state == "list":
        # ...
    elif state == "fetch_initial_keys":
        # 前面生成的密钥放在/var/lib/ceph/mon/ceph-{{ inventory_hostname }}
/keyring下
        # inventory_hostname 在 hosts 中一般写的是主机全名，如 x01.gz.cn
        hostname = socket.gethostname().split('.', 1)[0]
        # 碰到密钥文件无法生成的情况，如果是主机名中点号的问题，则对主机名不进行裁剪
          即可
        # hostname = socket.gethostname()
```

```
    user = "mon."
    keyring_filename = cluster + "-" + hostname + "/keyring"
    user_key = os.path.join("/var/lib/ceph/mon/", keyring_filename)
    rc, cmd, out, err = exec_commands(
        module, list_keys(cluster, user, user_key, container_image))
else:
    # ...

# ...
```

从前面的 mon.yml 文件中可以看到，在安装 ceph-mon 服务中所生成的 keyring 文件的全路径为/var/lib/ceph/mon/ceph-{{ inventory_hostname }}/keyring，这里 inventory_hostname 表示的是 hosts 中指定的主机名，一般笔者会写全名，和 /etc/hosts 中的主机名对应。而在 ceph_key 模块中，对于 state= "fetch_initial_keys" 的情况，user_key 会指定前面生成的 keyring 文件，而这里的 hostname 变量是对主机名进行裁剪，去掉了点号及其之后的字符串，所以导致在后面执行的 Ceph 命令出错，从而使相关的密钥文件无法生成。但是该任务执行时并没有报错，只是显示为绿色，比较难发现。解决方法也很简单，不对获取的主机名进行裁剪，保证能找到对应生成的 keyring 文件即可。这是笔者在天翼云科技有限公司实际部署 Ceph 集群时遇到的问题，希望能对读者有所帮助。

6.4.3 小结

本节从 ceph-mon 服务部署角度分析了 ceph-ansible 的源码，并找出了部署 ceph-mon 服务的全过程，最后实现了简化版部署 ceph-mon 服务的 Playbook。至此，本章的内容也全部讲完了，这场 Ansible 项目的探索之旅也将告一段落，期待我们青山不改，"江湖"再会。

推荐阅读